Developments in Agricultural Engineering, 2

Land and Stream Salin

OTHER

1. Con
by
19

TITLE IN THIS SERIES

rolled Atmosphere Storage of Grains
. Shejbal (Editor)
30 viii + 608 pp.

This volume is reprinted from Agricultural Water Management Vol. 4, nos. 1, 2, 3 (1981)
pp. 1–392

Acknowledgement
The typing of the text and preparation of the illustrations were arranged for by K.C.
Webster and P. Hulbert of the Public Works Department, Western Australia

Developments in Agricultural Engineering, 2

Land and Stream Salinity

An International Seminar and Workshop held in November 1980 in Perth, Western Australia

edited by

J.W. HOLMES
Flinders University of South Australia, Adelaide, South Australia

and

T. TALSMA
CSIRO Division of Forest Research, Canberra, A.C.T., 2600

ELSEVIER SCIENTIFIC PUBLISHING COMPANY
Amsterdam – Oxford – New York 1981

ELSEVIER SCIENTIFIC PUBLISHING COMPANY
Molenwerf 1
P.O. Box 211, 1000 AE Amsterdam, The Netherlands

Distributors for the United States and Canada:

ELSEVIER/NORTH-HOLLAND INC.
52, Vanderbilt Avenue
New York, N.Y., 10017

ISBN 0-444-41999-3 (Vol. 2)
ISBN 0-444-41940-3 (Series)

Printed in The Netherlands

LAND AND STREAM SALINITY SEMINAR AND WORKSHOP

Papers presented at the Land and Stream Salinity Seminar and Workshop, Perth, Western Australia, November 1980

Guest Editors : J.W. Holmes
T.Talsma

CONTENTS

Review of the workshop and its conclusions 3

Land and stream salinity in Western Australia
R.M. Hillman (Perth, Australia) 11

The influence of plant communities upon the hydrology of catchments
J.W. Holmes (Adelaide, Australia)
E.B. Wronski (Perth, Australia) 19

Transport of salts in soils and subsoils
E. Bresler (Bet Dagan, Israel) 35

Residence times of water and solutes within and below the root zone
P.A.C. Raats (Haren, RA, The Netherlands) 63

Analyses of solute distributions in deeply weathered soils
A.J. Peck, C.D. Johnston and D.R. Williamson (Wembley, Australia) 83

Transport of salts in catchments and soils
T. Talsma (Canberra, A.C.T., Australia) 103

Saline seep development and control in the North American Great Plains - Hydrogeological aspects
M.R. Miller, J.J. Donovan, R.N. Bergatino, J.L. Sonderegger and F.A.Schmidt (U.S.A.)
P.L. Brown (Fort Benton, U.S.A.) 115

Terrain, groundwater and secondary salinity in Victoria, Australia
J.J. Jenkin (Victoria, Australia) 143

Groundwater systems associated with secondary salinity in Western Australia
R.A. Nulsen and C.J. Henschke (Perth, Australia) 173

Role of solute-transport models in the analysis of groundwater salinity problems in agricultural areas
L.F. Konikow (Virginia, U.S.A.) 187
River basin hydrosalinity modeling
K.K. Tanji (California, U.S.A.) 207
Predicting stream salinity in South-Western Australia
I.C. Loh and R.A. Stokes (Perth, Australia) 227
The influence of stream salinity on reservoir water quality
J. Imberger (Nedlands, Australia) 255
Impact of water resource development on salinization of semi-arid lands
G.T. Orlob and A. Ghorbanzadeh (California, U.S.A.) 275
Dryland cropping strategies for efficient water-use to control saline seeps in the Northern Great Plains, U.S.A.
A.L. Black (North Dakota, U.S.A.)
P.L. Brown (Montana, U.S.A.) 295
Management of soil water budgets of recharge areas for control of salinity in South-Western Australia
R.H. Sedgley (Nedlands, Australia)
R.E. Smith (Colorado, U.S.A.)
D. Tennant (Perth, Australia) 313
South Australia's approach to salinity management in the River Murray
K.J. Shepherd (Adelaide, Australia) 335
The evolution of a regional approach to salinity management in Western Australia
B.S. Sadler and P.J. Williams (Perth, Australia) 353
Dryland management for salinity control
J. van Schilfgaarde (California, U.S.A.) 383

REVIEW OF THE WORKSHOP AND ITS CONCLUSIONS

1 OBJECTIVES

The papers in this special issue of Agricultural Water Management were presented at a Seminar and Workshop convened for the following purposes :

(a) To review the state of knowledge of processes involved in land and stream salinity and to consider its application in different environments,

(b) to identify gaps in research and development and to designate priorities that could be of significant benefit to salinity control,

(c) to identify alternative land and water management strategies for salinity control and to determine techniques for quantitative evaluation of alternative strategies, and

(d) to give a degree of emphasis in the above objectives to salinity problems in south-western Australia.

Other papers, not printed here, were presented in the poster format. The sessions of the Workshop were arranged to discuss five topics explicitly and these were :

(i) Transport of salt through the soil,

(ii) groundwater systems,

(iii) prediction of stream and surface reservoir salinity

(iv) soil and plant management to minimise mobilization of salts, and

(v) land and water management for salinity control.

2 CAUSES OF DRYLAND SALINITY

Dryland salinity problems have been reported from many parts of the world where there is, at least on a seasonal basis, a deficit of rainfall relative to potential evaporation. In such regions excess salts may be present in the surface soil layers but, wherever secondary (man-made) salinization occurs in non-irrigated areas, salts are more commonly present in the subsoil or in deeper strata. They may have accumulated from oceanic salt carried in rain, into a previously highly leached landscape (Western Australia) or be present initially in strata of marine origin (North American Great Plains). Whatever the origin, a subsoil salt store is a prerequisite for the development of secondary dryland salinity.

The occurrence of a dryland salt problem is almost invariably caused by a man-made disturbance of the hydrological balance. For example, it is commonly accepted that clearing of native scrub and trees in south-western Australia for an agricultural enterprise based on wheat-pasture rotation, or breaking of native sod on the North American Great Plains for wheat-long fallow rotation, has reduced the evapotranspiration component of the water balance. Holmes and Wronski (page 19) state that forests use more water, firstly by the evaporation of intercepted rain at a faster rate and secondly by creating a larger soil water deficit, to greater soil depths, than crops or pastures.

The additional water available on catchments converted to dryland farming is estimated to be only about 100 mm yr^{-1} and probably less for the generally open stands of native-forest vegetation. Independently, and differently assessed data from Western Australia suggest that the extra water available after clearing is 20 to 60 mm yr^{-1}, depending on annual rainfall. These small surpluses have generally led to the conclusion that it should be feasible to control dryland salinity by management of plants and crops on the recharge areas in order to promote a larger consumption of water.

3 TRANSPORT OF SALT THROUGH THE SOIL

The papers by Bresler (page 35) and Raats (page 63) showed that the theory of water and salt transport through soils and deeper aquifers is well developed and adequate for application to the problems of dryland salinity. Peck et al. (page 83) and Talsma's (page 103) papers contained examples of such applications, the paper of Peck et al. being particularly relevant to the regional problems of Western Australia.

However, the considerable variability of field observations, taken not only for purposes of investigation of local situations, but also for comparison with theoretical prediction of the phenomena, should give some warning that some of the theory (in particular the specification of boundary conditions) could be either inadequate, or that its application was difficult or even inappropriate. For example, the distribution of salts as a bulge in the pallid zone of the clay subsoil at many locations, may be caused by water flow paths that follow fissures and do not influence salt transport in the by-passed matrix of the clay. Such salt distributions, discussed by Peck et al. to explain the apparently stable "bulge-profiles" were not to be predicted by Raats' theory, which uses a transpiration function that distributes the intensity of root abstraction of soil water, together with a steady-state, convective solute drainage model. Furthermore, the theoretical prediction of the accumulation of salts at the ground surface should probably be reconsidered by taking thermal and salt concentration gradients into account. The assumption of steady-state conditions may sometimes be a poor assumption.

4 GROUNDWATER SYSTEMS

Examples of regional groundwater systems were given for three areas with dry-land salinity problems. Miller et al. (page 115) discussed the Great Plains, U.S.A., Jenkin (page 143) dealt with secondary salinity in the geologically much more diverse and complex terrain of Victoria, Australia, and Nulsen and Henschke (page 173) described the physical factors involved in the development of saline seeps in the wheatbelt of Western Australia. Konikow (page 187) briefly surveyed general, three-dimensional equations governing water and salt transport in groundwater and then solved a particular case of water contamination with a numerical solution of an approximate two-dimensional equation.

The first three papers all dealt with the causes, in recharge (upland) areas, and the consequences, in the discharge areas. In general, recharge areas are close to discharge areas, and geological- as well as soil-stratigraphy is continuous over the problem areas in the Great Plains and in the Western Australian wheatbelt. In the more complex situation in Victoria some discharge areas receive saline groundwater from distant, as well as nearby recharge areas. A point in common, for all areas, appears to be that the characteristic lengths of these systems exceed their depths by factors varying between 10 and 100. Hence solutions to groundwater flow problems, using the simplifying Dupuit-Forchheimer assumptions, should be valid.

In the Western Australian landscape there are typically two nearly-horizontal pathways for water movement; rapid response flow of water temporarily perched on the lateritic mottled or pallid zones of low hydraulic conductivity, and slow response flow of the deeper and more saline groundwater. The composition of the discharge at any time is then determined by the relative amounts and compositions of the two components. The physical properties of the pallid zone appear to be the key to understanding salt and water transport. Given the very small hydraulic conductivity of the dense kaolin matrix, preferred - and presumably dominantly vertical - pathways should be important.

5 PREDICTION OF STREAM AND RESERVOIR SALINITY

The prediction of temporal and long-term changes in surface water salinity with changing physical conditions in catchments was mainly discussed in terms of hydro-salinity models. Tanji (page 207) described ten models, developed in the U.S.A., as a basis for assessing the adequacy of available models to answer specific questions and for discussing criteria for selection of a given modelling approach.

Orlob and Ghorbanzadeh (page 275) addressed the effect on a specific irrigated basin, the San Joaquin Valley, of increased upstream water development. Loh and Stokes (page 227) discussed the effect of land use changes in Western Australia on river flow and salinity, while Imberger (page 255) applied a detailed reservoir simulation model to the Wellington Reservoir in order to evaluate the consequences of varying operating strategies.

Most, if not all, models were developed for site-specific problems and for specific objectives. None have strength in all phases - water or salt transport, soil phase, groundwater phase or streamflow - none lend themselves to universal application. The most useful purposes of models were seen to lie in obtaining a first, overall, approach to a system, to determine its sensitivity to specified parameters, to clarify data needs and to foster communication among researchers representing different disciplines.

Detailed modelling is often not feasible, because, in an initial appraisal, the data base for model evaluation and calibration is usually not adequate. Also, it is not immediately obvious how a detailed sub-model, as e.g. discussed by Orlob and Ghorbanzadeh for small-field drainage, can be satisfactorily linked to their basin model for which a much longer time-scale applies. The broad scale model of Loh and Stokes for stream salinity in Western Australia simulated the effects of catchment clearing very adequately. However, due to lack of data, the effects of reforestation cannot be predicted with confidence.

6 SOIL AND PLANT MANAGEMENT

Soil and plant management strategies to minimise mobilization of salts in the subsoil were discussed on the basis of data and concepts presented in the papers by Black et al. (page 295) for the Great Plains, and Sedgley et al. (page 313) for Western Australia. Both stressed plant management on recharge areas (see Section 2); Black et al. described successful cropping sequences, most of them including lucerne (alfalfa) in the rotation, that effectively used excess soil water that accumulated and drained under earlier wheat-long fallow farming systems. Sedgley et al. presented a systematic and largely conceptual approach to estimate the water balance terms of evapotranspiration and recharge into and beyond the root zone, and then discussed the few available field data on increased water use by cropping systems that differed from the traditional wheat-pasture rotation system.

The detailed and successful strategies designed for the Great Plains are not readily adaptable to the Western Australian problem area (especially not in the wheatbelt), since climatic conditions and basic soil physical and chemical properties in the two regions are quite different. For example a suitable lucerne variety for the Western Australian region is currently not available.

Partial reforestation with native tree species, currently favoured for the Western Australian water supply catchments (higher rainfall areas), is not readily acceptable in the drier wheatbelt since this does not, at present, provide the necessary economic incentive considered essential by both Black et al. and Sedgley et al.

Soil management in south-western Australian recharge areas, with predominantly sandy soils, would need to be directed towards decreasing surface infiltration, increasing water storage and retention, together with the closure of preferential flow paths in the clay subsoil. However little of this is practised or even contemplated.

7 LAND AND WATER MANAGEMENT

Management of land and water, mainly for downstream control of surface water salinity was discussed by Shepherd (page 335) for the lower Murray River in South Australia, by Sadler and Williams (page 353) for Western Australia, and was summarized by van Schilfgaarde (page 383) for irrigation basins in the U.S.A. together with an appraisal of the salinity problem in Western Australia.

Management and salinity control measures on the lower Murray River in South Australia are partly dictated by widespread upstream diversion of water for irrigation outside that State's control. There are legal and constitutional constraints upon South Australia's response to the problem. Measures adopted for stream-water salinity control include diversion of saline drainage and groundwater into off-stream evaporation basins, and conversion from furrow irrigation to more effective systems. These are expected to provide adequate control for the next 5 to 10 years.

A different situation exists in Western Australia. Here management is in the hands of a single Authority. Several of the small river basins regarded as candidates for future management have their main tributaries originating in a naturally saline environment. Clearing for agriculture, or for timber production, has long been recognised as a cause for salinity increases in downstream river water. While this has led, so far, to the spread of salt upon a relatively small area of agricultural land, such practices are now recognised to be harmful as well to water supply for urban and irrigation use. Sadler and Williams provide details of a design programme for catchment-wide planning to control river salinity, based on the method for predicting stream salinity as described by Loh and Stokes (Section 5). They draw attention to the complexity of trade-offs between river water quality and dryland agricultural production, and to the needs for research into design analysis.

Van Schilfgaarde stressed that, although we have what appears to be a sufficent understanding of the physical processes, the need remains to distinguish between various problem situations. Feasible solutions in one circumstance may not apply elsewhere. The constraints may be physical, economic or political. Where the objective is to control river salinity, control of the water balance at re-charge sites is favoured because control of saline seep areas would be harder and might not be viable in the long-term. Nevertheless this should not be ignored completely as an option.

8 CONCLUSIONS AND RECOMMENDATIONS

(i) The participants agreed that the theory and concepts regarding salt and water transport through soil and deeper aquifers were adequate to provide the necessary technical insight of the problems. However, application is to a large degree site-specific. Therefore much remains to be done to gather sufficient data for extensive application, despite what has already been achieved. Data acquisition and analysis is needed in the characteristics of rainfall, the characteristics of the root zone and water extraction patterns of several plant and tree species, and in the water transmission properties of the pallid zone subsoil.

(ii) There was general endorsement of the policy to seek solutions to salin-ity problems in the recharge areas, as opposed to the treatment of the saline discharge areas. This was seen to be more opportune and it possessed the ad-vantage that the surplus water could be used on the site, would be conserved and the costs of diversion would be saved.

(iii) Although the current policy of reforestation of water supply catchments is appropriate, this could be unacceptable in the agricultural (wheatbelt) areas, because it would result in significant loss of agricultural areas. As first priority therefore, the Workshop participants strongly recommended that the scope of investigation of plant management should be widened beyond the current agro-forestry solution. At present it may be difficult and not immediately att-ractive to the land holders to seek alternative crops rotations and new crops for new markets. The participants were not convinced that these difficulties and objections were insuperable.

(iv) The customary engineering solutions such as shallow and deep drainage to remove saline water should not be rejected. There is ample terrain variability for such solutions to be feasible at specific sites without the need for pumping. By themselves they may have only small impacts on the improvement of the water balance but, since the original disturbance in terms of excess water was small, such impacts may nevertheless be worthwhile. The disposal of the saline water from drains would have to be considered carefully at the planning and design stage.

(v) The participants concluded that the control of stream salinity would probably require engineering works on a scale much larger than that needed for mitigating land salinity. Stream salinity would need to be viewed on a catchment scale. Alternative strategies can be evaluated by available models, techniques and information.

(vi) No management strategy should be considered to be exclusive of others. In certain circumstances a combination of a number of approaches could offer the best solution.

(vii) To consider, and to help to implement these recommendations, the Workshop participants recommended the formation of a small advisory group. Furthermore, they suggested that community involvement and extension of technical information could play a strong part in gaining the acceptance of the research and management programmes by the public.

LAND AND STREAM SALINITY IN WESTERN AUSTRALIA

R.M. HILLMAN
Director of Engineering, Public Works Department, Western Australia.
West Perth, W.A. 6005.

1 INTRODUCTION

Dryland salinity has been a threat to the land and water resources of the south-west of Western Australia since the commencement of agriculture in this State. However it is only in recent years that the seriousness of the problem has become widely known and accepted within the community. The phenomenon is not restricted to Western Australia; it is found in countries throughout the world, but it is believed that in few other parts of the world is it so evident or so widespread, and nowhere else is it a cause for such general community concern. Because of this concern the phenomenon has attracted relatively great research effort in this State, and those working on the problem here are in a good position to exchange ideas and information with other concerned people from throughout the world. For this reason it is believed that Western Australia provides a suitable venue for the Seminar. The papers, which were presented by scientists from a number of different countries, were chosen for the authors' particular expertise in salinity and the relevance of that expertise to dryland salinity.

Dryland salinity is manifest in each of the other States of Australia. It is also a significant problem in the Great Plains region of North America. In Canada it occurs extensively in the prairie provinces of Manitoba, Saskatchewan and Alberta, and in the United States in the states of Montana, North and South Dakota (Miller et al., 1976). In total, approximately 0.8 million hectares (2 million acres) of farmland in this region is severely affected by saline seeps. Dryland salinity is also said to occur in South Africa and Thailand, and it probably exists - recognised or unrecognised - in other countries.

The serious effects of dryland salinity on farmlands and water resources in all these areas has stimulated extensive research in an effort to find practical solutions to this complex problem. Many of the phenomena associated with salt and water movement through the soil under dryland agriculture are similar to those associated with such movement under irrigation. It was felt that the bringing together of the group of engineers and scientists at the Seminar and Workshop would provide an opportunity for all aspects of the problem to be discussed.

2 HISTORY OF SALINITY IN WESTERN AUSTRALIA

By overseas standards agriculture in Western Australia is young. Last year the 150th anniversary of the first settlement by Europeans was celebrated. On the other hand the land mass is very old. Over thousands of years a degree of equilibrium had been reached in the salt cycle within the hydrological cycle. The introduction of agriculture has upset this equilibrium. The result is a dryland salinity problem in the south-west of the State which is of very serious proportions.

The area of Western Australia is about $2.5 \times 10^6 km^2$ and its present polulation is some 1.25×10^6 people. Most of the population is to be found in the south-west corner of the State (Fig. 1) within an area of less than about $200 \times 10^3 km^2$, and within which most of the State's agricultural development has taken place.

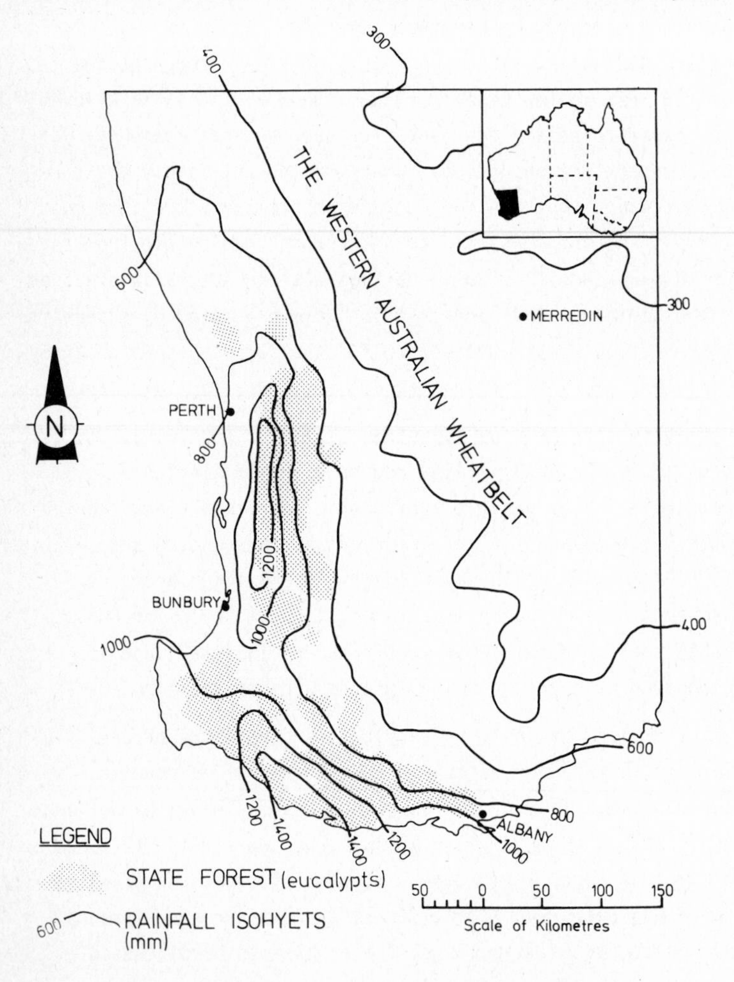

Fig. 1. The south-west region of Western Australia.

Settlement came late to Western Australia and for many years the population increased very slowly. People of European origin in the population numbered only about 180×10^3 by 1900. The growth of the agricultural industry, in parallel with the polulation, was at first slow.

Farming in the wheatbelt first commenced in the 500 to 600 mm rainfall region and has gradually developed eastwards into the lower rainfall regions as agricultural practices improved. By 1900 there were only 300 km^2 sown to wheat in the whole of the State. At present there are approximately $41 \times 10^3 km^2$. After the first seventy years of settlement, clearing in the susceptible areas of the wheatbelt would, no doubt, have provided evidence of salt leaching, but it certainly could not then have been considered a problem.

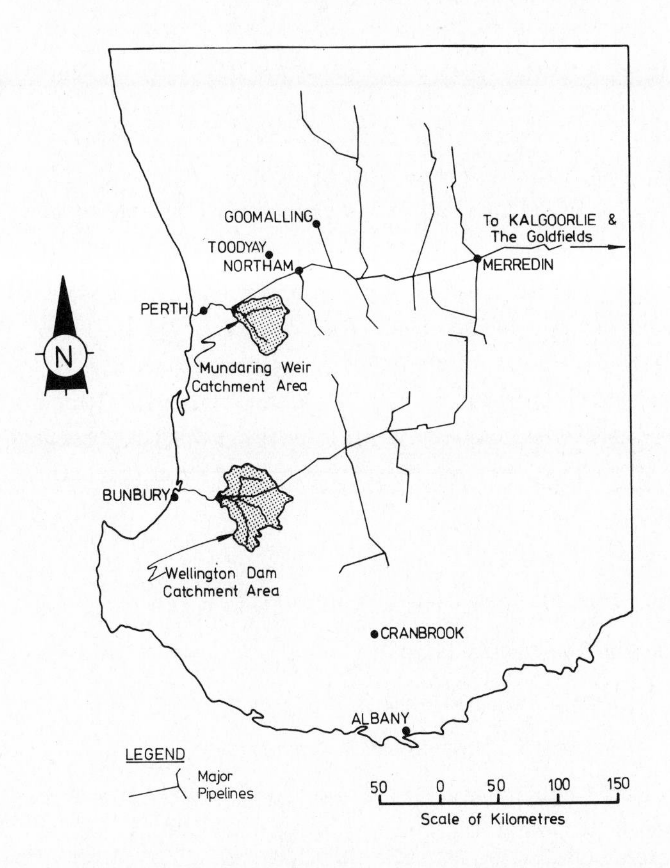

Fig. 2. Great southern towns, goldfields and agricultural water supply systems.

In this State salinity first came to be recognised as a problem in connection with the supply of boiler feed water for railway steam locomotives. A railway engineer, W.E. Wood was the first to publish data on the phenomenon. In a paper in the Journal of the Royal Society of Western Australia in 1924 he hypothesised as to the cause and cited examples he had found, or of which he had heard during the course of his work. Early examples were from Yorke Peninsula in South Australia in 1894; Northam and Toodyay in 1897 (Fig. 2); Goomalling in 1904 and Cranbrook in 1905. Many others followed.

In the early years of the century concern was felt for the apparent increase in salinity in the water stored in the reservoir behind the recently completed Mundaring Weir, some 40 km from Perth (Fig. 2). The water from this reservoir was to be pumped 584 km inland to supply the Kalgoorlie and eastern goldfields.

A sequence of below average rainfall years followed the completion of the weir in 1902. In an effort to improve the runoff, trees on an area of the catchment were ringbarked. As a result of this action there was an apparent increase in the salinity of water in the streams from the ringbarked area. In 1908 a decision was taken to cease ringbarking on unalienated land, to resume the alienated land whenever possible and to reforest the resumed lands. A recommendation was also made to scour from the bottom offtake valve of Mundaring Weir, but this was not accepted. This course of action and the scouring recommended have many parallels with the action now being taken on the Wellington Catchment.

In the years 1900 to 1930 agriculture expanded rapidly; over this period an estimated $50 \times 10^3 km^2$ was brought into production. With this expansion dryland salinity emerged as a significant problem to agriculture and started to receive more detailed attention. Expanding upon Wood's hypothesis, valuable contributions to the understanding of the process were made by L.J.H. Teakle, G.H. Burvill and others from the Department of Agriculture over the decade prior to the start of the second world war.

By 1929 it had been recognised that:

(i) dryland salinity was at least partly due to cyclic salt in rainfall,

(ii) it was most prevalent in low rainfall areas,

(iii) its occurrence was influenced by topography and soil type, and

(iv) its basic cause was the removal of the natural vegetation.

As more and more land was released for agriculture the likely impact of dryland salinity was frequently raised. However, the demand for new land was so great that the adverse effects were overlooked with the result that the release of large contiguous blocks of land was condoned and the complete clearing of all natural vegetation occurred over extensive areas.

Agriculture in the State went through a period of decline from about 1930 to the late 1940's. During this time as new salt patches developed dryland salinity continued to be regarded essentially as an agricultural problem. The problem of Mundaring Weir had been resolved. Railway operations were fairly flexible and adequate sources of boiler water supply could always be found.

A second major period of expansion of agriculture in the State started in the early 1950's and continued up to the mid-1960's. As a consequence soil salinisation increased and increased concern over the problem followed. Salt land surveys to gauge the full extent of affected land were carried out in 1955 and again in 1962, and, also in 1962, a major study of the cause of dryland salinity was undertaken within the Department of Agriculture.

In parallel with the agricultural expansion of the 1950's there was a demand for the provision of reticulated water supplies to the country towns and farms. These supplies are provided from the major surface storages of Wellington and Mundaring reservoirs or from local catchments. During this period there was growth in the south-west irrigation areas and also an expansion in mining and industry. The increased water demand gave rise to an increased awareness of the effect that increasing salinity was having on the water resources of the region. For the general public dryland salinity progressed from being regarded as a problem of importance only to farmers, to a matter of concern to all and of vital concern to the future development and prosperity of the State. The increased general public awareness of the problem has had two important results.

Firstly, it has brought about a widespread recognition of the inter-relationship between land uses and their effect on the quantity and quality of runoff. This has led to the introduction of more careful and sophisticated land use planning.

Secondly, the need for more research both fundamental and applied has become obvious and this has led to a substantial increase in activity in this area.

3 EFFECTS OF SALINITY IN WESTERN AUSTRALIA

The seriousness of the salinity problem in this State can be illustrated by quoting some figures of the effects on farming and of the effects on resources of water.

3.1 Effects on Farming

Dryland salinity has resulted in the loss of considerable areas of productive farmland. In the most recent 1979 salt land survey by the Department of Agriculture, (Henschke, 1981) it was found that there are 264×10^3 ha of once productive land in the south-west that are no longer suitable for agriculture. This figure represents 1.75% of all cleared land in the region. The average increase has been 7.8 x 10^3 ha yr^{-1} since 1955. In two shires the ratio of salt land to cleared land is greater than 6%, in five other shires between 4% and 5%. Many other shires are also seriously affected.

In most instances where land has been lost it was land with the best soil, the most productive land on the farm. The lost land capitalised as it was by clearing, fertilisation, fencing and other factors represents a considerable economic loss to individual farmers, to the State and the nation.

3.2 Effects on Water Resources

The south-west of Western Australia is an area with relatively limited water resources. It has been assessed that within the area, the surface water nominally available for diversion, was approximately 2.6 x $10^9 m^3$ yr^{-1} (Fig. 3). However, of this total 135 x $10^6 m^3 yr^{-1}$ or 5% are now saline (over 3000 mg L^{-1} TSS), 780 x $10^6 m^3 yr^{-1}$ or 30% are brackish (1000 - 3000 mg L^{-1} TSS) and 425 x $10^6 m^3 yr^{-1}$ or 16% are marginal (500 - 1000 mg L^{-1} TSS). This leaves only some 50% of the divertible surface water resource largely unaffected and still fresh.

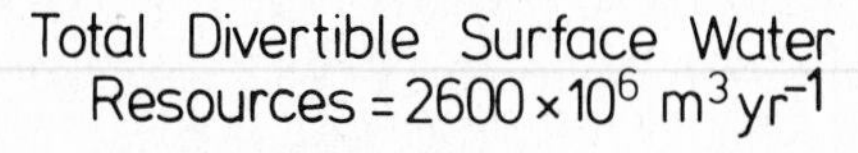

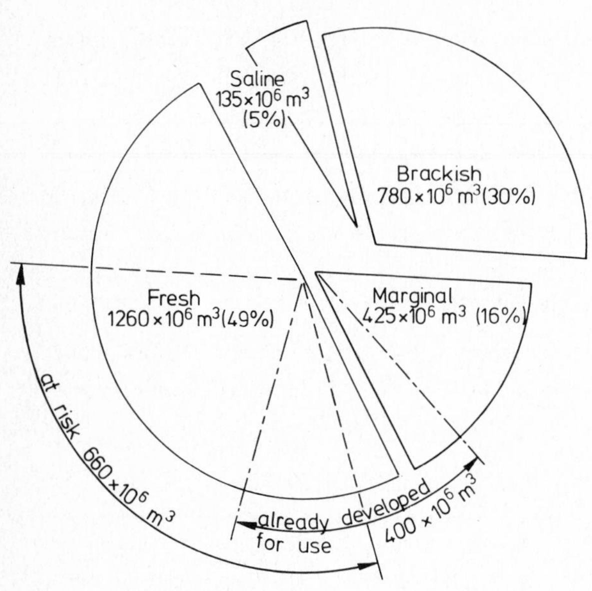

Fig. 3. Divertible surface water resources of the south-west of Western Australia.

Thus it can be seen that the leaching of the salts stored in the landscape has had a most serious effect on the quality of the water resources in the region. The importance of the remaining divertible resources which contain less than 1000 mg L^{-1} TSS is appreciated when it is recognised that 400 x $10^6 m^3 yr^{-1}$ or about 25% of the total remaining fresh and marginal resources are already used. In addition it has been estimated that a further 660 x $10^6 m^3 yr^{-1}$, i.e. about 40% of the totals in those categories, could be adversely affected by salinity if not carefully managed. Furthermore, environmental considerations may preclude the diversion of much of the nominally divertible resources.

To arrest and reverse the deterioration of the quality of these water resources, clearing controls have been introduced on five important river catchments. The first of these was on the Wellington Catchment on the Collie River (Fig. 2), where controls were introduced in 1976. In 1978 controls were introduced on four other catchments.

Wellington Reservoir on the Collie River is the water storage with the greatest yield in the south-west. The average yield of this reservoir is 100 x $10^6 m^3 yr^{-1}$. It provides a water supply for irrigation on the coastal plain and for town supply in the wheatbelt. It has been calculated that if clearing controls had not been introduced the average water quality in the reservoir would have reached 1700 mg L^{-1} TSS and that during dry cycles of weather the figure would have been much higher. Even with clearing controls salinity is expected to increase for some time because there is a considerable time lag between clearing and the full manifestation of the resulting salinity. The salinity of the water in the reservoir may reach an average of 1100 mg L^{-1} TSS. Similar conclusions could be drawn from studies of each of the other catchments over which clearing controls have been imposed.

4 RESEARCH ACTIVITIES

The realisation of the seriousness and complexity of the problem has generated considerable research in Western Australia over the last 10 years. This research is being undertaken at all levels, theoretical, fundamental and applied. Broadly, theoretical and fundamental research is being undertaken by the Universities and the Commonwealth Scientific and Industrial Research Organisation. State Government Departments have concentrated on applied research. The number of organisations involved is quite large but research has benefited from the informal coordination and co-operation which has developed amongst the researchers. The helpful co-operation which has developed has had a major influence on the enthusiasm and calibre of the research which is being undertaken in Western Australia.

The availability of basic data is always a problem when such complex and widespread problems are involved. The careful installation of gauging weirs, piezometers and other equipment is now beginning to bear fruit and will be invaluable to the research effort in the coming years.

5 CONCLUSION

It is hoped that in this short introductory paper some indication has been given of the magnitude of dryland salinity in Western Australia. The following papers, which were presented at the Seminar and discussed in Workshop sessions, cover a range of research activities and management strategies which are being undertaken to combat stream salinity. It is hoped the papers and discussion presented in these proceedings will benefit the many people who are working in the field of dryland salinity.

6 REFERENCES

Henschke, C.J., 1981. The 1979 Saltland Survey. J. Agric. West. Aust. (in press).

Miller, M.R., Van der Pluym, H., Holm, H.M., Vasey, E.H., Adams, E.P. and Bahls, L.R., 1976. An overview of saline-seep programs in the States and Provinces of the Great Plains. In: "Regional Saline Seep Control Symposium Proceedings". Bull. No. 1132, pp. 4-17. (Montana State Univ., Bozeman, Montana).

Wood, W.E., 1924. Increase of salt in soil and streams following the destruction of the native vegetation. J. Roy. Soc. West. Aust. 10: 35-47.

THE INFLUENCE OF PLANT COMMUNITIES UPON THE HYDROLOGY OF CATCHMENTS

J.W. HOLMES,
School of Earth Sciences, Flinders University of South Australia,
Adelaide, South Australia.

E.B. WRONSKI,
C.S.I.R.O. Division of Land Resources Management,
Perth, Western Australia.

ABSTRACT

Holmes, J.W. and Wronski, E.B., 1981. The influence of plant communities upon the hydrology of catchments. Agric. Water Manage., 1981.

This paper reviews some of the present scientific knowledge about the water yield from catchments that vary in their land use. Afforested catchments yield less runoff because the evapotranspiration from them is larger than from catchments not in forest. The evaporation of rain water held upon the wet foliage of trees proceeds at a faster rate than would the transpiration if the leaves were dry. Since the interception store can be 2 mm and is often more than this in dense forests, the interception loss can probably account for the observed yields from afforested catchments of up to 200 mm yr^{-1} less than from (say) grassland, where rainfall is >1 m yr^{-1}. The soil water deficit that deep-rooted, perennial plants can create is also likely to be larger than that caused by annual pastures and cereal crops. Although it is hard to find reliable data for comparative purposes, it is likely that a soil water deficit of 250 mm each year could be created by evergreen forest stands and 180 mm each year by annuals, in the climatic regions of southern Australia where the rainfall is 700 mm yr^{-1} or more. These figures themselves would imply a yield of 70 mm yr^{-1} less from afforested land, in runoff of surface water, or recharge of groundwater.

1 INTRODUCTION

Our knowledge of the processes that determine the water yields of catchments has been aided by the concept of the water balance. The stream discharge that may be measured at a gauging station at the defined exit of a catchment should be equal to the difference between the precipitation upon the catchment and the sum of evaporation and groundwater discharge. A trend to change of the stream discharge, as may be observed after many years of record, must signify some change in one or more of the individual components of the balance equation that is expressed in equation (1), namely:

$$P = E + SD + UD + \Delta S. \quad (1)$$

Here P is the precipitation; E is the evaporation; SD is the surface drainage (run-off); UD is the underground discharge not available for measurement in the stream; and ΔS is the change in soil moisture and groundwater storage.

All parameters can be conveniently expressed in mm yr^{-1}, but the yearly cycle may, in fact, be too short for some hydrological purposes.

Precipitation can show long-term trends although its variability from year to year, without trend, is much more important from an economic point of view. Evaporation is often nearly equal to precipitation so that the three other terms on the right hand side of equation (1) are often only about 5 to 10% of the water budget. Therefore, if the average evapotranspiration from a catchment should vary because land use has changed, it is to be expected that a change in runoff from the catchment should be observed. Unfortunately, detection of such change may be delayed for a long time by undetected change in soil and groundwater storage.

Langford and O'Shaughnessy (1977) reviewed world-wide evidence about the effects that forest changes can have upon catchment hydrology. Thinning and particularly clearfelling of the forest has often caused much greater runoff than before the cutting. Bethlahmy (1974) reported that runoff in a Colorado catchment was increased substantially after the trees (*Picea engelmannii*) had been killed by an attack of bark beetles. Langford (1976) described the hydrological aftermath of the 1939 bushfires in Victoria. Five years after the destruction of the mature-age mountain ash forests the regrowth of seedling trees was extremely dense with thousands of stems per hectare. Streamflow was observed to increase immediately after the fire when there was little vegetation upon the ground. But it soon began to decline as the regrowth increased in density and leaf-area-index. Eventually runoff became 24% less than from comparable catchments that had not been burned, during the 20 years of observation 1945 to 1965. These results were subsequently confirmed on three other groups of catchments, also mainly in mountain ash (*Eucalyptus regnans*), which had an average water yield of 25% less than their yield estimated as if they had not been burned in 1939, for the four water years 1972/73 to 1975/76 (Langford et al., 1980). The mean annual rainfall on these catchments is about 1400 mm yr^{-1}.

Besides the contrasts that have been noted in the runoffs from land in varying degrees of afforestation, there are also interesting comparisons to be drawn between forest and grassland. For example, Clarke and Newson (1978) described the differences in streamflow of the upper catchments of the Rivers Wye and Severn, which rise in the mountains of Wales. Table 1 is a brief extract from their paper.

TABLE 1

Precipitation (P), Streamflow (Q) and loss by evapotranspiration (P-Q) (Mean of 1974 to 1976) data of the Wye and Severn catchments.

	P mm yr^{-1}	Q mm yr^{-1}	P-Q mm yr^{-1}
Wye (grassed)	2210	1788	421
Severn (afforested)	2237	1552	685
(Data from Clarke and Newson, 1978)			

The catchment of the Wye is prediminantly in rough hill pasture, whereas the catchment of the Severn has a forest cover upon about two thirds of its area.

Similar contrasts between natural grass cover and eucalyptus forest (*Eucalyptus grandis*, Hill ex Maiden) upon steep hill catchments of the Drakensberg escarpment have been reported by van Lill et al. (1980). This well-designed experiment, which was conducted over the years 1956 to 1977, led the authors to state that the streamflow from the afforested catchment was between 300 and 380 mm yr^{-1} less than it would have been over the years of observation, if this catchment had been retained in the natural pasture. The annual rainfall in that part of South Africa (25°S) is about 1200 mm and possesses a strong monthly maximum about mid-summer. The reduction in streamflow produced by the afforestation could be observed to start only two years after tree planting, and it reached a steady value apparently only four years after the trees were established at an initial density of 1370 stems ha^{-1}.

Holmes and Colville (1970a,b) measured the recharge of groundwater in a karstic terrain of South Australia where surface discharge was an insignificant component of the regional runoff. They found that the infiltration to the groundwater beneath grass was 63 mm yr^{-1} on average, for the years 1960 to 1965, whereas there was no groundwater recharge beneath plantation forest (*Pinus radiata*), established upon a similar soil. Allison and Hughes (1972) confirmed these results by the technique of interpreting the tritium hydrology of the groundwater. There was so little tritium remaining in water sampled from the water table under the forests, when compared with the tritium concentration in water beneath grassland, that it appears unlikely that significant recharge could ever occur beneath mature pine forest in the Mount Gambier district, where the mean annual rainfall is about 720 mm yr^{-1}. In fact, the water was so old beneath forest that the regional groundwater discharge from the extensive grasslands must be the source of the water beneath the forests, which themselves allow no rainfall excess to infiltrate past their root zone.

Many other experimenters have observed the propensity of forests to consume more water annually, by evapotranspiration, than do other kinds of plant communities. In the interests of brevity, we refer the reader to the comprehensive list of references given in the paper by Langford and O'Shaughnessy (1977).

2 EVAPOTRANSPIRATION RATES FROM WET AND DRY FOLIAGE

There is much evidence that forest cover makes streamflow diminish if it is compared to the streamflow to be expected from similar catchments not in forest. It is logical to argue that the lesser stream discharge has been caused by a larger evapotranspiration of the forests. However, the micro-meteorological results required to substantiate that argument have been hard to get. The research work is financially expensive and success has depended upon taking the techniques available to the limit of their capability.

The Bowen ratio, β (=H/LE) (Bowen, 1926) is a useful parameter to employ in describing variations of the evapotranspiration rate (LE). Equation (2) expresses the simplest concept of the energy budget when minor terms can be neglected, namely:

$$R = H + LE, \tag{2}$$

where H is the sensible heat transfer.

Here the available energy, R, is assumed to be the net radiation measured above the experimental site in question. However, the purpose of this Section is to present some evidence that advected sensible heat can be an important source of energy and it makes water evaporate at a fast rate when it is held as water droplets upon the foliage. At a symposium held in 1965 in Pennsylvania, of which all papers were subsequently published as the book "Forest hydrology" (Sopper and Lull, 1967), Rutter (1967) summarised some of his work upon Scots pine and made the startling claim that "the rate of evaporation of intercepted water is, on average, about 4 times as great as the transpiration rate in the same environmental conditions". The actual rates were stated to be 1 to 3 mm day^{-1} (in southern England) in winter and 7 to 10 mm day^{-1} in summer. These rates exceed the net radiation above the ground surface several-fold. Rutter stated that... "the additional energy required is obtained from the air". Although he did not elaborate that statement further, the sources of the energy he refers to are distributed within the atmosphere and at the surface of the globe. They include air subsidence on a synoptic scale, latent heat transfer and its release following condensation and cloud formation, surface heating by global radiation upon the continents as well as local scale heating of surfaces that are not afforested. In fact, these are some of the main processes that keep the atmosphere generally in a state of deficit with respect to saturated water vapour pressure.

Rutter began his field experiment about 1957. The actual evaporation from the forest, which was near Crowthorne (Berkshire), exceeded the estimated evaporation from an open water surface in a manner that is summarised in Table 2. It should be noted that the data of Table 2 are not necessarily in conflict with a four-fold enhancement of evaporation when the canopy is wet because the periods of time indicated in the Table are approximately 10 months when most of the water loss would occur from the dry canopy through transpiration.

TABLE 2

Evaporation from a plantation of *Pinus sylvestris*.

Year	*E_o mm	E_F/E_o
1957	410	1.13 ± .09
1958	380	1.07 ± .05
1959	540	0.98 ± .04
1960	430	1.15 ± .07
1961	410	1.20 ± .08
1962	470	1.06 ± .03

It can be seen that the actual evaporation from the plantation forest was generally larger than the evaporation estimated to have been lost from an open water surface for the same period. Assuming, as Rutter did, that evaporation from grass is often about 0.75 E_o, the conclusion is that actual evaporation from the forest could exceed actual (potential) evaporation from grass by about 1.5. Rutter was cautious in claiming any general validity of his experimental results until further studies could place them in the context of the energy budget of the ground surface and its plant communities.

Rutter observed the water balance of his forest and deduced the large evaporation loss because his rain-gauge measurements required that it be so. Moore (1976) attempted a direct measurement of evaporation rates from a forest canopy and their range of variation when the canopy was wet and dry. His experiment was in, and above a plantation forest of *Pinus radiata* near Mount Gambier, South Australia. He had 16 days of each condition, chosen from a much longer set of days through the period August to October, which is part of the rain season in that district.

*E_o is the estimated evaporation from a free water surface, following Penman (1956) and E_F is the actual evaporation of the forest, for the periods May to December, approximately, each year, derived from soil moisture and precipitation measurements. (Data from Rutter, 1964).

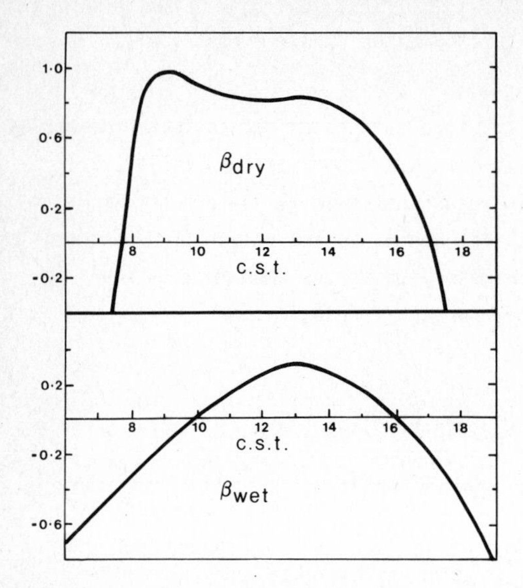

Fig. 1. Variation in the Bowen ratio of a *Pinus radiata* forest, diurnally and when it was wet or dry. (Data from Moore, 1976).

Fig. 1, from Moore's data, shows the contrast of Bowen ratio, β (=H/LE), for wet and dry forest canopies, through the hours of daylight. The experimental method was to measure the sensible heat and latent heat fluxes independently by the eddy correlation technique, which was first developed by CSIRO. The mean Bowen ratios, weighted with respect to contemporary net radiation above the forest were 0.124 for the wet and 0.74 for the dry foliage and these may be taken to be the best values of Bowen ratio to be applied to the daily energy budget, for purposes of partitioning the available energy when advected, sensible heat is relatively small.

A numerical example to show the significance of these results may be helpful. Suppose that the available energy at Moore's forest site at the beginning of August was about 4.4 MJ m^{-2} for a 24-hour period, equivalent to the energy required to evaporate 1.8 mm of water at about 20°C. If the forest canopy were wet, the evaporation rate would be 1.60 mm day^{-1} but if the canopy were dry it would be 1.04 mm day^{-1}. The ratio of evaporation rates, wet to dry, is seen to be 1.55, a result that alternatively we could have derived directly from an application of the expression for evaporative flux in terms of net radiation and the Bowen ratio, namely:

$$LE = R_n/(1+\beta). \tag{3}$$

Wronski continued the experimental work, begun by Moore, to attempt an interpretation of processes of forest hydrology. He measured the amount of rain that could be held upon the foliage, and also assessed it by suitable interpretations of the field evaporation rates. In plantation forest the amount should be very dependent upon the leaf-area-index (LAI). In his thesis he stated (Wronski, 1980) ..."the way intercepted water is distributed in a canopy plays a large part in determining the rate at which it is evaporated".

The maximum repeatable interception store that Wronski could measure at the Noolook forest site was 1.8 mm. His direct measurements of Bowen ratio there were about 1.1 for the dry forest and 0.2 when the canopy was wet. His results therefore confirmed Moore's data but there is some further work of interpretation of the details that promises to be enlightening. The Noolook forest was a younger and lower forest than the Mount Gambier forest, though both had closed canopies and about 2000 stems ha^{-1}.

In Britain during the 1970's, the Institute of Hydrology undertook a large research programme on forest hydrology for purposes of assessing water yields from afforested land. Stewart (1977) analysed some of the data of the Thetford forest project (in East Anglia). He had numerous observations of available energy (R) and the mean temperature (T) and specific humidity (q) profiles for 20-minute periods, above the forest canopy, from which he selected periods when the canopy was fully wet. Table 3 shows a selection of his derived results.

TABLE 3
Mean values of energy fluxes above the fully-wet canopy of a pine forest in East Anglia.

Available energy (R) Wm^{-2}	30	68	171	240	352	827
Latent Heat Flux (LE_I) (evaporation rate of intercepted water) Wm^{-2}	83	96	190	228	293	675
Sensible heat flux (H) Wm^{-2}	-53	-28	-19	12	59	152

(Data after Stewart, 1977. The apparent surface resistance was between -10 s m^{-1} and 10 s m^{-1}).

It may be observed from Table 3, that the sensible heat flux was directed downwards (negative values) when the available energy flux was less than about 180 Wm^{-2}. Stewart used the Bowen ratio for the partitioning function to be applied to the available energy, as :

$$\beta = \frac{C_p}{L} \frac{dT/dz}{dq/dz} \tag{4}$$

Therefore, the balance of $R = H + LE_I$ which is satisfied by the columns of Table 3 cannot indicate the precision of the method. In fact one could entertain some doubt about the precision since the author had to use extremely small increments or decrements of observed T and wet-bulb temperatures, that demanded the utmost in good technique. The values of the fluxes in Table 3 are means of up to 29 individual observations but the largest energy budget shown was from one isolated observation.

Rutter and his associates in the Institute of Hydrology have also approached the problem by computing the evaporation rate as predicted by the Monteith-Penman equation (Monteith, 1965). This modelling method (Rutter et al., 1971, 1975; Gash and Morton, 1978) does predict a large evaporation rate from a wet canopy because the stomatal resistance item in the equation is made zero. By it the rate of evaporation of water from the wet canopy is calculated by applying the formula for evaporation from a free water surface:

$$E_p = \frac{\Delta R_n + \rho C_p \ VPD/r_a}{L(\Delta+\gamma)} \tag{5}$$

where R_n is the net radiation above the forest canopy (Wm^{-2}); VPD is the vapour pressure deficit also measured above the forest canopy (m bar); Δ is the slope of the saturated vapour pressure curve at the air temperature above the forest (m bar K^{-1}); γ is the psychrometric constant (m bar K^{-1}); L is the latent heat of vaporisation of water at the evaporating temperature (J kg^{-1}); ρ is the density of the air (kg m^{-3}); C_p is the differential heat capacity of the air at constant pressure (J $kg^{-1}K^{-1}$) and r_a is the aerodynamic resistance to turbulent transfer of water vapour and should have a value appropriate to the roughness of the forest, not that of a plane water surface (s m^{-1}). Use of equation (5) in the forest situation can lead to evaporation rates that are up to 3 times the calculated transpiration rates derived by including in an equation like (5) a term for the stomatal resistance.

Wronski (1980) refined this approach considerably by attempting to allow for the distribution of water upon the foliage and the progressive drying downwards from the top-most branches. We cannot describe his calculating scheme here but Table 4 gives a summary of some of these results, which have yet to be published.

It may be observed that the Bowen ratios, calculated from the data of Table 4, are 0.55 and 0.20 respectively for a dry and wet canopy, assuming a stomatal resistance of 4.0 s cm^{-1} at the top of the canopy, and 0.86 and 0.13 for a stomatal resistance of 6.0 s cm^{-1}. These values may be compared with the measured values of 0.74 and 0.12, that Moore (1976) obtained near Mount Gambier during the same time of year (about August) and measured values of about 1.1 and 0.2 that Wronski himself got at the Noolook forest.

TABLE 4

Daily rates of transpiration from a dry canopy, and evaporation and transpiration from a wet canopy (simulated calculation using measured parameters for input). Available energy assumed to be 2.3 mm day^{-1}.

	Stomatal resistance at the top of the crown, s cm^{-1}.		
	4.0	6.0	8.0
Transpiration from dry canopy	1.48	1.24	1.08
Total evapotranspiration from wet canopy	2.03	1.91	1.84
Evaporation component	1.43	1.47	1.49
Transpiration component	0.60	0.44	0.35

(Data after Wronski, 1980)

We have probably written enough to justify the conclusion that the evaporation rate when the canopy of a forest is wet can exceed by 1.5 to 3 times the rate of transpiration that would have prevailed if the canopy were dry. Such a contrast is hard to make experimentally and some workers have approached it by comparing forest and grassland on the assumption that the evaporation rate from grassland is hardly affected by its wetness and, wet or dry, is reasonably close to the dry-weather transpiration rate of forests. Indeed Holmes and Colville (1968) drew attention to the differences in cumulative evaporation from grassed and afforested land. Table 5 shows some of their results, which were obtained by water balance techniques.

TABLE 5

Contrasts in cumulative evapotranspiration from grassland and afforested land (*Pinus radiata*)

	5/5/63 - 26/9/63	12/5/64 - 4/11/64	1/5/65 - 15/10/65
Net radiation (in mm of water equivalent)	160	267	228
Evapotranspiration from grassland (mm)	110	240	212
Evapotranspiration from afforested land (mm)	260	420	305

(Data after Holmes and Colville, 1968)

The period of observation during 1964 was exceptionally wet for the district and the large evapotranspiration from the forests then is consistent with an enhanced rate of evaporation from the wet canopy. The period of observation during 1963 was drier than usual. In fact, there may have been occasions when soil moisture deficit could have limited the transpiration rate.

Section 3 of this paper, which follows, presents a discussion of soil moisture deficit and the way it must relate to the hydrology of catchments and their plant communities.

3 EVAPOTRANSPIRATION RATES AND SOIL WATER DEFICIT

The climate of a large part of southern Australia has such a prolonged dry season in the summer that the soil becomes very dry at its end. Crops such as wheat and barley, and pastures composed of annual species, taper off their transpiration rates as maturity, grain and seed production and senescence progress, even if there should be abundant soil moisture available. Nevertheless, conditions are often very dry by the middle of December. Unlike the annual plants, perennials cannot avoid the drought and have to endure it. When the neutron moisture meter became available for field experiments (about 1956), the depletion of soil water reserves could be studied intensively.

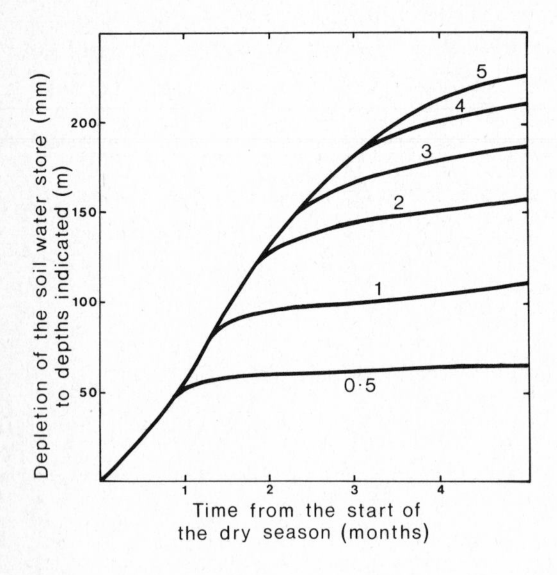

Fig. 2. Depletion of the soil water store as a function of time and depth in the soil (generalised data to facilitate discussion of the effects of different plant communities).

Fig. 2 depicts, in a generalised way, how plant roots extract water from the soil. The total deficit to a depth of 5 m in the soil profile would depend upon the characteristics of the plant community. The root density, in particular, would determine the intensity of the drying of the soil in the shallower layers, such as that one from the surface to 0.5 m. Drying at the greatest depth would depend upon the penetration of roots, for it is believed that the soil water content changes only very slowly if the distance of travel to an absorbing root surface is greater than about 0.1 m.

The experiments on the water budget of grassland and pine forest, referred to above (Holmes and Colville, 1970a,b) yielded data about soil water deficit that is summarised in Table 6. It may be observed that the soil profile beneath pasture dried to a depth of about 1.8 m seasonally and the soil water deficit so created was about 180 mm. The pasture was comprised of perennial rye grass (*Lolium perenne, L.*), subterranean clover (*Trifolium subterraneum, L.*), soft brome grass (*Bromus mollis, L.*), heron's bill (*Erodium botrys,* (Cav.) Bertol.), barley grass (*Hordeum leporinum,* Link), capeweed (*Cryptostemma calendula,* (*L.*), Druce) and Yorkshire fog (*Holcus lanatus, L.*).

TABLE 6

The soil water deficit that was recharged each year (in mm) and the approximate depth of drying by roots in the soil (in m, in parentheses)

		Year			
	1961	1962	1963	1964	1965
(Forest sites)					
1 duplex: sand over clay	-	-	208 (3.3)	514 (3.9)	242 (3.3)
2 duplex: sand over clay	-	-	414 (5.4)	811 (>4.8)	346 (4.8)
3 duplex: sand over clay	-	-	163 (2.1)	474 (>4.8)	308 (4.5)
4 deep sand	-	-	127 (3.3)	232 (>4.2)	187 (>4.2)
5 deep sand	-	-	126 (3.0)	227 (6.3)	163 (4.5)
6 deep sand	-	-	141 (3.3)	231 (5.4)	164 (3.6)
Grassland site	135 (1.5)	190 (1.8)	192 (1.8)	190 (1.8)	160 (1.5)

(Data from field experiments of Holmes and Colville, 1970a,b).

Beneath forest (*Pinus radiata*, , D. Don) the soil dried and wetted up in a manner that was dependent upon both soil conditions and canopy density. Forest sites 1, 2 and 3 were in the Penola forest where mature-age trees had been thinned to a stand density of 450 ha^{-1}. The trees there were about 40 m high. Sites 4, 5 and 6 were in the Mount Gambier forest, where the trees were about 15 m high, canopy had closed and the density was 2200 ha^{-1}.

It may be observed that the soil water deficit (SWD) that was recharged at Sites 4, 5 and 6 in 1963 was less than the SWD at the grassland site. It was greater than that at the grassland site in 1964 and about the same in 1965. Those three years were drier than, much wetter than and about normal for the district. The influence of interception of rain and its evaporation from the foliage of the forest is very conspicuous in these data.

At Sites 1, 2 and 3 the subsoil clay at about 0.9 m was rather impermeable and a temporarily-perched water table was often observed. The very large change in soil water content at Site 2 was caused by inflow of "foreign" water. Nevertheless, the whole of the soil water increment was dissipated by transpiration and evaporation and none by deep infiltration to the country water table, which remained at about 8 m. There was also no drainage through the soil profile at Sites 4, 5 and 6. By contrast, the drainage at the grassland site was 40, 134 and 72 mm in the years 1963, 64 and 65, and the water table fluctuated between 1 and 2.4 m below ground surface.

Schultz (1971) showed that the effect of fallowing soils for cereal production in South Australia was to conserve between 40 and 120 mm of soil water, depending upon soil type, that otherwise would have been lost by evapotranspiration during the season December to May. We do not believe it is an exaggeration to propose that SWD could vary in the range 50 to 300 mm annually, on the one site if it were occupied by cereal-fallow, annual grassland or pine forest.

There is, unfortunately, not much information of this kind about eucalyptus forests in Australia.

Williams and Coventry (1979) reported that they had observed SWD to be in the range 220 to 110 mm in the root zones of an open woodland community in northern Queensland, which was made up of yellow jack (*Eucalyptus similis*), bloodwood (*E. dichromophloia*), iron bark species (*E. whitei* and *E. crebra*) together with acacias and other non-eucalypts. Table 7 shows values of SWD observed by Holmes (1960) at four sites in mallee heath in South Australia. Two sites in lucerne that had replaced the mallee communities, were also studied.

The mean annual rainfall in that region is about 510 mm and it has a strong winter incidence. However, one of the years (1956) during which the experiment was conducted happened to be exceptionally wet. The soil water deficits shown are likely to be much larger than usual and yet there appeared to be insignificant drainage to the water table, which was present at a depth of about 7 m.

TABLE 7
The maximum observed soil water deficit that was created by mallee heath and lucerne on deep sands in South Australia. (In mm).

Site	Soil water deficit
1	350
2	377
3	345
4	473
Lucerne	
1	450
2	273

(Data from Holmes, 1960).

Butcher (1979) reported some measurements of soil water depletion beneath native woodland on the coastal plain north of Perth, W.A. The sites were occupied variously by jarrah (*E. marginata*), other eucalyptus species, *Banksia attenuata* and other banksias. He did not draw attention to any noticeable differences in SWD corresponding to different woodland communities. A deficit of about 250 mm seems to have been created each year to a depth of 5 m on deep sandy profiles.

4 CONCLUDING COMMENTS

The enhanced rates of evaporation of water, present as droplets upon the leaves after rain, are hard to measure directly. It is probably impossible to develop calculating schemes that would be universally valid. The factors that combine to produce the fast rate of evaporation include the amount of advected heat (i.e. the vapour pressure deficit) available for the period of wetted foliage, the interception store which determines the duration of the period, the distribution of water upon the foliage which determines the pattern of drying downwards from the top-most branches, the leaf-area-index of the community and the fetch to the site concerned if leading-edge advection (the oasis effect) is a large contributor to the energy budget.

However, complexity in detail should not allow pessimism to prevail about future success in this research. Our understanding of the problem has enlarged greatly in the last 20 years. It is probably time to write that we now know the principles of physics and meteorology that are involved. Quantitative prediction of the effects of changed land use may not yet be possible to the desired precision, but we know what must be measured to gain that precision.

Not much has been written in this paper about the climatic regime of rainfall. Its significance should be apparent. If, for example, the precipitation in the wettest month is 100 mm and the characteristic amount per wet day is 3.3 mm, the trees are likely to be wet nearly all the time. On the other hand, if the amount of rain per wet day were 10 mm, the foliage might be wet for only one day in three. The interception loss would therefore be larger in the former example. Perhaps this consideration gives some reason for the apparent lack of significant change in the hydrology of river basins in eastern and north-eastern Australia, where rainfall characteristically occurs as heavy downpours.

The experience in Western Australia, which the papers for this International Seminar are meant to augment, suggests that there can be a delay of some decades before hydrologic change is obvious. Relatively crude observations may not be sufficiently sensitive to detect changes in stream discharges that are not merely the result of seasonal variability. However, trends in groundwater levels that are caused by changed amounts of deep infiltration and trends in concentrations of salts as tracers could be quicker indicators. There are numerous unpublished sets of records held by Government departments in Western Australia that would repay further examination.

Aston (1979) has recently compared the relative effectiveness of eight tree species in retaining water upon the leaves. Six eastern Australian eucalypts, a *Pinus radiata* tree and an *Acacia longifolia* were examined with the results shown in Table 8.

TABLE 8
Relative effectiveness of interception of sprayed water by eight small trees.

Order of decreasing interception
Pinus radiata
Eucalyptus pauciflora
Acacia longifolia
E. maculata
E. cinerea
E. mannifera, subsp. *maculosa*
E. dives
E. viminalis

(Data after Aston, 1979).

The *Pinus radiata* tree was five times as effective in intercepting and holding water droplets as was the *Eucalyptus viminalis* tree.

5 ACKNOWLEDGEMENTS

The research work of Holmes, Moore and Wronski was supported in part by grants from the ARGC, AWRC, WRFA, RCDF (Reserve Bank) and Dept. of Agriculture and Fisheries, S.A. We thank the Dept. of Woods and Forests, S.A. for approval to experiment in their commercial forests and for the cooperation of their officers. CSIRO Division of Forest Research made facilities available to us at the Mount Gambier Regional Laboratory.

6 REFERENCES

Allison, G.B. and Hughes, M.W., 1972. Comparison of recharge to groundwater under pasture and forest using environmental tritium. J. Hydrol.,17: 81-96.

Aston, A.R., 1979. Rainfall interception by eight small trees. J. Hydrol., 42: 383-96.

Bethlahmy, N., 1974. More streamflow after a bark beetle epidemic. J. Hydrol., 23: 185-89.

Bowen, I.S., 1926. The ratio of heat losses by conduction and by evaporation from any water surface. Phys. Rev., 27: 779-87.

Butcher, T.B., 1979. Management of *Pinus pinaster* plantations on the Swan coastal plain for timber and water yield. Aust. Water Resour. Council, Tech. Pap. 42: 60 pp.

Clarke, R.T. and Newson, M.D., 1978. Some detailed water balance studies of research catchments. Proc. R. Soc. Lond., A363: 21-42.

Gash, J.H.C. and Morton, A.J., 1978. An application of the Rutter model to the estimation of the interception loss from Thetford forest. J. Hydrol., 38: 49-58.

Holmes, J.W., 1960. Water balance and the water table in deep sandy soils of the upper south-east, South Australia. Aust. J. Agric. Res., 11: 970-88.

Holmes, J.W. and Colville, J.S., 1968. On the water balance of grassland and forest. Trans. 9th Congr. Int. Soil Sci. Soc., Adelaide, 1: 39-46.

Holmes, J.W. and Colville, J.S., 1970a. Grassland hydrology in a karstic region of southern Australia. J. Hydrol., 10: 38-58.

Holmes, J.W. and Colville, J.S., 1970b. Forest hydrology in a karstic region of southern Australia. J. Hydrol., 10: 59-74.

Langford, K.J., 1976. Change in yield of water following a bushfire in a forest of *Eucalyptus regnans*. J. Hydrol., 29: 87-114.

Langford, K.J. and O'Shaughnessy, P.J., 1977. Some effects of forest change on water values. Aust. Forestry, 40: 192-218.

Langford, K.J., Moran, R.J. and O'Shaughnessy, P.J., 1980. The North Maroondah experiment pre-treatment phase comparison of catchment water balances. J. Hydrol., 46: 123-45.

Monteith, J.L., 1965. Evaporation and environment. Symp. Soc. Expt. Biol., 19: 205-34.

Moore, C.J., 1976. Eddy flux measurements above a pine forest. Quart. J.R. Met. Soc., 102: 913-18.

Penman, H.L., 1956. Evaporation: an introductory survey. Neth. J. Agric. Sci., 4: 9-29.

Rutter, A.J., 1964. Studies on the water relations of *Pinus sylvestris* in plantation conditions. II. The annual cycle of soil moisture changes and derived estimates of evaporation. J. App. Ecol., 1: 29-44.

Rutter, A.J., 1967. An analysis of evaporation from a stand of Scots Pine. In: Forest Hydrology, Sopper, W.E. and Lull, H.W. (Eds.). Pergamon, pp. 403-17.

Rutter, A.J., Morton, A.J. and Robins, P.C., 1975. A predictive model of rainfall interception in Forests II. Generalization of the model and comparison with observations in some coniferous and hardwood stands. J. App. Ecol., 12: 367-80.

Rutter, A.J., Kershaw, K.A., Robins, P.C. and Morton, A.J., 1971. A predictive model of rainfall interception in forests I. Derivation of the model from observations in a plantation of Corsican pine. Agric. Meteorol., 9: 367-84.

Schultz, J.E., 1971. Soil water changes under fallow-crop treatments in relation to soil type, rainfall and yield of wheat. Aust. J. of Exp. Agric. and Animal Husbandry, 11: 236-42.

Sopper, W.E. and Lull, H.W., 1967. Forest hydrology. Proc. Sem. Pennsylvania State Univ., Sept. 1965. Pergamon.

Stewart, J.B., 1977. Evaporation from the wet canopy of a pine forest. Water Resour. Res., 13: 915-21.

Van Lill, W.S., Kruger, F.J. and Van Wyk, D.B., 1980. The effect of afforestation with *Eucalyptus grandis* (Hill ex Maiden) and *Pinus patula* (Schlecht et Cham) on streamflow from experimental catchments at Mokobulaan, Transvaal. J. Hydrol., 48: 107-118.

Williams, J. and Coventry, R.J., 1979. The contrasting soil hydrology of red and yellow earths in a landscape of low relief. In: The hydrology of areas of low precipitation. IAHS Publ. No. 128, pp. 385-95.

Wronski, E.B., 1980. Hydrometeorology and water relations of *Pinus radiata*. Ph.D. Thesis. Flinders Univ. of South Australia, 317 pp.

TRANSPORT OF SALTS IN SOILS AND SUBSOILS

E. BRESLER
Division of Soil Physics, Institute of Soils and Water, The Volcani Center, Bet Dagan, Israel

ABSTRACT

Bresler, E., 1981. Transport of salts in soils and subsoils. Agric. Water. Manage., 1981.

Factors affecting transport of salts in saturated-unsaturated dryland soils are reviewed. Simultaneous movement of water and salts occurring in homogeneous soils is discussed first. Governing equations describing combined diffusion-convection transient transport and miscible displacement of salts, as well as nonsteady water flow are given. Effects of salinity on soil water transmission rates are described. The governing equations and boundary conditions appropriate to three processes of the dryland hydrologic cycle: infiltration, redistribution, and evaporation, are formulated in a manner suitable for mathematical modeling. Numerical solutions are obtained. One-dimensional, vertical profiles of non-interactive salts are described for homogeneous, bare, fallow soil. Effects of physico-chemical interactions between solution and soil matrix on transport of interactive anions and cations are incorporated. Vegetation factors and their effects on salt dynamics and distribution are also considered. Consideration of salt transport in saturated-unsaturated heterogeneous fields concludes the paper. Statistics of field profile concentration and salt dynamics in field scales are given.

1 INTRODUCTION

The imbalance between transported incoming and outgoing salt causes salinization of soils and subsoils which result in increasing the salinity of streamflows and agricultural land. This salinization is a serious dryland environmental problem. Salt transport is affected by a combination of several soil-water-salt-plant factors. To estimate the magnitude of the hazard posed by salinity, it is important to understand and identify the processes that control salt movement from the soil surface through the root zone to the groundwaters and streamflows. Knowing these processes makes it possible to develop optimum management schemes for environmental control with the purpose of preventing groundwaters, streamflows and farmland salinization.

This paper reviews the factors affecting the movement of salts in saturated-unsaturated soils and subsoils under dryland conditions. It first describes the transport occurring in homogeneous field soils. Important factors affecting salt transport in homogeneous soils are quantitatively formulated in a manner suitable for mathematical modeling. Numerical solutions using finite difference or finite element methods are obtained. Consideration of salt transport in saturated-unsaturated heterogeneous field soils concludes this paper.

2 SALT MOVEMENT IN HOMOGENEOUS SOILS

2.1 Combined Convective-Diffusion Transport

In dealing with the simultaneous transfer of solute and water, one usually assumes that the transport of solute is governed by convection (viscous movement of the soil solution) and diffusion (thermal motion within the soil solution). The diffusion process is described by Fick's law. The macroscopic convective transport of a solute is usually described by an equation that takes into account two modes (or components) of transport: (a) the average flow velocity, and (b) mechanical dispersion (resulting from local variations in flow velocities). This mechanical dispersion effect is similar to diffusion in the sense that there is a net movement of solute from zones of high concentration to zones of low concentration. It is commonly agreed, therefore, that an equation similar to Fick's equation provides a good first-order description of the dispersion component of convective flow, provided that the diffusion coefficient D_p is replaced by a coefficient of mechanical dispersion (D_h), considered proportional to the first power of the average velocity:

$$D_h(V) = \lambda|V| \tag{1}$$

where λ is the dispersivity and $|V|$ is the absolute value of the average flow velocity.

The three components of solute transport, molecular diffusion and the two modes of convective flow, occur simultaneously in natural soils. Generally, the coefficients of mechanical dispersion and of molecular diffusion can be included in a combined coefficient, the so-called "hydrodynamic dispersion coefficient." Effects of molecular diffusion on overall dispersion become more important as the average flow velocity becomes smaller.

Assuming that the soil is an inert porous body makes it simpler to simulate the system theoretically. In this case, the joint effects of diffusion and convection can be described by:

$$J_s = - [\theta D_h(V)+D_p(\theta)]\frac{dc}{dx} + V\theta c = - \theta D(V,\theta) \frac{dc}{dx} + qc \tag{2}$$

where c is salt concentration in the soil solution; D is the hydrodynamic dispersion coefficient; θ is the volumetric water content; x is the flow direction; and q is the volumetric water flux. The right hand terms in equation (2) are only approximate, because the macroscopic quantities D, V, θ, and c are actually spatial averages. Nevertheless, equation (2) is useful in predicting solute transport through soils. An expression for one-dimensional transient conditions can be derived from a consideration of continuity, or mass conservation. This states that the rate of change of solute within a given soil element must be equal to the difference between the amounts of solute that enter and leave that element. By equating the difference between outflow and inflow to the amount of salt that has accumulated in the soil element, for the case of one-dimensional vertical flow, one obtains the expression (Bresler, 1973a):

$$\frac{\partial}{\partial t}(Q+\theta c) = \frac{\partial}{\partial z}\left[\theta D(V,\theta)\frac{\partial c}{\partial z}\right] - \frac{\partial(qc)}{\partial z} + S' \quad . \qquad (3)$$

Here, t is time; Q is the local concentration (positive or negative) of solute in the "adsorbed" phase (meq cm^{-3} soil), usually depending on both θ and c; S' is any solute loss (sink) or gain (source) due to salt uptake, sorption, precipitation, or dissolution; and z is the vertical space coordinate (considered to be positive downward). Equation (3) applies when the solute either does not interact (neglecting Q and S') or does interact chemically with the soil, including when there is loss or gain of salt inside the flow system. Mathematical solutions of equation (3) subject to specific initial and boundary conditions, may be developed for problems involving salinization under a wide range of soil conditions.

2.2 Miscible Displacement in Soils

Most work on miscible displacement phenomena in soils has been limited to steady state water flow with constant flow velocities and water contents (e.g., Biggar and Nielsen, 1967). As such, these studies provide a means of determining hydrodynamic dispersion coefficients, evaluating macroscopic flow velocities, and giving physical explanations for mixing phenomena which occur when salts flow through soils. For an inert system (Q and S' of equation (3) being negligible) undergoing steady state unidirectional flow of water, constant θ and V, equation (3) becomes :

$$\partial c/\partial t = D(\partial^2 c/\partial z^2) - V(\partial c/\partial z) \quad . \qquad (4)$$

To illustrate a general case of mixing during miscible displacement, let us consider an elution curve obtained when a salt-free soil solution is displaced through a column of soil by a solution containing an inert (non-interacting) solute of concentration C_o at pore water velocity V and water content θ. The fraction of this solute in the effluent at time t can be designated as c/C_o. Plots of c/C_o versus pore volumes of effluent (ratio of volume of effluent to

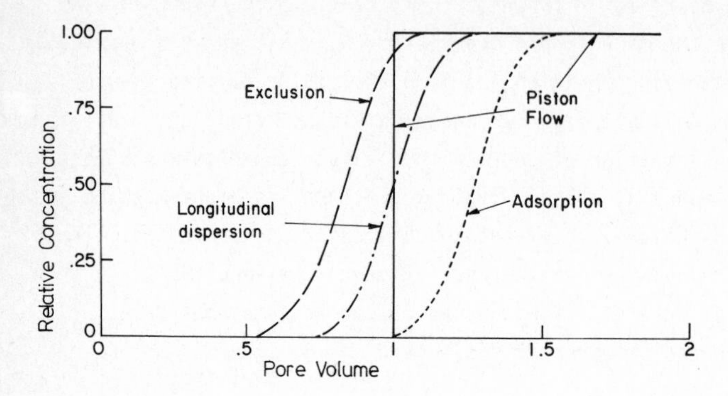

Fig. 1. Schematic breakthrough curves (BTC) for various miscible displacement conditions.

volume of solution contained in the soil column) are commonly called breakthrough curves (BTC). If piston displacement were operative, no mixing would occur between the displacing and displaced solutions, and a vertical line would represent solute "breakthrough" (Fig. 1). A sigmoid shape of elution breakthrough curve, on the other hand, indicates mixing (longitudinal dispersion). Shifting of the curve to the left indicates exclusion from, or bypass of a significant portion of the soil solution. Displacement to the right indicates adsorption or retention of the solute by the soil. Solving equation (4) for a semi-infinite column subject to the initial conditions, $c = c_n$ when $z > 0$ and $t = 0$, and to the boundary conditions $c = C_o$ when $z = 0$ and $t > 0$ and $c = c_n$ when $z \to \infty$ and $t > 0$, yields:

$$\frac{c - c_n}{C_o - c_n} = \frac{1}{2}\left[\mathrm{erfc}\left(\frac{z-Vt}{\sqrt{4Dt}}\right) + \exp\left(\frac{Vz}{D}\right) \mathrm{erfc}\left\{\frac{z+Vt}{(4DT)^{\frac{1}{2}}}\right\}\right] \tag{5}$$

where erfc is the complementary error function. Experimental data that fit the breakthrough curve calculated from equation (5) can also be used in an inverse manner to identify the dispersivity parameter λ which is necessary for modeling purposes (Bresler and Laufer, 1974).

2.3 Nonsteady (Transient) Water Flow

For one-dimensional vertical solution flow in the z direction, the condition of conservation of matter is that:

$$\frac{\partial(\rho\theta)}{\partial t} = -\frac{\partial(\rho q)}{\partial z} \tag{6}$$

where ρ is the density of the solution (generally assumed constant). Eliminating q by substituting the Darcy equation:

$$q = V\theta = - K(\theta)\ dH/dz \tag{7}$$

into the conservation equation (6) we obtain the general nonsteady (transient) one-dimensional (vertical) water flow equation:

$$\frac{\partial\theta}{\partial t} = \frac{\partial}{\partial z}\left[K(\theta)\frac{\partial h}{\partial z} + K(\theta)\right] . \tag{8}$$

Here, $K(\theta)$ is the unsaturated hydraulic conductivity function; $h = h(\theta)$ is the soil water pressure head function; and H is the hydraulic head (sum of pressure head and the gravitation head z). As h and θ are already interrelated by the soil water retentivity function, $h(\theta)$, equation (8) can be rewritten either in terms of h or θ. In cases where hysteresis in soil water retentivity is particularly important and $h(\theta)$ is not a single-valued function, the h-based version of equation (8) is:

$$C_w(h)\frac{\partial h}{\partial t} = \frac{\partial}{\partial z}\left[K(h)\frac{\partial h}{\partial z} + K(h)\right] \tag{9}$$

where C_w is the differential water capacity defined by $C_w\ (h|w) = \partial\theta/\partial h$ (a function depending on h and on the wetting history, w, of the system).

2.4 Effects of Salinity on Soil Water Transmission Rates

Macroscopic solution flow under isosalinity (and isothermal) conditions is well described by Darcy's equation (7) which assumes that the hydraulic gradient H is the only driving force causing water flow. Dynamic changes of soil solution salt concentration due to mass movement of salt or due to water content fluctuations, however, may create an additional driving force due to osmotic gradients. In addition, variations in salt concentration and composition affect the hydraulic conductivity and soil water retentivity functions.

2.4.1 Effects of salt concentration gradients

The negatively charged surfaces of soil particles tend to exclude anions and hence, because of electroneutrality, accompanying cations as well. Even when solution in larger soil pores is interconnected by thin films of solution, soluble salts tend to be excluded from the films and to remain in the bulk solution of the larger pores. Such an electric restriction can be viewed as similar to the geometrical restrictions of semi-permeable membranes.

If salt is completely excluded from the film, the system behaves as a perfect osmotic membrane. In such a case the specific flux of solution (q) in response to the hydraulic gradient dH/dx, is just the same as if an osmotic pressure gradient, $d\Pi/dx$, of equal magnitude has been applied. Under these conditions, the hydraulic conductivity K is valid for both hydraulic and osmotic pressure gradients. Thus, Darcy's equation (7) must be modified to the form:

$$q = - K \left[\frac{dH}{dz} - \frac{1}{\rho g}\frac{d\Pi}{dz}\right] . \tag{10}$$

The osmotic pressure can be estimated from Van't Hoff's law:

$$\Pi = \phi RTC \tag{11}$$

where ϕ is the osmotic coefficient of the electrolyte; R is the universal gas constant; T is the absolute temperature; and C is the sum of the molar concentrations of all anions and cations in the equilibrium solution. When the salt is only partially restricted, however, isothermal water transmission is described by:

$$q = - K\left(\frac{dH}{dz} - \frac{\sigma}{\rho g}\frac{d\Pi}{dz}\right) \tag{12}$$

in which σ is the osmotic efficiency coefficient or macroscopic reflection coefficient. The osmotic efficiency coefficient varies between 0 to 1 and represents the degree of semipermeability of the soil. When σ is 0, water transmission is not affected at all by salt concentration gradients (the classical Darcy equation is valid). When $\sigma = 1$ osmotic gradients are as effective as hydraulic gradients in moving water (equation (10)). The greater the restriction of solute relative to solvent (water), the greater the effect of solute gradients on soil water flow.

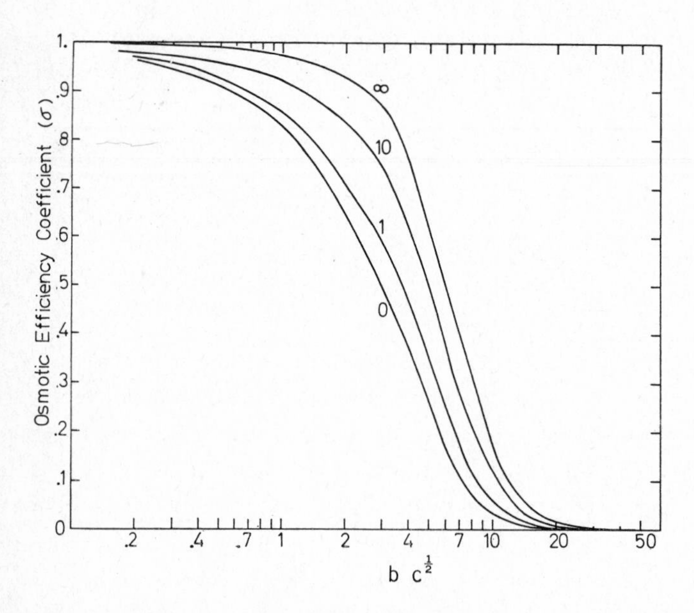

Fig. 2. Theoretical osmotic efficiency coefficient (σ) as a function of $b(\theta)c^{\frac{1}{2}}$ for four monovalent to divalent cationic ratios (numbers labeling the lines).

Fig. 2 gives estimates of σ calculated by Bresler (1973b) as a function of film thickness $b(\theta)$, equilibrium solution concentration c and ratio of concentration of monovalent to divalent cations in the equilibrium solution. Diffuse double layer theory (Bresler, 1970) was used to obtain salt exclusion data with surface charge density Γ_s being estimated from cation exchange capacity (CEC) divided by the specific surface area of the soil. Fig. 2 verifies experimental findings that the reflection or osmotic efficiency coefficient becomes a more important factor in transport of soil solution as the solution becomes more diluted, the water film thinner (the soil drier), or as larger proportions of monovalent cations are present in the soil.

2.4.2 Effects of salts on soil hydraulic parameters

(i) Hydraulic conductivity function $K(\theta)$. Swelling of soil clay particles in a confined system causes the size of large soil pores to decrease. Dispersion and movement of clay platelets further block soil pores. As swelling and dispersion of soil particles are highly affected by solution salt concentration and its composition, as given by Na to Ca ratio, low hydraulic conductivity which results from such geometric restrictions must, in turn, be affected by soil solution concentration and composition. This is predicted by the diffuse double layer theory for mixed electrolyte systems (Bresler, 1972). Russo and Bresler (1977a,b) tested effects of mixed Na-Ca salt solutions on unsaturated hydraulic conductivity $K(\theta)$ for a loamy soil from Gilat, Israel.

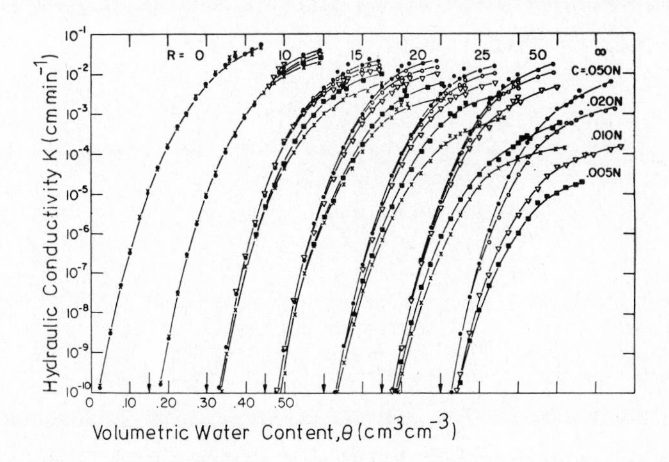

Fig. 3. Hydraulic conductivity (K) as a function of volumetric water content (θ) and solution concentration (C) for seven cationic ratios $[R=Na^+/(Ca^{2+})^{\frac{1}{2}}]$. Note that the point θ = 0.0 has been shifted for successive sets of curves, as indicated by the arrows along the abscissa.

Their results (Fig. 3) show that K(θ) is independent of solution concentration for a Ca-saturated system. For mixed Na-Ca systems, on the other hand, hydraulic conductivity functions are highly dependent on the composition and concentration of the soil solution, in addition to its dependence on soil water content. For any given θ, K(θ) decreased either as the soil solution salt concentration decreased or as the proportion of sodium in the soil solution increased. The negative effect of a combination of high R value (Na to $Ca^{\frac{1}{2}}$ ratio) and low soil solution concentration decreased with decreasing degree of water saturation.

(ii) Soil water retentivity function h(θ). The double-layer theory for mixed electrolyte systems (Bresler, 1972) predicts that the spacing between adjacent clay platelets increases as the value of R increases or as C decreases, for a given pore water suction. This, in turn, results in an increase in the amount of water retained by the clay as the pore water suction decreases. For a constant-volume system, such as a confined soil, changes in the volume of the clay mass are at the direct expense of the quantity and distribution of soil pores. Since the swelling of clay decreases as the pore water suction increases, the amounts of water retained (and the decreases in hydraulic conductivity) become smaller as the soil water suction increases.

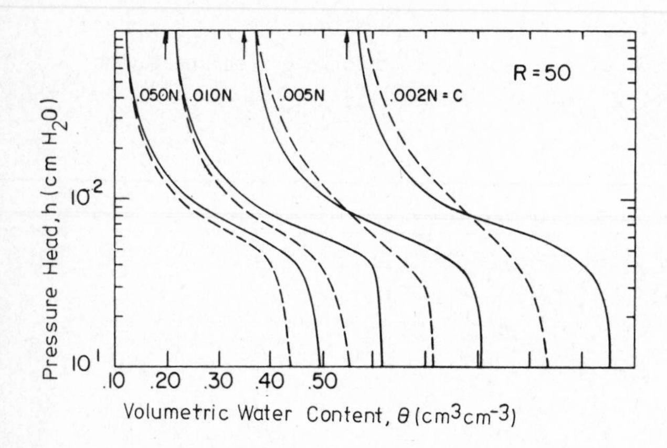

Fig. 4. Soil water suction head (h) as a function of soil water content (θ) for 4 values of solution concentration (C). Computed results (solid lines) are compared with measured curves (dashed lines). Arrows indicate that point θ = 0.10 is shifted and data are translated along θ axis. (After Russo and Bresler, 1980).

From diffuse double layer theory the distance 2d between adjacent quasi-crystals can be calculated as a function of h,R, and C, using electrical potential values for mixed systems. The volume of water which is retained by the clay fraction per unit volume of soil can be obtained from:

$$\theta(h,R,C) = \frac{\rho_b S_p\{d(h,R,C) + d_o[N_p(h,R,C)-1]\}}{N_p(h,R,C)} \quad (13)$$

in which S_p is the specific surface area of the soil; ρ_b its bulk density; d_o is the average half spacing between platelets in the quasi-crystals (4.5 Å); and N_p is number of clay platelets per quasi-crystal (tactoid). A comparison between data calculated from equation (13) and experimental results are given in Fig. 4.

2.4.3 Modified solution flux equation

Equation (7) is valid as long as $K(\theta)$ and $h(\theta)$ are unique. Because of the dynamic changes of salt concentration and composition in the field and the creation of driving forces arising from osmotic gradients, however, in addition to the effect on $h(\theta)$ and $K(\theta)$ functions result of soil swelling and dispersion, the flux of soil solution is more accurately expressed as:

$$q = - K(\theta)K^r(\theta,R,C)\left[\frac{\partial h(\theta,R,C)}{\partial z} - \frac{\sigma(\theta,R,C)}{\gamma}\frac{\partial \Pi(C,R)}{\partial z} - 1\right] \quad (14)$$

in which K^r is the relative hydraulic conductivity $(0 < K^r \leq 1)$ defined by $K^r = K(R,C,\theta)/K(\theta)$; with $K(\theta)$ being the hydraulic conductivity function of Ca-saturated soil (taken as the standard $K(\theta)$ for stable soil conditions); C is the molar concentration of all ions in the soil solution $(C = \sum_m c_m)$; R denotes the ionic composition; and γ $(=\rho g)$ is the specific weight of water pressure.

2.5 Salt Dynamics and Distribution in Fallow Soils

2.5.1 Non-interactive solute

The models to be described in this section are applicable mainly to fallow field conditions or when the effect of vegetation may be neglected for other reasons. Neither salt precipitation nor dissolution, nor uptake by plant roots, are assumed to take place. Solute interaction with the soil matrix is also assumed negligible, as are the effects of salt fluctuation on water flow, and on adsorption and exclusion of ions. The soil is considered to be non-sodic, and both horizontally and vertically homogeneous. Isothermal conditions are assumed. Such a model is restricted to considerations of variations in total salt concentration, since neither salt precipitation nor dissolution are allowed to take place. One-dimensional horizontally homogeneous soil (with respect both to water content and salt concentration) can be approximated by flooding a leveled soil or by a soil wetted at a moderate rate by rainfall (or sprinklers) or upward movement from shallow water table. One-dimensional wetting and salinity patterns can be simulated by considering water and salt to enter over the total field surface, or total field subsoil. One-dimensional evaporation and one-dimensional drainage are also assumed.

(i) Governing equation and boundary conditions for water flow. The appropriate partial differential equation governing one-dimensional vertical isothermal hysteretic water flow in inert soils under fallow conditions is equation (9), which can be approximated by a finite difference or finite element technique, subject to appropriate boundary conditions, and can be solved numerically with the aid of a computer. As we restrict ourselves to inert solute and porous material (non-interactive solute) no coupling coefficients need to be considered in the governing equation (9).

To formulate suitable boundary conditions, especially at the soil surface, three water flow processes must be considered: (1) infiltration, (2) redistribution or drainage, (3) evaporation. In the absence of water ponding at the soil surface, the soil mass can gain water by infiltration; neither gain nor lose water during redistribution; or lose water to the atmosphere by evaporation. While the potential (i.e., maximum possible) rate of evaporation from a soil depends only on atmospheric conditions, the actual flux across the soil surface is limited by the ability of the soil to transmit water from below in the soil profile. Similarly, if the potential rate of infiltration (e.g. the rainfall intensity) exceeds the infiltrability of the soil, part of the water may be lost by surface runoff. Here, again, the potential rate of infiltration is controlled by external conditions, whereas the actual water flux depends on soil hydraulic conditions, including the influence of antecedent soil moisture. Thus, except for the case of redistribution without evaporation, the exact boundary conditions to be assigned for water flux at the soil surface ($z = 0$) cannot be predicted a priori. Instead, a numerical solution must be sought by maximizing the absolute value of the specific water flux while maintaining the correct sign (i.e., positive flux for infiltration or negative flux during evaporation) subject to the requirements:

$$|q(0,t)| = |K(\theta)\frac{\partial h}{\partial z} + K(\theta)|_{z=0} \leqslant |IR(t)| \tag{15a}$$

$$h_d \leqslant h \leqslant 0 \tag{15b}$$

where $IR(t)$ is the prescribed potential surface flux as a function of time; $q(0,t)$ is the specific flux of water at the surface; and h_d is the minimum allowed pressure head at the soil surface. Boundary conditions at the soil-air interface ($z = 0$) are as follows:

(1) during the infiltration process:

$IR(t) > 0 \quad h(0,t) \leqslant 0 \quad t > 0$ (15c)

(2) during the redistribution process:

$IR(t) = -K(\theta)\ (\partial h/\partial z) + K(\theta) = 0 \quad t > 0$ (15d)

(3) during the evaporation process:

$IR(t) < 0 \quad h(0,t) \geqslant h_d \quad t > 0$ (15e)

The value of h_d is usually taken as the air-dry pressure head. It may also be calculated as a function of time according to the equilibrium conditions between soil water pressure head and atmospheric vapor, using the formula:

$$h_d(t) = (RT/Mg)\ \ln\ [\overline{RH}(t)] \tag{16}$$

where R is the universal gas constant; T is absolute temperature; M is the molecular weight of water; g is the acceleration due to gravity; and $\overline{RH}$ is the relative humidity of the air.

The lower geometric boundary of the soil profile at the depth $z = Z$ should always be chosen such that it is below both the root zone and the wetting front. In the case of drainage to the groundwater or vertical upward movement from the groundwater, the bottom boundary ($z = Z$) is taken at the water table where atmospheric pressure (assigned a value of zero) is maintained. Thus, mathematically:

$\partial h/\partial z = 0$ or $h(z,t) = 0 \qquad z = Z \qquad t \geqslant 0$ (15f)

In addition to the surface and bottom boundary conditions, it is necessary to specify initial conditions. These are the predetermined water pressure profile $h_n(z)$ throughout the soil domain, i.e.:

$h(z,0) = h_n(z) \qquad 0 \leqslant z \leqslant Z \qquad t = 0$ (15g)

From numerical solution of the governing equation (9) subject to the pertinent boundary conditions and including hysteresis, values of pressure head, water content, and water flux as a function of time and depth (i.e., $h(z,t)$, $\theta(z,t)$ and $q(z,t)$ are obtained. These can be used to solve for distribution of non-interacting solutes in fallow soils.

(ii) Governing equation and boundary conditions for solute flow. The governing differential equation for transient one-dimensional vertical diffusive and convective flow of an inert solute under isothermal fallow conditions is given by equation (3). For zero values of Q and S' this expression becomes:

$$\frac{\partial(c\theta)}{\partial t} = \frac{\partial}{\partial z}\left[\theta D(\theta\ ,\ V)\ \frac{\partial c}{\partial z} - qc\right] \tag{17}$$

To solve equation (17), the values of $D_h(z,t)$ and $D_p(z,t)$ must also be known. With $\theta(z,t)$ already known from solution of equation (9) using equation (15), the value of $D_p(z,t)$ may be calculated from Olsen and Kemper (1968). Similarly, once

$q(z,t)$ is known, one can calculate $V(z,t)$ from the macroscopic definition of the average velocity $V = q/\theta$. Knowing V, values of D_h can be determined from the $D_h(V)$ relationship given in equation (1).

The boundary conditions, appropriate to distribution of a solute in fallow soils, that must be satisfied at the soil-air interface ($z = 0$) and at any time $t > 0$ during infiltration, redistribution and at evaporation are given as:

$$J(0,t) = -\{D_p[\theta(0,t)] + \theta D_h[V(0,t)]\}\frac{\partial c}{\partial z} + q(0,t)c(0,t) \quad (18a)$$

Furthermore, 1) during infiltration, surface solute flux must equal the product of water flux and solute concentration of the infiltrating water i.e.:

$$J(0,t) = q(0,t)\ C_0(t) \quad (18b)$$

2) during redistribution or drainage, surface solute flux as well as surface water flux must remain at zero, i.e.:

$$J(0,t) = 0 \quad q(0,t) = 0,\ \text{so}\ \ \partial c/\partial z|_{z=0} = 0 \quad (18c)$$

and 3) during evaporation, solute flux at the soil surface remains equal to zero, because salt is not a volatile substance, i.e.:

$$J(0,t) = 0 \quad . \quad (18d)$$

Boundary conditions at the bottom of the soil domain and initial conditions are:

$$t \geqslant 0 \qquad z = Z \qquad \partial c/\partial z = 0 \quad (18e)$$

$$t = 0 \qquad 0 \leqslant z \leqslant Z \qquad c(z,0) = c_n(z) \quad (18f)$$

where $c_n(z)$ is the predetermined initial salt concentration profile.

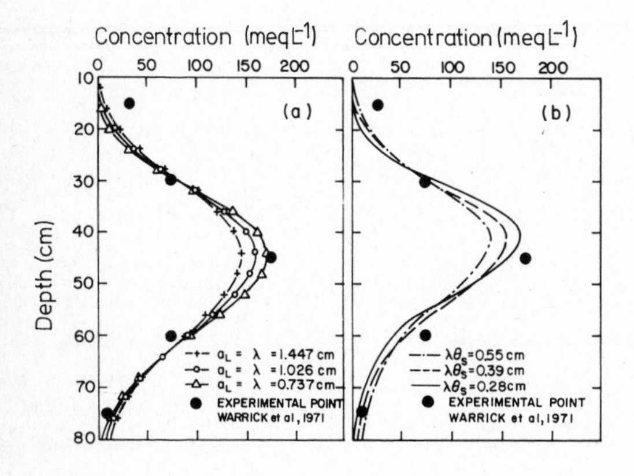

Fig. 5. Field distribution of chloride at t = 9 hours after initiation of irrigation (Warrick et al., 1971) compared with (a) results of the finite element model of Segol (1977) and (b) results of the finite difference model of Bresler (1973a).

Salt concentration distribution results after 9 hours of infiltration are presented in Fig. 5 for three values of the dispersivity λ. For comparison with experimental data, results obtained with the finite element method of Segol (1977) are given in Fig. 5a, whereas those obtained from the finite difference method of Bresler (1973a) are plotted in Fig. 5b. The experimental points of Warrick, Biggar and Nielsen (1971) are indicated on each figure. Good agreement exists between either of the numerical models and the field data, although the point of maximum concentration is located somewhat more accurately by the finite element model.

Good agreement was also obtained between calculated and observed values of chloride during redistribution and evaporation in the laboratory (Fig. 6). This indicates that the described models are reliable and suited for the analysis and prediction of one-dimensional transient transport of non-interacting solute in fallow soils.

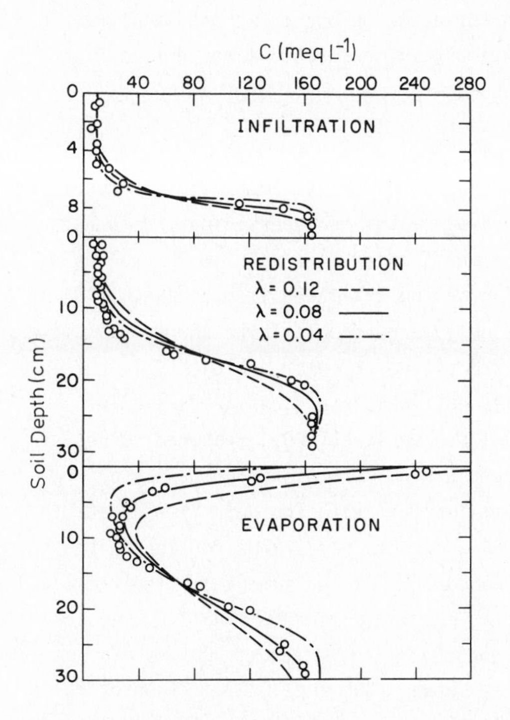

Fig. 6. Theoretical (continuous lines) and measured (open circles) chloride distribution for infiltration, redistribution, and evaporation and three values of λ.

2.5.2 Solute interacting with soil

To take into account the various physico-chemical interactions which occur between ions and the soil matrix, the effects of osmotic gradients and of anion exclusion or cation adsorption and solute flow must also be considered. The procedure in this case must use the continuity equation (6), along with the specific solute flux, such as given by equation (14). For solute flow models, governing equation (3), with or without the term S', must be applied. In evaluating the value of Q in equation (3) a distinction must be made between anions and cations. For cations, the value of Q is always positive due to attraction of cations to negatively charged soil particles. Conversely, the value of Q is always negative for anions, because of anion exclusion from the vicinity of negatively charged soil particles.

(i) Distribution of anions. Anions and cations on the external surfaces of the exchanger phase can be assumed distributed according to the theory of the planar diffuse double-layer. If the solid surface has a known density of charge, then the exclusion of a particular anion (Γ^-) can be calculated from the equilibrium salt concentration (c) and the thickness of the solution between any two charged platelets, $b(\theta)$. Since ions on the exchange phase are assumed to be in equilibrium with the soil solution, the amount of excluded anions per cm^{-3} of soil may be estimated from:

$$Q(c,\theta) = \Gamma^-[b(\theta),c]\, A_{ex}\, \rho_b = \theta_{ex}(\theta,c)\, c \quad . \qquad (19)$$

Here, Γ^- is the calculated anion repulsion or negative adsorption (a function of $b(\theta)$ and c); A_{ex} is specific surface area of the soil participating in anion exclusion; and θ_{ex} is the equivalent volume of anion-free solution per unit volume of bulk soil. The value of $Q(c,\theta)$ may be estimated from equation (19) if Γ^- is calculated (for external surfaces only) from double-layer theory, assuming either a symmetric or a nonsymmetric mixed system (Bresler, 1970).

The theoretical relationship $Q(c,\theta)$ as obtained from equation (19) using calculated values of Γ^- from equation (8) of Bresler (1970), measured values of ρ_b, and estimated values of A_{ex}, is given in Fig. 7. This figure shows a comparison between calculated and experimental anion exclusion data (for a $CaCl_2$ system) as reported by (a) Mokady, Ravina and Zaslavsky (1968) for kaolinitic and illitic clay soil; (b) Thomas and Swoboda (1970) for a montmorillonitic clay soil; and (c) Krupp, Biggar and Nielsen (1972) for a montmorillonitic clay loam soil. Fig. 7 shows the relative anion exclusion concentration (f) as a function of $b\, c^{\frac{1}{2}}$, where the latter varies between 200 and 20 000. Theoretical f data are obtained from the relationship $f_{calculated} = \Gamma^-/(bc)$. To calculate f for experimental data, values of b, A_{ex}, ρ_b, and either the exclusion volume (v_{ex}) or the exclusion water content (θ_{ex}) must be known. Then:

$$f_{measured} = v_{ex}/(bA_{ex}) = \theta_{ex}/(bA_{ex}\rho_b) \quad . \qquad (20)$$

It is clear from equation (19) that for anions Q in equation (3) can be replaced by $-\theta_{ex}c$. This enables the replacement of $\theta(z,t)$ by:

$\theta'(z,t) = \theta(z,t) - \theta_{ex}(z,t)$ which largely facilitates numerical computations.

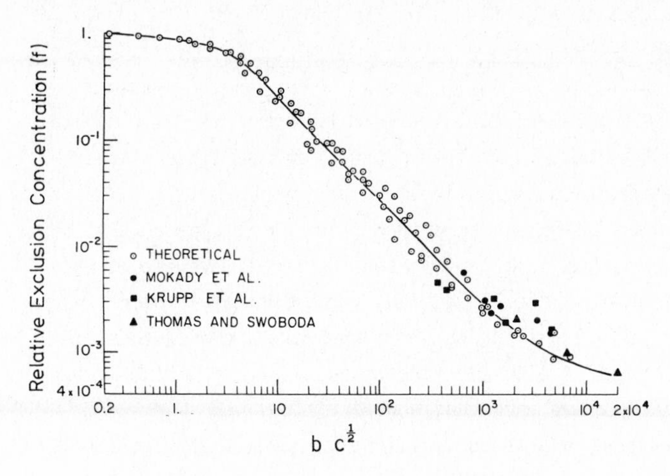

Fig. 7. Relative exclusion concentration (f) as a function of b $c^{\frac{1}{2}}$.

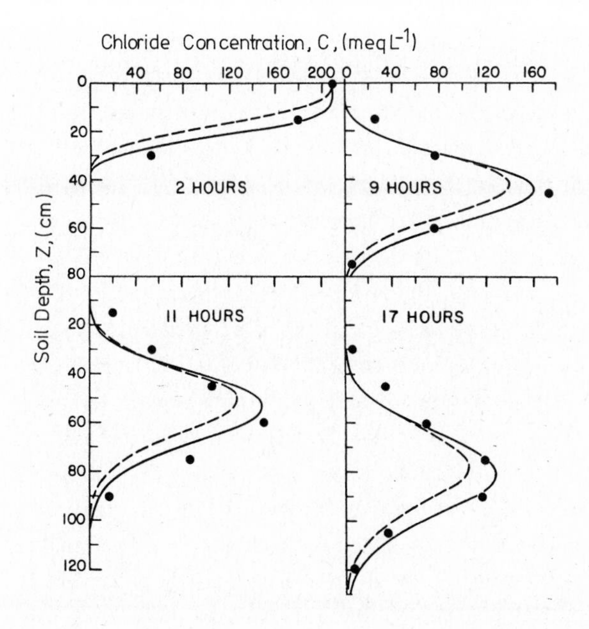

Fig. 8. Computed chloride distributions (solid and dashed lines) as compared with field data (black circles) for four infiltration times. Dashed lines denote modeling data for a non-interacting solute, whereas solid lines denote a model in which anion exclusion has been considered.

A comparison between results calculated from numerical solution of equation (3) which considers σ and θ_{ex}, and Warrick et al. (1971) data for 2, 9, 11, and 17 hours of infiltration is given in Fig. 8.

Figs. 2 and 7 were used to model $\sigma(c, \theta)$ and $Q(c, \theta)$ or $\theta_{ex}(c, \theta)$, respectively Values of $Q(z,t)$ or $\theta_{ex}(z,t)$ were calculated from equation (19). Values of $\lambda\theta_s$ = 0.55 cm were assigned. Similar results were obtained for $\lambda\theta_s$ = 0.28 and 0.39. A slight improvement is evident in the agreement between theoretical predictions and measured chloride distributions.

In analyzing the effect of ion-soil interaction (Fig. 8) it appears that salt concentration gradients are not an important factor in normal movement of the soil solution. It remains likely, therefore, that the slight differences between the pairs of calculated lines in Fig. 8 may be more a consequence of anion exclusion effects than of osmotic flow processes. Only about 10% of the chloride moving in this profile was excluded, however. Thus, the agreement between calculated and measured data was only slightly improved by incorporating anion exclusion estimates. From analysis of other experiments one can also conclude that salt concentration gradients and anion exclusion are normally relatively minor factors in describing the movement of soil solution.

(ii) Mixed cationic (Na/Ca) systems. Mixed cation systems differ from previously discussed anionic systems in two main respects. First the value of Q in equation (3) governing solute flow is always positive, due to positive adsorption to the solid surface (as compared to $Q < 0$ and $Q = 0$ for the aforementioned cases of anions and non-interacting solutes, respectively). Second, due to the pronounced effect of the sodium to calcium ratio on the $K(\theta)$ and $h(\theta)$ functions, the partial differential equation which governs transport of water is obtained by combining equations (14) and (6) instead of equations (7) and (6). The parameters Q, K and h of the governing equation are affected by interactions between the cations and the unsaturated soil matrix. As before, it will be assumed that cations on the "exchanger phase" are in equilibrium with the contiguous solution, with exchange processes being virtually instantaneous. Anions and cations on the "external" surfaces of the "exchanger phase" will be assumed distributed according to the planar diffuse double layer theory. If the CEC and the specific surface area of the soil (S_p) are known, the surface charge density (Γ_s) of soil particles can be estimated and then used to calculate the distribution of diffuse-layer cations. This calculation is needed for modeling the functional relationships of $Q(\theta,R,C)$, $h(\theta,R,C)$ and $K^r(\theta,R,C)$. These in turn must be known in order to solve the governing water and solute flow equations and to eventually obtain the distribution of each cationic species m. Numerical solutions for the mixed cationic system can again be performed using conventional algorithms, as for the cases of flow involving non-interacting solutes.

As instantaneous equilibrium between solution and exchange phase is assumed during most transport models, it is possible to adapt the diffuse double-layer (DDL) approach to estimate Na-Ca exchange equilibria. Knowing the anion distribution as before and exchange equilibria, the value of Q for both Na and Ca can be obtained.

A comparison between the DDL-based exchange model and the mass action and Gapon-based models for mixed Na-Ca exchange equilibria in soils is given in Fig. 9. As most types of exchange-equilibria models give similar results, Magdoff and Bresler (1973) used the simple Gapon-type equation to predict the time-dependent distribution of ESP in a profile of initially sodic soil (initial ESP = 22) irrigated with water containing various concentrations of $CaCl_2$. Their results showed that such a modeling procedure is able to distinguish between treatments and to rank them in the same order of Na-replacement efficiency as for the experimental results. Moreover, for some of the treatments there was fairly good quantitative agreement between results calculated by the model and actual soil ESP profiles.

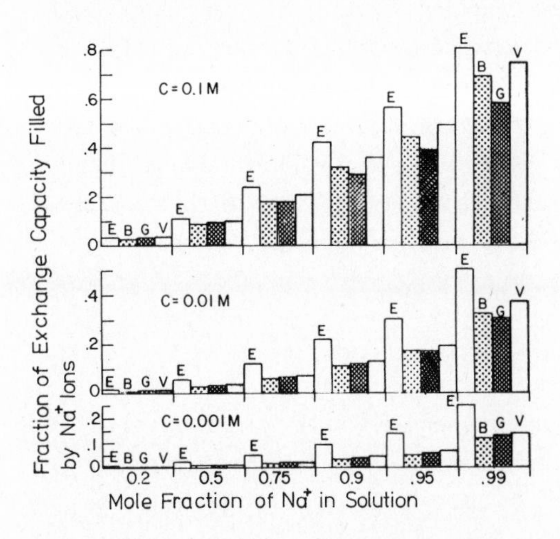

Fig. 9. Relationships between adsorbed ions and ions in the contiguous solution for clay in equilibrium with a mixed Na-Ca solution, as calculated by four different methods: (a) Erikson model, denoted by E; (b) Bresler model, denoted by B; (c) Gapon equation, with K_G = 0.146, denoted by G; and (d) Vanselow equation, with K_{Ca}^{Na} = 4.0, denoted by V.

2.6 Salt Dynamics and Distribution in Soils Under Plant Growing Conditions

As the amount of salt uptake by plant roots is generally small compared to changes in total solute concentration, it is reasonable to assume that salt uptake can be neglected in the presence of plants, analogous to the case of fallow soils. Assuming also that other solute sinks and sources are not present in the soil, the term S' in equation (3) can safely be neglected. Water uptake by plant roots is usually a dominant factor affecting simultaneous water and salt transport, however, and must therefore be considered when calculating salt distribution under plant growing conditions.

2.6.1 Models for water extraction by plant roots

Water uptake by plant roots can be represented by a volumetric, macroscopic water extraction term accounting for water loss during transpiration. This term S is simply added to the continuity equation (6). In one, vertical, dimension this leads to:

$$\partial\theta/\partial t = -(\partial q/\partial z) - S(z,t) \quad . \tag{21}$$

Estimation of salt distribution is carried out as before, after solving the governing water flow equation (21) numerically. This is provided, of course, that the required water extraction parameters can be estimated either empirically or theoretically.

A major difficulty in solving equation (21) to obtain values of $\theta(z,t)$ and $q(z,t)$ stems from the unknown form of the function $S(z,t)$. As a result of lack of physical understanding, there has been a tendency to describe water uptake by macroscopic equations which, by analogy to Darcy's equation assume the rate of water uptake by roots to be proportional to soil hydraulic conductivity $K(\theta)$ and to the difference between total pressure head at the root-soil interface (i.e., $\psi_r^t = h_r + \Pi_r/\rho g$) and corresponding pressure head $\psi^t = h + \Pi$ of the soil. According to this approach the sink term S entering into equation (21) can be expressed as:

$$S(z,t) = - [1/b(z,t)]\ K[\theta(z,t)]\ [\psi_r^t(z,t) - h(z,t) - \Pi(z,t)/\rho g] \tag{22}$$

where S is the volumetric rate of water uptake per unit bulk volume of soil. The coefficient of proportionality 1/b represents the geometry of the flow path. If one is willing to continue the analogy to Darcy's law, it is reasonable to assume that 1/b is proportional to the specific area of the soil-root interface (total surface area of roots per unit bulk volume of soil) and inversely proportional to the impedance (thickness divided by the hydraulic conductivity) of the soil-root-interface. Note from equation (22) that b must have units of length squared.

The root extraction term S in equation (22) also includes the effect of soil osmotic pressure on plant water uptake. When the osmotic pressure head Π is low, it may not be possible for the plant to extract enough water to meet potential transpiration and growth reduction may occur. In fact, actual transpiration will be less than potential transpiration if both h and Π are relatively low, for these two water pressure terms have additive effects on water extraction patterns.

Similar to the case of potential evaporation from fallow soils, the potential rate of transpiration by plants (E_p) is also dependent on atmospheric conditions. The potential rate of transpiration is assumed to be equal to the maximum possible rate of water extraction by all roots of the crop per unit horizontal area of the soil. This quantity can be calculated according to Feddes et al. (1974) from:

$$E_p = \frac{\delta(R_n - G)/L + \rho_a C_p (e_z^* - e_z)/r_a - \delta R_n e^{-0.39LAI}}{(\delta + \gamma)L} \tag{23}$$

where R_n is net radiation flux; δ is the slope of the saturation vapor pressure curve; LAI is leaf area index of the crop; G is heat flux into the soil; ρ_a is density of moist air; C_p is specific heat of air at constant pressure; e_z^* and e_z are unsaturated and saturated water vapor pressures, respectively, at elevation z and ambient temperature; γ is psychrometric constant; L is latent heat of vaporization; and r_a is resistance to vapor diffusion through the air layer around leaves.

The approach to be used in solving both for solute flow and water flow during the infiltration, redistribution and evapotranspiration stages is virtually identical to that discussed earlier for the fallow-field case. The major modification to be made occurs during the evapotranspiration stage, because water flow changes due to the presence of plants significantly affect solute distribution. During this stage, the rate of water uptake by plants depends both on atmospheric conditions (with the potential rate of transpiration given by equation (23)) and on below-ground (soil) conditions as before.

With this approach, different crops of unequal rooting depth can be treated simultaneously. In addition, the rooting depth can be allowed to vary with time simply by assigning new values to the function b(z) at any given time step. One can approximate changes in root development during the growing period, provided only that b as a function of plant growth stage and time is known for any depth.

To apply water extraction models to the question of salt distribution profiles under plant growing conditions, one must first examine the agreement between model calculations and actual fluctuations in soil water content over time. In this respect, the dimensionless scaling factor $\alpha_r = \lambda_r/\lambda^*$ applies (in which λ_r is the microscopic value of a selected parameter for a particular soil sample and λ^* is some reference value of λ). Peck et al. (1977) used α as a single parameter which then allowed them to approximate $h(\theta)$ and $K(\theta)$. As a consequence of their scaling theory, the pressure head h_r and the hydraulic conductivity K_r at a given water content θ and location r in the field is related to the respective average values h^* and K (averaged over an entire field) by :

$$h_r = h^*/\alpha r$$

$$K_r = \alpha_r^2 K^* \tag{24}$$

Warrick et al. (1977) found that the distribution of α_r is approximately log-normal, with some 70 to 95% (depending on the soil series and on the data used for obtaining α_r estimates) of α_r values in the range 0.2 to 2.5.

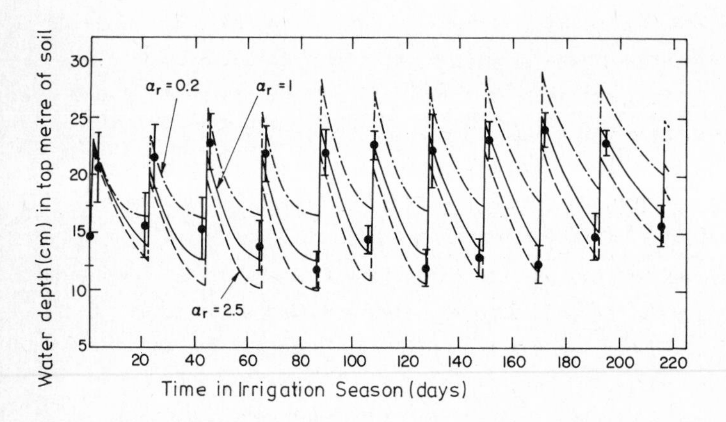

Fig. 10. Computations of changes with time in soil water content for the 1 m rooting depth of a citrus orchard (continuous lines), compared to measured data (black circles with vertical bars) for three α values. Vertical bars denote standard deviations of the measured data.

A comparison of measured and computed water contents in soil profiles from z = 0 to z = 100 cm, and of changes with respect to time for the entire irrigation season, is given in Fig. 10. The experimental data represent average values (the circles) and standard deviations (the vertical bars) obtained from six replications with three to four repeated measurements in each instance with the neutron probe method. Results of model computations (the continuous solid and broken lines) were obtained from the water extraction model employing three values of α_r (i.e. α_r = 0.2, α_r = 1, and α_r = 2.5) to account for the main effect of spatial variability in field soils. When considering the spatial variability of measured values (as indicated by the size of the vertical bars) and of calculated soil hydraulic parameters (as indicated by the distance between the three lines in Fig. 10) there appears to be good agreement between field water content data and time-course water content data calculated from the root extraction field model. In general, average observed values are close to values computed with α_r = 1. The range of standard deviations in the field data is generally within the range of modeling results computed with α_r = 0.2 and with α_r = 2.5.

The generally good agreement between computed and measured time variations in field water content (Fig. 10) indicate that the water extraction model should be applicable to the simulation of field salt distribution data under plant growing conditions. Experimental chloride profiles measured in an irrigated citrus field are compared in Fig. 11 with computed data. In this case, $q(z,t)$ and $\theta(z,t)$ were calculated using the scaled α_r = 1, and 4 values of λ. The variability of the field measurements with which the modeling comparison is being made is illustrated by the standard deviation of the measurements, made at 30 cm intervals, as given by the horizontal portions of the rectangles drawn in Fig. 11.

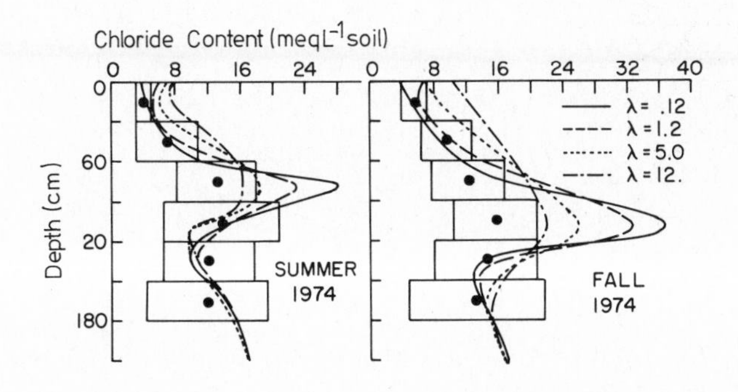

Fig. 11. Measured data and results computed from the model (solid, dashed, dotted, and dot-dashed curves) with α_r = 1 and using four different values of dispersivity (λ in cm).

The degree of accuracy that one obtains in estimating salt distribution profiles under crop growing conditions depends primarily on factors such as (a) proper choice of a mathematical model to adequately describe the physical system at hand. This includes proper specification of boundary conditions; (b) accuracy and stability of the numerical procedure, and (c) accuracy in the estimation of soil and plant parameters used for the computations, including treatment of their field variability. The most critical factor is the last one.

Because of the large effects of field heterogeneity (section 3) and since difficulties involved in obtaining all necessary soil-water-plant-salt measurements required for the above-mentioned models, a simplified modeling procedure seems appropriate. Such a model was described by Bresler (1967). This model entails numerical solution of equation (17) with $D(\theta,V) = 0$, and with the time increment Δt being set equal to the time interval between successive water applications (j-1 and j). Furthermore, only downward water and solute flow was considered, with such flow assumed to take place in the range of water contents ($\bar{\theta}$) between saturation and the assumed "field capacity" for that soil. In addition the amount of water passing any soil depth $z \neq 0$ at t^j (i.e. $q(z)\Delta t^j$) was estimated as the difference between the amount of water applied when runoff does not take place ($\bar{Q}^j = \bar{q}(0,j)\Delta t^j$) and the water consumption by the crop from the soil surface down to soil depth z in time interval Δt^{j-1}.

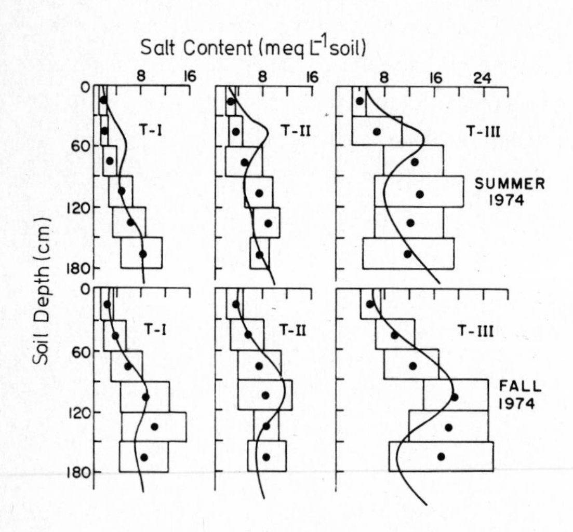

Fig. 12. Measured data (as in Fig. 11) and results computed from a simple model with $\alpha_r = 1$ (solid curves).

A comparison of field results for the chloride ion, as obtained from citrus irrigation experiments, with those calculated according to the more crude model is shown in Fig. 12. Note that in this figure the experimental data of T - III (averages and standard deviations) are identical to those of Fig. 11. The calculations producing the lines in Fig. 12 were performed using the assumption that $\bar{\theta} = 0.33$, with $\bar{S}_1^j$ and $\bar{Q}^j$ being calculated by the previously described water extraction model. The model was subject to the experimental boundary and initial conditions as actually measured in the field.

Examination of the results in Figs. 11 and 12 suggests that little is gained by the more sophisticated model (Fig. 11) compared to the more approximate one (Fig. 12). Salt distribution results were not appreciably affected by the larger time increments, or by the use of average values for the amount of leaching water and for the water content $\bar{\theta}$ at which leaching takes place.

In view of the results given in Figs. 11 and 12 it seems that both models described above constitute useful tools for the purpose of approximately estimating and predicting the accumulation and leaching of salts under plant growing conditions in a homogeneous field. As such, they may be applied as part of management-oriented models for control of soil salinity in the field, or for economic analysis of salinity in agriculture.

3 SALT DISPERSION AND DISTRIBUTION IN HETEROGENEOUS FIELDS

So far we have described water and solute transport by using macroscopic quantities which vary in a deterministic manner, obey physical and chemical laws, and are expressed in the form of partial differential equations. To solve these equations we have usually assumed that the soil is a homogeneous porous medium and that the flow parameters are uniform throughout the entire field. In reality, however, fields - unlike small laboratory soil columns - are seldom homogeneous, they generally display large variations so that their hydraulic properties vary from place to place in the field, i.e., they are heterogeneous. We can therefore regard an actual field as a collection of homogeneous vertical columns differing in their soil properties. As a result, the solute transport through the soil, under a given set of boundary conditions, will differ from profile to profile depending on the local soil properties.

A recent contribution to the development of models with statistically independent hydraulic parameters and random infiltration rate is the work of Dagan and Bresler (1979) and Bresler and Dagan (1979). In our work the problems of vertical solute transport caused by steady infiltration in unsaturated soils, is considered. To arrive at a simple model the following assumptions have been adopted: (i) the flow is vertical; (ii) the soil properties do not change along any vertical profile but vary considerably in the horizontal plane; (iii) this variation can be described in statistical terms; and (iv) the statistics are homogeneous. In addition, a few simplifying assumptions have been forwarded for a particular, but representative case; (v) the hydraulic conductivity-water content relationship $K(\theta)$ has a simple analytical structure; (vi) the randomness of K stems from the stochastic nature of the saturated conductivity K_s which depends quadratically on the scaling parameter (equation (24)) which in turn has a lognormal distribution (i.e. $\Upsilon = \ln\alpha$ is normal with mean m_Y and variance σ_Y); (vii) the flow is generated by steady recharge applied on the surface at a deterministic rate R; (viii) the water flow is steady so that the infiltration velocity V and the soil water content θ do not change with depth and time; (ix) the dimensionless concentration C is initially zero throughout the profile and is equal to unity at the soil surface ($z = 0$) for any infiltration time t.

3.2 Piston Flow Profiles

3.2.1 Concentration distribution for steady leaching

In field problems in which pore scale dispersion D is insignificant, an approximation of practical value can be described by neglecting D in equation (4) and solving the piston flow equation for any $V = q/\theta$ in a closed form by the method of characteristics. Thus, $C = (c-c_n)/(c_n-C_0)$ is constant for:

$$(dz/dt)_C = V(z,t) \tag{25}$$

and for steady uniform flow C is propagated along fronts of $z = Vt =$ constant. For the boundary and initial conditions leading to equation (5) the value of C from (25) is given by:

$$C(z,t) = H(Vt-z) \tag{26}$$

where H is the Heaviside step function, (i.e., $H(x) = 0$ for $x < 0$, $H(x) = 1$ for $x \geqslant 0$). Equation (26) represents a sharp front separating $C = 1$ and $C = 0$ which moves downwards at constant velocity V, whereas in equation (4) dispersion "smears" this front over a region of certain length.

For a particular case of uniform soil in the vertical direction with steady rate of deterministic recharge IR and steady gravitational flow with constant θ and V in each profile, two situations may occur: (a) for that portion in the field in which $IR \geqslant K_s$ (i.e., $IR \geqslant K_s^{*}\exp(2Y)$) :

$$V = K_s/\theta_s = K_s^{*}\exp(2Y)/\theta_s \tag{27}$$

(b) for $IR < K_s$ (i.e. $IR < K_s^{*}\exp(2Y)$) the profile is unsaturated and :

$$V = K(\theta)/\theta = IR/\theta = IR\exp(2\beta Y)/\theta_s(IR/K^{*})^{\beta} \tag{28}$$

Here, we adapt for $K(\theta)$ the relationship :

$$K/K_s = (\theta/\theta_s)^{1/\beta} \tag{29}$$

where θ_s and K_s are values at saturation. We assume that θ_s and β are deterministic and constant and the randomness due to field variability is associated entirely with K_s.

Finally, for K_s we assume that there exists a scaling parameter, α (equation (24)) which has a lognormal distribution, i.e.:

$$f(Y) = \frac{1}{\sigma_Y\sqrt{2\Pi}} \exp\left[-\frac{(Y-m_Y)^2}{2\sigma_Y^2}\right] \tag{30}$$

where $Y = \ln\alpha$, with $m_Y = -0.616$ and $\sigma_Y = 1.16$ for Panoche soil. Because of the random nature of K_s, and therefore of V, the concentration $C(z,t)$ satisfying equation (26) is also a random function $C(z,t;\,K_s)$ of the parameter K_s. Hence, C cannot be predicted deterministically as in laboratory column experiments, but only in terms of its conditional probability which depends on K_s through equations (27) and (28).

We cannot, therefore, answer the question, "what is the magnitude of C in the field at x,y,z,t" but only, "what is the probability of C taking values between A and A + dA, or the probability that $C \leqslant A$, at a depth z and time t".

By the various simplifications adopted so far, we have reduced the entire statistical description of $C(z,t)$ to the frequency function of K_s, or V, which is given in turn by the three statistical constant parameters K_s^*, m_Y and σ_Y. Hence, V depends on the deterministic parameters K_s^*, β and θ_s and is a random function of the variable Y. The cumulative probability distribution of V can be computed from :

$$P(V) = \int_{-\infty}^{Y} f(Y)\, dY \qquad (31)$$

with $f(Y)$ given by equation (30). In the simple piston flow model, in which C depends on V through equation (26), for given values of z and t $C = 1$, or $C = 0$ depending on whether $V > z/t$ or $V < z/t$, respectively. The cumulative probability of C is, therefore, given by :

$$P(C = 1-\varepsilon) = P(V \leq \frac{z}{t}); \quad P(C = \varepsilon) = 1-P(V \leq \frac{z}{t}); \quad (\varepsilon \to 0) \quad . \qquad (32)$$

In particular, the average concentration over the field is given by the first moment :

$$\overline{C}(z,t) = \int_0^1 C\, f(z,t;C)dC = \int_0^1 C \frac{dP}{dC}\, dC = CP\big|_0^1 - \int_0^1 PdC = 1-P(C) = 1-P(V) \qquad (33)$$

where $P(V)$ is calculated from equation (31) using equation (30) and $V = z/t$. The value of calculated $\overline{C}$ exhausts the statistical information on the solute concentration C, it represents the ratio, for a given t and at a fixed z, between the area in the x,y plane for which $C = 1$ and the total area.

For convenience in applying the model to a real heterogeneous field, the average concentration function $\overline{C}(z,t;\ K_s^*,\ \theta_s,\ \beta,\ m_Y,\ \sigma_Y,\ IR)$ is cast in a dimensionless form by adapting the following variables: $\xi = z\theta_s/t\ K_s^*$ and $r = IR/K_s^*$. Fig. 13 gives calculations of $C(\xi)$ with K_s^*, θ_s, m_Y and σ_Y of Panoche soil (Warrick et al., 1977) with a value of $1/\beta = 7.2$ (Bresler et al., 1978). The three curves in Fig. 13 represent three values of deterministic $IR = r \times K_s^*$. For the concentration profile which lies between $\xi = 0$ and $\xi = r$ solute transport is controlled by the saturated flow beneath the ponded area. For the zone in which $\xi > r$ the $C(\xi)$ profile is controlled by the unsaturated flow. The three $C(\xi)$ curves of Fig. 13 differ mainly in the location of the breaking points at $\xi = r$ and in the spreading of the solute which is larger for larger values of r. Fig. 13 can be applied in order to estimate the percentage of the field which has been leached to a depth z after a time t of leaching. For example, for Panoche soil with $r = 1$ ($IR = 0.25$ cm hour^{-1}) 30% of the field is leached to a depth $z = 0.5$ m after a time of $t = 100$ hours has elapsed from the beginning of the leaching process.

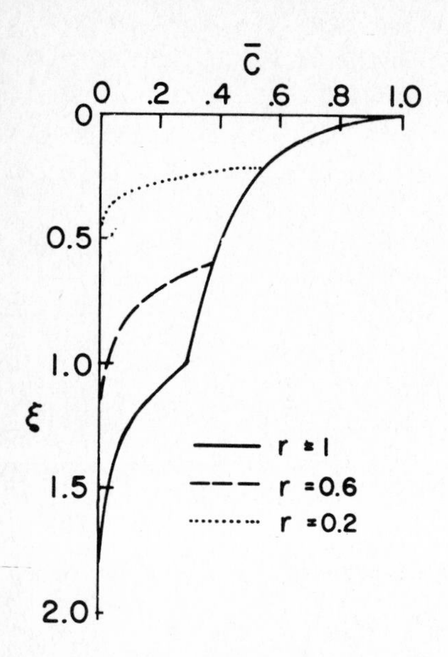

Fig. 13. The average concentration profile ($\overline{C}$) as a function of $\xi = z\theta_s/tK_s^*$ for three values of $r = IR/K_s^*$.

3.2.2 Solute distribution for transient leaching

Equation (9) governing nonsteady water flow can be solved for V at any location i in the field provided that $K^i(\theta)$ and $h^i(\theta)$ are known. Substituting $V^i(z,t)$ into equations (25) and (26) gives $C^i(z,t)$ at any location i = 1,2, ..., N in the field. Using the same semi-infinite boundary conditions for solute flow as before (equation (5)) average concentration profiles over the entire field $\overline{C}(z,t)$ can be calculated. Such $\overline{C}(z,t)$ calculations have been performed for a field in Israel by counting the fraction of sites of the N = 30 sites in the field through which the salt concentration front, separating the zone where C = 1 from the zone where C = 0, have been passed the depth z in time t. These values of $\overline{C}$ (Fig. 14) represent precisely the ratio for a given t and at a fixed z, between the area in the field for which C = 1 and the total area. The results demonstrated in Fig. 14 are the basis of application of water and salt flow models to leaching of salts in heterogeneous field soils. From the data of Fig. 14 one can obtain the length of time for the leaching process to be complete to a given depth and for a given portion of the entire field. For example, 50% of the field has been leached to a depth of .25 m after 4.2 hours of leaching at a rate of 0.015 m h^{-1}. This time is equivalent to 0.062 m of leaching water. At the same time and water quantity, only 14% of the field has been leached to .35 m. Similarly, the leaching process is completed throughout the entire field to a depth of .35 m only after about 11 hours of irrigation and the amount of water needed for this complete process is as much as .50 m.

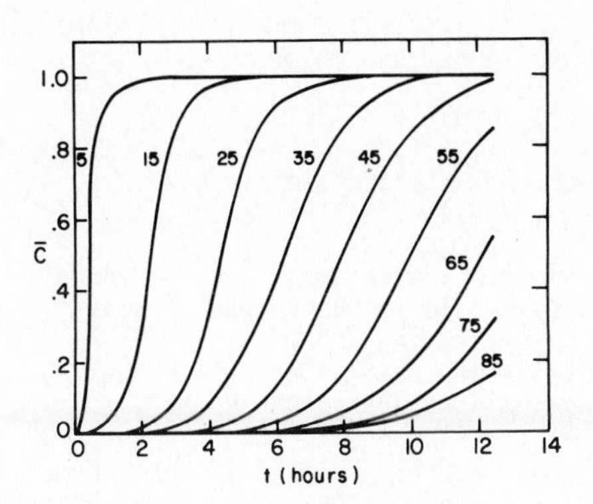

Fig. 14. Average concentration ($\bar{C}$) as a function of time (t) and depth z in cm (the numbers labeling the curves).

4 REFERENCES

Biggar, J.W. and Nielsen, D.R., 1967. Miscible displacement and leaching phenomena. In: Irrigation of Agricultural Lands (Editor R.M. Hagen), Agron. 11: 254-274. Am. Soc. Agron: Madison, Wis.

Bresler, E., 1967. A model for tracing salt distribution in the soil profile and estimating the efficient combination of water quality and quantity under varying field conditions. Soil Sci.,104: 227-233.

Bresler, E., 1970. Numerical solution of the equation for interacting diffuse layers in mixed ionic system with non-symmetrical electrolytes. J. Colloid Interface Sci.,33: 278-283.

Bresler, E., 1972. Interacting diffuse layers in mixed mono-divalent ionic systems. Soil Sci. Soc. Am. Proc.,36: 891-896.

Bresler, E., 1973a. Simultaneous transport of solute and water under transient unsaturated flow conditions. Water Resour. Res.,9: 975-986.

Bresler, E., 1973b. Anion exclusion and coupling effects in non-steady transport through unsaturated soils. I. Theory. Soil Sci. Soc. Am. Proc.,31: 663-669.

Bresler, E. and Laufer, A., 1974. Anion exclusion and coupling effects in non-steady transport through unsaturated soils. II. Laboratory and numerical experiments. Soil Sci. Soc. Am. Proc.,38: 213-218.

Bresler, E., Russo, D. and Miller, R.D., 1978. Rapid estimate of unsaturated hydraulic conductivity function. Soil Sci. Soc. Am. J.,42: 170-172.

Bresler, E. and Dagan, G., 1979. Solute dispersion in unsaturated heterogeneous soil at field scale: II. Applications. Soil Sci. Soc. Am. J.,43: 467-472.

Dagan, G. and Bresler, E., 1979. Solute dispersion in unsaturated heterogeneous soil at field scale. I. Theory. Soil Sci. Soc. Am. J., 43: 461-467.

Feddes, R.A., Bresler, E. and Neuman, S.P., 1974. Field test of a modified numerical model for water uptake by root systems. Water Resour. Res.,10: 1199-1206.

Krupp, H.K., Biggar, J.W. and Nielsen, D.R., 1972. Relative flow rates of salt and water in soil. Soil Sci. Soc. Am. Proc.,36: 412-417.

Magdoff, F. and Bresler, E., 1973. Evaluation of methods for reclaiming sodic soils with $CaCl_2$. In: Physical Aspects of Soil Water and Salt in Ecosystems. Ecological Studies, Springer Verlag: Berlin 4: 441-452.
Mokady, R.S., Ravina, J. and Zaslavsky, D., 1968. Movement of salt in saturated soil columns. Israel J. Chem.,6: 159-165.
Olsen, S.R. and Kemper, W.D., 1968. Movement of nutrients to plant roots. Adv. Agron.,30: 91-151.
Peck, A.J., Luxmoore, R.J. and Stolzy, J.L., 1977. Effects of spatial variability of soil hydraulic properties in water budget modeling. Water Resour. Res., 13: 348-354.
Russo, D. and Bresler, E., 1977a. Effect of mixed Na/Ca solutions on the hydraulic properties of unsaturated soils. Soil Sci. Soc. Am. J.,41: 713-717.
Russo, D. and Bresler, E., 1977b. Analysis of the saturated-unsaturated hydraulic conductivity in a mixed Na/Ca soil system. Soil Sci. Soc. Am. J., 41: 706-710.
Russo, D. and Bresler, E., 1980. Soil-water-suction relationships as affected by soil solution concentration and composition. In: A. Banin and U. Kafkafi (Editors), Agrochemicals in Soils, Pergamon Press, pp. 287-297.
Segol, G., 1977. A three-dimensional Galerkin finite element model for the analysis of contaminant transport in saturated-unsaturated porous media. In: W.G. Gray and G.F. Pinder (Editors), Finite Elements in Water Resources, Pentech Press, London, pp. 2.123-2.144.
Thomas, G.W. and Swoboda, A.R., 1970. Anion exclusion effects on chloride movement in soils. Soil Sci.,11: 163-166.
Warrick, A.W., Biggar, J.W. and Nielsen, D.R., 1971. Simultaneous solute and water transfer for an unsaturated soil. Water Resour. Res.,7: 1216-1225.
Warrick, A.W., Mullen, G.J. and Nielsen, D.R., 1977. Scaling field-measured soil hydraulic properties using similar media concepts. Water Resour. Res., 13: 355-362.

RESIDENCE TIMES OF WATER AND SOLUTES WITHIN AND BELOW THE ROOT ZONE

P.A.C. RAATS
Institute of Soil Fertility, Postbus 30003, 9750 RA Haren (Gr.), The Netherlands

ABSTRACT

Raats, P.A.C., 1981. Residence times of water and solutes within and below the root zone. Agric. Water Manage., 1981.

The progress of a parcel of water or solute in the course of time can be determined by integrating its speed along its path. This basic information can be used to describe the fate of collections of parcels of water forming a surface or occupying a region and to formulate input/output relationships characterizing transport across a region. It is shown that within the root zone the speed of the water or solute depends primarily on the components of the overall water balance, the average water content, and the distribution of the water uptake. Particular attention is given to recent attempts to infer water uptake from salinity data. Transport to drains, ditches, or streams induced by an input distributed uniformly over the soil surface is discussed in detail. If the ratio of the half-spacing between drains, ditches, or streams and the depth to the impermeable layer is larger than about five, then i) the isochrones are horizontal, except close to the outlets, and ii) the transit time density distribution is approximately exponential, i.e., the system approximates an apparently well-mixed system. Methods for determining transit time density distributions for more complicated flow patterns are discussed briefly. Estimates are also given for the retardation due to adsorption, for the influence of reactions, and for, the often small, influence of dispersion.

I INTRODUCTION

Traditionally the main concern of research related to water management has been to determine the quantities of water being transported and the distribution of pressure head and water content. But lately the quality of water is of at least as much interest. One possible approach to the management of water quality is to split the problem in two parts:

* find the space-time trajectories of parcels of water;

* determine the changes in quality of these parcels.

The space-time trajectories of parcels of water can be calculated from:

$$t-t_o = \int_{s_o}^{s} v^{-1} ds, \qquad (1)$$

where $t-t_o$ is the time it takes for a parcel to travel from s_o to s; and v is the speed along its path. The speed v can, in principle, be determined by solving the appropriate flow equation.

The second objective has many aspects:

* the change in the concentration of a parcel of water due to evaporation at or near the soil surface and due to selective uptake of water by plant roots;

* the gain or loss of solutes by parcels of water as a result of diffusive and dispersive mixing with their surroundings;

* the retardation of solutes relative to the water resulting from adsorption;

* changes due to precipitation or dissolution and due to decay or production.

At any point in the soil, the balance of mass for the water may be written as:

$$\partial\theta/\partial t = -\nabla.(\theta \underset{\sim}{v}) - \lambda T, \qquad (2)$$

where t is the time; ∇ is the vector differential operator; θ is the volumetric water content; $\underset{\sim}{v}$ is the velocity of the water; T is the rate of transpiration; and λ is the spatial distribution function for the uptake of the water. The flux, $\theta\underset{\sim}{v}$, of the water is given by Darcy's law:

$$\theta\underset{\sim}{v} = -k\nabla h + k\nabla z, \qquad (3)$$

where h is the tensiometer pressure head; k is the hydraulic conductivity; and z is the position in the gravitational field.

Also at any point in the soil, the balance of mass for a solute may be written as:

$$\frac{\partial}{\partial t}\theta c = -\nabla.\underset{\sim}{F}_s - \frac{\partial}{\partial t}\mu_a - \frac{\partial}{\partial t}\mu_f - \lambda_s N, \qquad (4)$$

where c is the concentration of the solute in the aqueous phase; $\underset{\sim}{F}_s$ is the flux of the solute; μ_a and μ_f are the densities of the solute per unit volume in the adsorbed and immobile phases; N is the rate of uptake by plant roots; and λ_s is the uptake distribution function. The flux $\underset{\sim}{F}_s$ is assumed to be the sum of a convective component $\theta\underset{\sim}{v}c$ and a diffusive component $-\mathcal{D}\nabla c$:

$$\underset{\sim}{F}_s = \theta\underset{\sim}{v}c - \mathcal{D}\nabla c. \qquad (5)$$

Combining equations (4) and (5) and using (2) gives:

$$\frac{\partial}{\partial t}c + \underset{\sim}{v}.\nabla c = \frac{\partial}{\partial t}c\Big|_{\text{parcel of water}}$$

$$= \left\{ \lambda Tc + \nabla . \mathcal{D} \nabla c - \frac{\partial}{\partial t} \mu_a - \frac{\partial}{\partial t} \mu_f - \lambda_s N \right\} / \theta . \tag{6}$$

On the left hand side of equation (6) appears a material time derivative, i.e., a time derivative following the motion of a parcel of water. On the right hand side appear five possible causes for change of the concentration of a parcel of water. They are all proportional to the dryness θ^{-1}, implying that for a given cause the absolute value of the rate of change of c is largest in sands, intermediate in loams, and smallest in clays.

In using equation (1) to describe the space-time trajectories of parcels of water and equation (6) to describe the change of the concentration of these parcels, I emphasize one of many possible approaches to the analysis of simultaneous transport of water and solutes. My aim is to demonstrate that this approach gives good qualitative insight in the fate of solutes within the root zone and in the region between the water table and the drains, ditches or streams.

2 THE ROOT ZONE

2.1 Depth-time Trajectories for Downward Flow of Parcels of Water

Within the root zone the time averaged velocity of the water at depth z is found by integrating equation (2) (Raats, 1975):

$$v = \theta v / \theta = \left\{ R+I-E-T \int_0^z \lambda dz \right\} / \theta = \left\{ D/T + \int_z^\infty \lambda dz \right\} T/\theta , \tag{7}$$

where R is the rate of rainfall; I is the rate of irrigation; E is the rate of evaporation from the soil surface; T times the integral of λ from o to z represents the cumulative rate of uptake above depth z; and D = R+I-E-T is the rate of drainage. Below the root zone $\int_0^z \lambda dz = 1$ and the velocity v approaches the constant value:

$$v = v_\infty = (R+I-E-T)/\theta = D/\theta . \tag{8}$$

Introducing equation (7) into equation (1) gives:

$$t - t_i = \int_{z_i}^{z} v^{-1} dz = (\theta/T) \int_{z_i}^{z} \left\{ D/T + \int_z^\infty \lambda dz \right\}^{-1} dz . \tag{9}$$

Equation (9) describes depth-time trajectories of parcels of water starting at time t_i at depth z_i. Below the root zone the trajectories approach straight line asymptotes with a slope equal to the velocity D/θ defined by equation (8). The intercept of these asymptotes with the z-axis is given by:

$$d_i = z_i + \int_{z_i}^{\infty} (1 - v_\infty / v) dz . \tag{10}$$

Introducing (7) into (10) shows that d_i can be expressed entirely in terms of z_i, D/T and λ:

$$d_i = z_i + \int_{z_i}^{\infty} \left\{ \int_z^{\infty} \lambda dz \right\} \left\{ D/T + \int_z^{\infty} \lambda dz \right\}^{-1} dz. \tag{11}$$

If $\lambda = o$ for $z \geqslant \delta$ then δ may be regarded as the rooting depth. If δ is finite then equation (9) implies:

$$t_\delta - t_i = \int_{z_i}^{\delta} v^{-1} dz = (\theta/T) \int_{z_i}^{\delta} \left\{ D/T + \int_z^{\delta} \lambda dz \right\}^{-1} dz, \tag{12}$$

while equation (11) reduces to:

$$d_i = \delta - v_\infty (t_\delta - t_i). \tag{13}$$

If $z_i = z_o = 0$ then the time interval $t_\delta - t_i = t_\delta - t_o$ represents the residence time in the root zone. Fig. 1 shows a graphical interpretation of equation (10). At any depth the integrand $(1 - v_\infty / v)$ is the fraction of the flux which will be taken up below that depth.

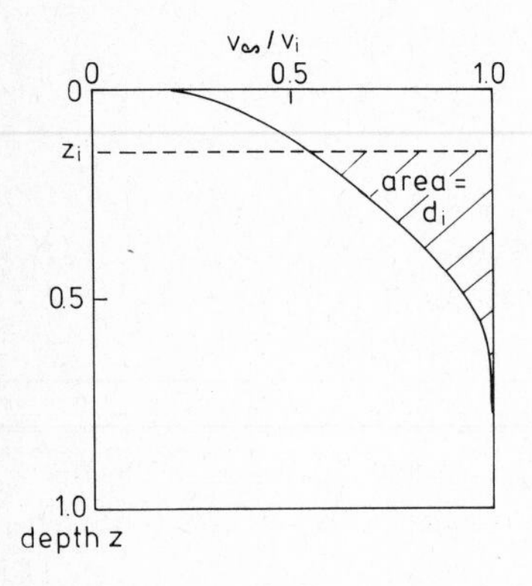

Fig. 1. Graphical interpretation of equation (10)

The second terms on the right hand sides of equations (10) and (11) represent the cumulative displacement induced by water adsorbed by roots at some distance below the soil surface. A large rooting depth induces a relatively rapid leaching of the soil solution over a large depth even if the rate of drainage is small. The role of uptake of water by plant roots in restricting the salinity near the soil surface was already clearly understood by Hilgard and Loughridge (1895, 1906). Fig. 2 shows the distribution in March at the end of the wet season. Most roots of the native spring growth of herbs and flowers were found in the top 40 cm. Near the end of the dry season in September the distribution of the salts had hardly changed from that shown in Fig. 2.

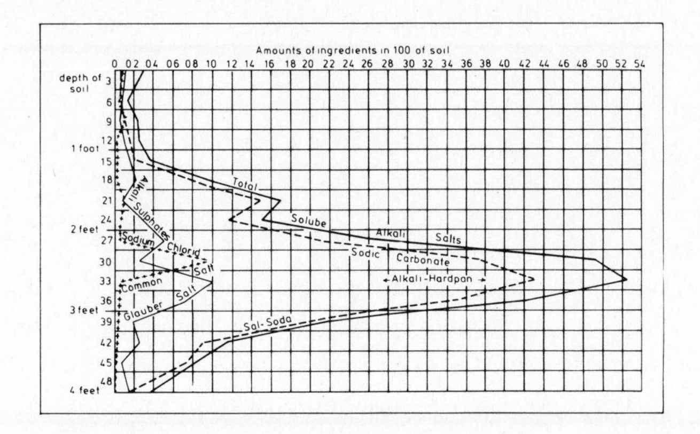

Fig. 2. Distribution of salts in March 1895 for unirrigated, black alkali land of Tulare Experiment Substation, California (after Hilgard and Loughridge, 1895, 1906).

Ancient Mesopotamian farmers may already have profited from water extraction over large depth. Jacobson and Adams (1958) wrote:

"In spite of almost proverbial fertility of Mesopotamia in antiquity, ancient control of the water table was based only on avoidance of overirrigation and on the practice of weed-fallow in alternate years. As was first pointed out by J.C. Russel, the latter technique allows the deep-rooted *shoq (Proserpina stephanis)* and *agul (Alhagi maurorum)* to create a deep-lying dry zone against the rise of salts through capillary action. In extreme cases longer periods of abandonment must have been a necessary, if involuntary, feature of the agricultural cycle. Through evapotranspiration and some slow draining they could eventually reduce an artifically raised water table to safe levels".

2.2 Change of the Solute Concentration of Parcels of Water

If only changes in solute concentration due to selective uptake of water by plant roots are considered then equation (6) reduces to:

$$\partial c/\partial t + v\partial c/\partial z = \left.\frac{dc}{dt}\right|_{parcel} = \frac{\lambda T}{\theta} c. \tag{14}$$

Thus the rate of increase of the concentration is proportional to the rate of uptake of water λT, the dryness θ^{-1}, and the current concentration of the parcel of water. Integration of equation (14) gives:

$$c_{parcel} = c_i \exp \int_0^t (\lambda T/\theta)dt,$$

$$= c_i \exp \int_{z_i}^{z} (\lambda T/\theta v)dz,$$

$$= c_i \exp \int_{z_i}^{z} \lambda \left\{D/T + \int_z^\infty \lambda dz\right\}^{-1} dz, \qquad (15)$$

where c_i is the concentration at time t_i.

The dependence on the initial solute distribution enters in equations (9) and (15) through (t_i, z_i, c_i). Setting $z_i/\delta = 0$ in equation (9) and replacing in the resulting equation = by < defines the region in which the distribution of the concentration at time t is determined by the time course of the concentration at the soil surface for times $t > t_i$:

$$t-t_i < \int_0^z v^{-1} dz = (\theta/T) \int_0^z \left\{ D/T + \int_z^\infty \lambda dz \right\}^{-1} dz. \qquad (16)$$

In particular, if the concentration at the soil surface has a constant value c_o then in the region defined by the inequality (16) a time-invariant distribution of the concentration will have been reached. An explicit expression for this concentration profile valid for an arbitrary uptake distribution function λ is obtained by setting $c_i = c_o$ and $z_i = 0$ in equation (15):

$$c = c_o \exp \int_0^z \lambda \left\{ D/T + \int_z^\infty \lambda dz \right\}^{-1} dz. \qquad (17)$$

The steady profile for some ratio $(D/T)_i$ and uptake distribution λ_i may serve as an initial state for a transition to another ratio D/T and another uptake distribution λ. The space time trajectories will be given by:

$$t-t_o = (\theta/T)_i \int_{z_o}^{z_i} \left\{(D/T)_i + \int_{z_i}^\infty \lambda_i dz\right\}^{-1} dz$$

$$+ (\theta/T) \int_{z_i}^{z} \left\{D/T + \int_z^\infty \lambda dz\right\}^{-1} dz, \qquad (18)$$

and combining equations (15) and (17) will give:

$$c/c_o = \exp \left\{ \int_o^{z_i} \frac{\lambda_i}{(D/T)_i + \int_z^{\infty} \lambda_i dz} dz \right.$$

$$\left. + \int_{z_i}^{z} \frac{\lambda}{(D/T) + \int_z^{\infty} \lambda dz} dz \right\} \tag{19}$$

More generally, gradual changes of D/T and λ could be treated in a similar manner by incremental extensions of equations (18) and (19):

$$t - t_o = \sum_{n=1}^{N} (\theta/T)_n \int_{z_{n-1}}^{z_n} \left\{ (D/T)_n + \int_{z_n}^{\infty} \lambda_n dz \right\}^{-1} dz, \tag{20}$$

$$c/c_o = \exp \sum_{n=1}^{N} \int_{z_{n-1}}^{z_n} \lambda_n \left\{ (D/T)_n + \int_z^{\infty} \lambda_n dz \right\}^{-1} dz. \tag{21}$$

2.3 Results for Specific Uptake Distributions

The theory presented above applies to any uptake distribution function λ. In the literature various aspects of the theory have been worked out in detail for two special cases:

1. Step function uptake distribution:

$$\lambda = \delta^{-1}, \quad o < z < \delta, \tag{22}$$

$$\lambda = 0, \quad z \geqslant \delta, \tag{23}$$

where δ is the rooting depth. This assumption was used in a pioneering paper by Gardner (1967), in a recent review by Parlange (1980), and in an analysis of supply of water and nutrients in soilless culture (Raats, 1980c).

2. Exponentially decreasing uptake distribution:

$$\lambda = \delta_e^{-1} \exp - z/\delta_e \tag{24}$$

where δ_e corresponds to the depth of an equivalent, uniform root system with the same rate of uptake at the soil surface and rate of transpiration T. Elsewhere I have presented in detail various implications of equation (24). Rawlins (1973) and Jury et al. (1977) used equation (24) for $0 < z < \delta$ and equation (23) for $z \geqslant \delta$.

Fig. 3 shows depth-time trajectories for the water under an orange tree based upon equations (9) and (24) and data that will be discussed later on (cf. Fig. 15 of Van Schilfgaarde, 1977). Fig. 4 shows steady salinity profiles calculated from equations (17) and (24) for leaching fractions L = 0.2 and 0.05

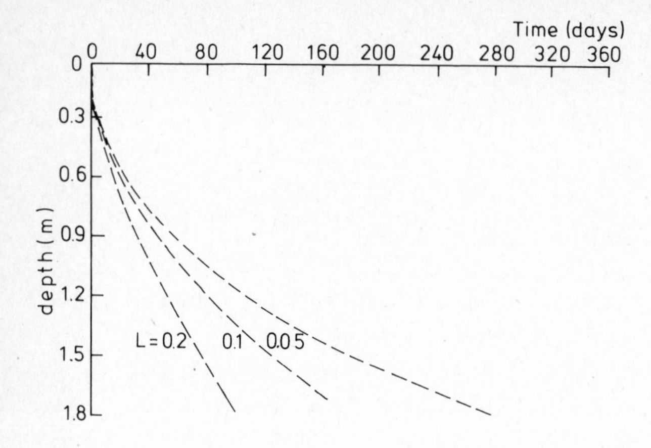

Fig. 3. Depth-time trajectories of parcels of water under an orange tree, based on equation (24) with δ_e = 0.4 m for L = D/(R+I) = 0.05, 0.1, and 0.2.

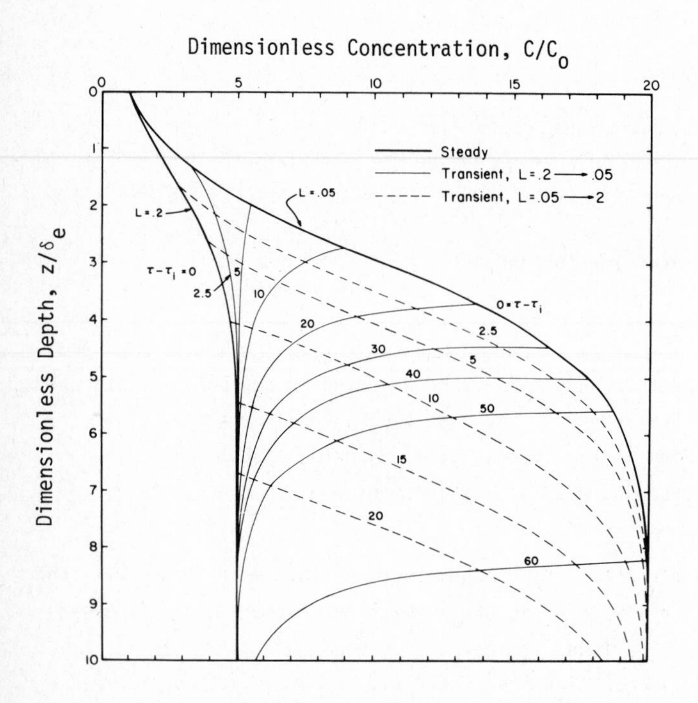

Fig. 4. Steady salinity profiles for L = D/(R+I) = 0.05 and 0.2, and transient salinity profiles for various dimensionless times $\tau = \frac{E + T}{\theta \delta_e} t$ (after Raats, 1975).

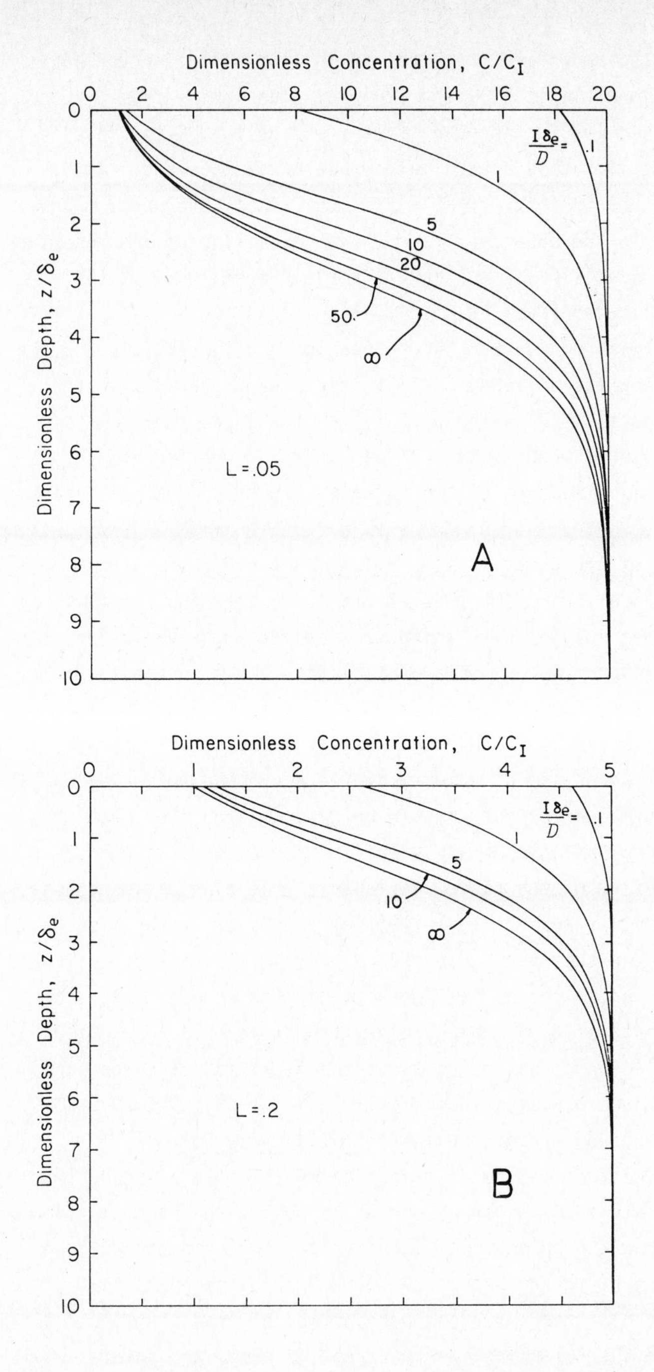

Fig. 5. Influence of dispersion upon steady salinity profiles
A : L = 0.05, B : L = 0.2.

(Raats, 1974a, b; 1975). Also shown are transient salinity profiles at various times calculated from equations (18), (19), and (24) for transitions from $L = 0.05$ to 0.2 and the reverse (Raats, 1975). In going from $L = 0.05$ to 0.2 the "old" salinity profile experiences virtually a parallel downward displacement. In going from $L = 0.2$ to 0.05, the salinity profile at later times develops a bulge, whose front becomes steeper in the course of time. The salinity profiles reported by Peck (1975, 1977) have shapes similar to these transient profiles, but their origin is perhaps more complicated (Peck et al., 1981).

Diffusion and dispersion will counteract the steepening due to selective uptake of water. Figs. 5a and b show the influence of diffusion and dispersion upon steady salinity profiles (Raats, 1977). The salinity profiles are functions of the leaching fraction L and the Peclet number $I\delta_e/D$. For a given leaching fraction, the Peclet number is a measure of the relative importance of convective and diffusive transports.

2.4 Lognormal Distribution of the Concentration Below the Root Zone

The one-dimensional, vertical transport model formulated above implies that, at any time, the concentration at any depth is uniform. However, wide variations of the concentrations of individual samples taken at a certain depth in the lower part of the root zone have been reported independently by Oster and Wood (1977) and by Wierenga and Sisson (1977). In both cases the distribution of log c was found to be normal; in other words the distribution of c was found to be lognormal. It is well known that lognormal distributions can be generated by a process in which the change of the random variable at any step of the process is a random proportion of the previous value of the random variable. This type of genesis of lognormal distributions was first formulated in 1903 by the Dutch astronomist Kapteyn and is now known as the theory of proportionate effect (Aitchison and Brown, 1957). If $\lambda T/\theta$ is a random variable, then according to equation (14) the change of c with a given change of t will be a random proportion of the previous value of c. Given the likely random nature of λ, θ, and T, it is not surprising that the distribution of c tends to being lognormal.

2.5 Inference of the Distribution of Water Uptake from Salinity Data

The uptake of the water does affect the distribution of the solute. Roughly, as water is taken up and solute is excluded by the plant roots, the concentration of the solute increases. Thus the space-time distribution of the solute will in part reflect the distribution of the uptake of the water. This suggests that the distribution of the uptake of the water can perhaps be inferred from the distribution of the solute. In the following the transport of water and solutes will be assumed to be one-dimensional (see also Raats, 1980b). For one-dimensional transport, solving equation (5) for the water flux θv gives:

$$\theta v = F_s/c + D\frac{\partial}{\partial z}\ln c. \tag{25}$$

Introducing equation (25) into equation (2) and solving for λT gives:

$$\lambda T = -\frac{\partial}{\partial t}\theta - \frac{\partial}{\partial z}F_s/c - \frac{\partial}{\partial z}D\frac{\partial}{\partial z}\ln c. \tag{26}$$

Integration of the mass balance for the solute expressed in equation (4) gives an expression for F_s:

$$F_s = F_{so} - \int_{z_o}^{z}\left\{\lambda_s N + (\theta c + \mu_a + \mu_f)\right\}dz, \tag{27}$$

$$= F_{so} - \frac{d}{dt}\int_{z_o}^{z}(\theta c + \mu_a + \mu_f)dz - N\int_{z_o}^{z}\lambda_s dz. \tag{28}$$

Equations (27) and (28) simply show that the flux of solute at depth z is equal to the flux of solute at depth z_o minus the time rate of change of storage between z_o and z. Introducing equation (28) into equation (26) gives:

$$\lambda T = -\frac{\partial}{\partial t}\theta - F_{so}\frac{\partial}{\partial z}c^{-1} - \frac{\partial}{\partial z}D\frac{\partial}{\partial z}\ln c$$

$$+\frac{\partial}{\partial z}c^{-1}\left\{\frac{d}{dt}\int_{z_o}^{z}(\theta c + \mu_a + \mu_f)dz + N\int_{z_o}^{z}\lambda_s dz\right\}. \tag{29}$$

If the flow is steady and the effects of dispersion, adsorption, precipitation and uptake are negligible, then equation (29) reduces to:

$$\lambda T = -F_{so}\frac{d}{dz}c^{-1}. \tag{30}$$

According to equation (30), the rate of water uptake may be calculated as the product of the salt flux F_{so} and the negative of the slope of the dilution profile, $-dc^{-1}/dz$. Gardner (1967) appears to have been the first to realize this. He wrote: "Equation (5) (= (30) above) gives us a relation between the water uptake pattern with depth and the concentration distribution. Since it is easier to measure the concentration than to measure the flux directly, the concentration gradient may give a better measure of w (= λT above) than the divergence of the flux density. Furthermore, the lower limit of the water uptake can be ascertained from the depth at which the concentration becomes constant".

It appears that equation (30) was not noticed for the following seven years (Raats, 1974a, b). Oster et al. (1974) applied the method to brome grass under high frequency irrigation in outdoor lysimeters. The cumulative water uptake distribution estimated from chloride data was 60, 80 and 90% for depths of 15, 30, and 45 cm, respectively. Evaporation losses in the 0-1 cm depth interval accounted for about half of the applied water. Plots of the log of the cumulative uptake as a function of depth were approximately linear. This meant that the distribution of the rate of uptake could be approximated by equation (24).

For two different lysimeters the rooting depth parameters δ_e turned out to be 8.5 cm and 9.6 cm, respectively. Assuming a dispersion coefficient of 0.05 $cm^2\,day^{-1}$, the third term on the right hand side of equation (29) had a negligible effect on the estimate of δ_e. Fig. 5 shows that if the leaching fraction is small and the Peclet number is rather small, then the influence of dispersion will be noticeable. If electrical conductivity data are used as a basis, then dissolution/precipitation described by the last term on the right hand side of equation (12) also needs to be considered. In the lysimeters the sum of the mineral equilibria and diffusion corrections to the rate of uptake was zero to the 15 cm depth. At greater depths the mineral equilibria correction was dominant and increased the calculated rate of uptake by as much as 30%.

The steady state distribution of chloride was also used to estimate the distribution of the water uptake under an orange tree (Van Schilfgaarde, 1977; Dirksen et al., 1979). The cumulative relative water uptakes were 64, 86, 93, 97, and 98%, respectively for depths of 0.3, 0.6, 0.9, 1.2, and 1.5 m, respectively, corresponding roughly to an equivalent rooting depth δ_e of 0.4 m. This information can in turn be used to calculate depth-time trajectories of parcels of water. Assuming θ = 0.5, T = 7 mm day^{-1}, and δ_e = 0.4 m, Fig.3 shows such time courses for leaching fractions 0.05, 0.1, and 0.2 (cf. Fig. 15 of Van Schilfgaarde, 1977).

In a laboratory study of space/time distributions of matric and osmotic potentials of daily irrigated alfalfa, Dirksen et al. (1980) estimated the distribution of the water uptake on two different days from hydraulic data and from the salt flux and salinity sensor readings on another day. The agreement between the two estimates was good. Between 80 and 90% of the uptake occurred above 0.50 m.

Jury et al. (1978a, b, c) used soil salinity sensor and chloride data to estimate fractional water uptake above a depth of 5 cm and in the layer 0-20 cm, respectively, in a lysimeter experiment with wheat and sorghum. Corrections for precipitation were made by using the chemical equilibrium model of Oster and Rhoades (1975) and calculating the electrical conductivity of the resulting mixed salt solutions by the method of McNeal et al. (1970). It turned out that 50% or more of the water was evaporated or was taken up within 5 cm from the soil surface.

Thus far, only one-dimensional flows have been discussed in this section. To infer anything about the distribution of the water uptake from the distribution of the salinity, one must have separate information about the flow pattern. For example, it can be shown that the proper generalization of equation (30) to multi-dimensional flows is given by :

$$\lambda T = -(A_o/A)\, F_{so} \frac{\delta c}{\delta s}^{-1}, \tag{31}$$

where $\delta/\delta s$ is the directional derivative along a streamline; A is the cross sectional area of a stream tube; and the subscript o indicates a reference point along the same stream tube. A logical next step would be to consider experiments involving localized irrigation or uptake patterns under trees and sufficiently detailed measurements of the salinity distribution and the flow pattern so that equation (31) can be used.

3 THE SATURATED ZONE

3.1 The Input-Output Relationship

Steady, multi-dimensional convective transport of solutes was discussed in detail in two recent papers (Raats, 1978a, b), the first paper dealing with the general theory, the second paper with specific flow problems. The time required for a parcel of water to move from one point to another along a streamline can be determined and this basic information can then be used to describe collections of parcels of water forming a surface. For any geometry and boundary conditions, the cumulative transit time distribution function, q, is defined as the fraction of the stream tubes with transit times smaller than τ. The transit time density distribution is defined as the derivative of q with respect to τ and may be regarded as the transfer function, $T[\tau]$, for the flow system:

$$T[\tau] = dq/d\tau, \tag{32}$$

where the square brackets denote functional dependence. The general relationship between the input I and the output O can be written as:

$$O[t] = \int_{t_o}^{t} T[\tau]\, I[t-\tau]\, d\tau + \text{a contribution of solutes present at time } t_o. \tag{33}$$

The function q can be determined by measuring the concentration of an ideal tracer in the output following a step change of the concentration in the input. The function $dq/d\tau$ can be determined by measuring the output resulting from a pulse distributed uniformly in the input.

To determine the travel time density distribution one must use equation (1) to calculate the transit times along the different stream lines. For steady flow, equation (1) is equivalent to the kinematical result:

$$\tau = t - t_\alpha = (\theta v)_\alpha^{-1} A_\alpha^{-1} \int_{s_\alpha}^{s} \theta A ds, \tag{34}$$

where A is the cross-sectional area of an infinitesimal stream tube and the subscript α refers to the input surface. If the flow pattern and the distribution of θ are known then the right hand side can be evaluated by graphical and numerical procedures. Other methods are based on an equation resulting from substituting Darcy's law into equation (1) and transforming from integration with respect to s to integration with respect to the total head H:

$$\tau = t - t_\alpha = - \int_{H_i}^{H} (\theta/k) \, |\nabla H|^{-2} \, dH. \tag{35}$$

For a few problems the right hand side of equation (35) has been evaluated analytically. If the distributions of θ, k, and H are known from an analytical or a numerical solution of the flow problem or from measurements, then the right hand side of equation (35) can always be evaluated numerically.

3.2 Apparently well mixed systems

The flow pattern shown in Fig. 6 is induced by a uniform input at the water table. The region in which the flow occurs is assumed to be a rectangle. A drain,

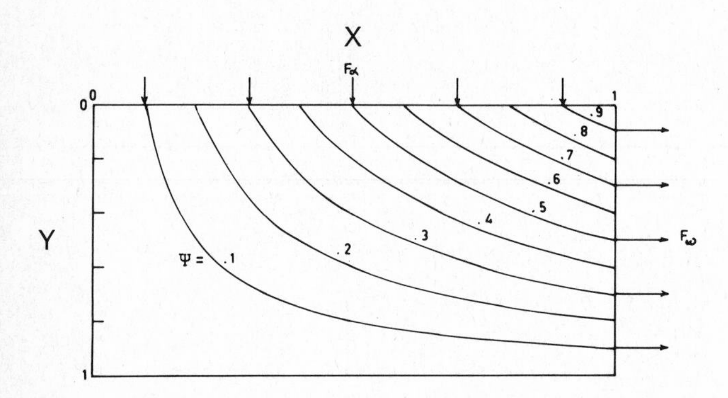

Fig. 6. Flow pattern with uniformly distributed input and output.

ditch, or stream is assumed to be located on the right hand side. For the time being the output is assumed to be uniformly distributed. On the left hand side is the midplane between drains, ditches, or streams. The bottom is assumed to be impermeable. In other words, at steady state the input at the water table is equal to the output at the drain, ditch or stream.

The turnover time $\bar{\tau}$ of the system is equal to the volume $-XY$ of water in the flow system divided by the flux $F_\alpha X$:

$$\bar{\tau} = \frac{\theta XY}{F_\alpha X} = \frac{\theta Y}{F_\alpha} . \tag{36}$$

The turnover time is the characteristic time for convective transport of solutes through the system. The horizontal and vertical components of the velocity are given by:

$$v_x = x/\bar{\tau} \tag{37}$$

$$v_y = (Y - y)/\bar{\tau}. \tag{38}$$

Both components are inversely proportional to the turnover time $\bar{\tau}$. They are shown in Figs. 7 and 8. The horizontal component of the velocity increases linearly from zero at the midplane to its maximum at the drain, ditch or stream.

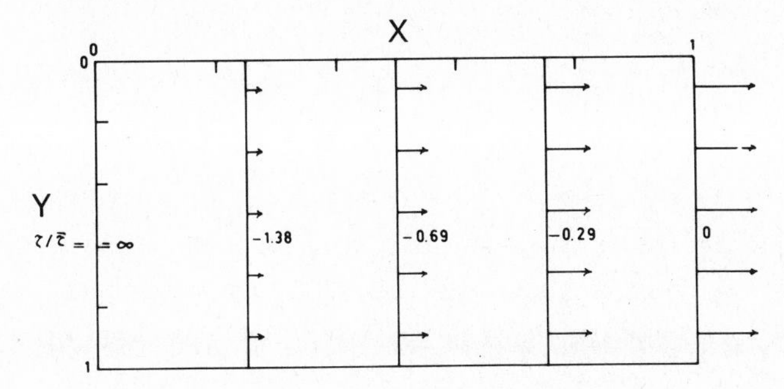

Fig. 7. Horizontal component of the velocity and associated vertical isochrones.

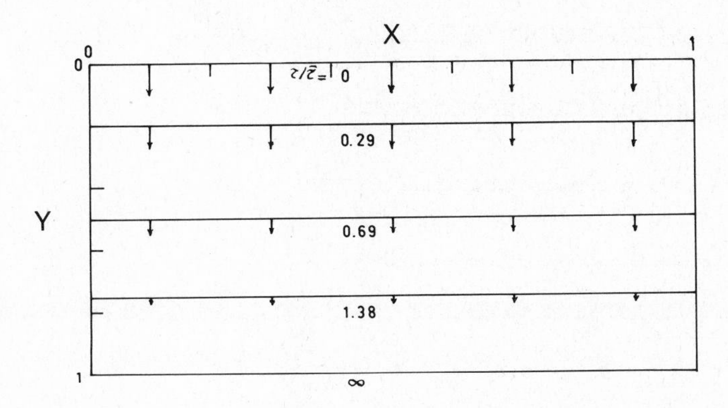

Fig. 8. Vertical component of the velocity and associated horizontal isochrones.

The vertical component of the velocity decreases linearly from its maximum at the soil surface to zero at the impermeable base. Since for this particular flow the vertical component of the velocity at a certain depth does not depend on the distance from the drain, ditch, or stream, parcels of water situated in a horizontal plane will jointly move down and remain in a horizontal plane. Integration of equation (37) shows that a parcel introduced into the system at $x = x_0$ will arrive at a drain, ditch, or stream located at $x = X$ after a time interval τ given by :

$$\tau = \bar{\tau} \ln X/x_0. \tag{39}$$

The implications of equation (39) can be best understood by considering a pulse of solutes in the input and the resulting output. Fig. 9 shows the distribution of the fraction of the solute remaining in the flow system at successive times. The remainder of the solute is uniformly distributed. It is as if such a band of solute is elastic and is being stretched uniformly. The output is largest at time zero and decreases exponentially with time. In other words, the transit time density distribution is given by :

$$dq/d\tau = \bar{\tau}^{-1} \exp -\tau/\bar{\tau}, \tag{40}$$

and the cumulative transit time distribution is given by :

$$q = 1 - \exp - \tau/\bar{\tau}. \tag{41}$$

This means that the system behaves as an apparently well mixed system. It is as if there is a steady flux $F_\alpha X$ of water through a perfectly stirred reservoir with a volume — XY. Of course, in reality the model is based on piston displacement and the transit-time density distribution is entirely dictated by the flow pattern shown in Fig. 6.

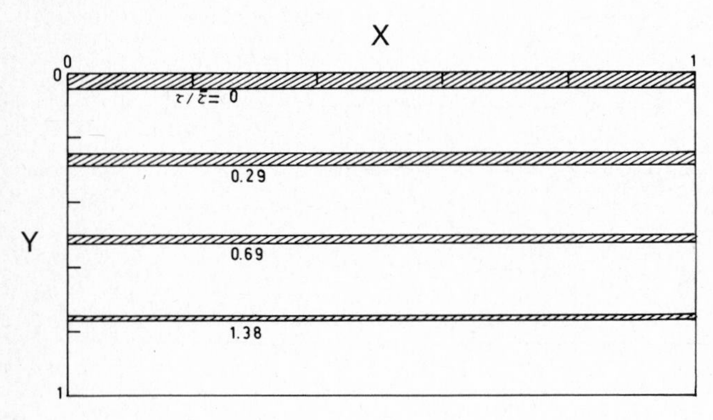

Fig. 9. Fate of a horizontal layer of solute.

It appears that an exponential distribution of arrival times was considered by Eriksson (1958). It was used later without further justification in theoretical discussions of input/output relationships (e.g. Eriksson, 1971; Nir and Lewis, 1975), and in attempts to fit experimental data on tritium in river basins (Eriksson, 1963), leaching of solutes from a laboratory model (Peck, 1973) and the chloride balance of some farmed and forested catchments in south-western Australia (Peck and Hurle, 1973). The relationship between the exponential travel time density distribution and the flow pattern shown in Fig. 6 gradually emerged in papers by Eldor and Dagan (1972), Ernst (1973), and Gelhar and Wilson (1974). Elsewhere I have discussed in detail implications and limitations of this important special case of the general theory (Raats, 1977, 1978b, 1980a; see also Rice and Raats, 1980). Here I note only that for $X/Y < 5$ the deviations from the flow pattern shown in Fig. 6 due to convergence to the drain, ditch, or stream must be taken into account (cf. Ernst, 1973 and Raats, 1978b, Fig. 7).

3.3 More Complex Flow Problems

Some of the literature on transit time density distribution for more complicated geometries was discussed elsewhere (Raats 1977, 1978b). Other examples, related to regional flow problems, are given by Nelson (1978). Here I will give just one example. If in the flow problem discussed in detail in subsection 3.2 the depth $Y \to \infty$, then the turnover time $\bar{\tau}$ is no longer useful. Ernst (1973) showed that in the limit $Y \to \infty$ the transit times are distributed according to:

$$\tau/\bar{\tau}_\infty = (1-x/X) \cot \frac{\pi X}{2X}, \tag{42}$$

where:

$$\bar{\tau}_\infty = \theta X/F_\alpha. \tag{43}$$

The theory discussed thus far applies to convective transport of a solute distributed uniformly over the input surface, and not subject to diffusion and dispersion, to adsorption, and to production or decay. For the flow pattern shown in Fig. 6, Eldor and Dagan (1972) have shown that diffusion and dispersion have only a very small influence upon the distribution of the solute in the horizontal direction and thus any resulting deviations from the transit time density distribution defined by equation (40) would be expected to be small. Linear adsorption can be accounted for by multiplying characteristic times such as $\bar{\tau}$ and $\bar{\tau}_\infty$ defined by equations (36) and (43) by a retardation factor $(\theta+k)/\theta$, where k is the adsorption constant (Raats, 1980a). If the adsorption is nonlinear, then the speed of the parcels of solute depends on the concentration and a detailed numerical calculation will generally be necessary, except for one aspect. For nonlinear adsorption with so-called favorable adsorption isotherms concentration profiles along streamlines will have a tendency to be compressed to shocks on the upstream side and the c-profile will have a tendency to be stretched on the downstream side. For unfavorable isotherms the reverse will occur. If c_α and c_∞ are the concentrations on the two sides of the shock then the average adsorption capacity over the range c_α to c_β serves as a retardation factor for the velocity of shock waves.

To accommodate non-uniform distribution of solute over the input surface and linear production or decay, the transfer function T [τ] is generalized to (Raats 1978a, 1980a):

$$T = dq/d\tau . r . \exp(\alpha t), \qquad (44)$$

where r is the input density distribution and α is a rate constant. Linear production and decay with a rate constant(> 0 for production, and <0 for decay) is accounted for by a factor exp αt.

4 CONCLUDING REMARKS

The principles governing the transport of solutes in the natural environment are quite well understood. Nevertheless quantifying a particular field situation remains a difficult task. Fortunately, in many cases it is possible to estimate the residence time within and below the root zone from the mass balance and flow pattern for the water and some information about the interaction of the solute with its environment.

5 REFERENCES

Aitchison, J. and Brown, J.A.C., 1957. The lognormal distribution. University Press, Cambridge.

Dirksen, C., Oster, J.D. and Raats, P.A.C., 1979. Water and salt transport, water uptake, and leaf water potential during regular and suspended high frequency irrigation of citrus. Agric. Water Manage., 2: 241-256.

Dirksen, C., Raats, P.A.C. and Shalhevet, J.S., 1980. Interaction of alfalfa with matric and osmotic soil water potentials nonuniform in space and time. II. Daily irrigation. Soil Sci. Soc. Am. J.,

Eldor, M. and Dagan, G., 1972. Solutions of hydrodynamic dispersion in porous media. Water Resour. Res., 8: 1316-1331.

Eriksson, E., 1958. The possible use of tritium for estimating groundwater storage. Tellus, 10: 472-477.

Eriksson, E., 1963. Atmospheric tritium as a tool for the study of certain hydrological aspects of river basins. Tellus, 15: 303-308.

Eriksson, E., 1971. Compartment models and reservoir theory. Ann. Rev. Ecol. Syst.,2: 67-84.

Ernst, L.F., 1973. De Bepaling van de Transporttijd van het Grondwater bij Stroming in de Verzadigde Zone. Nota ICW 755, 42 pp.

Gardner, W.R., 1967. Water uptake and salt-distribution patterns in saline soils. In: Isotope and Radiation Techniques in Soil Physics and Irrigation Studies, IAEA Proceedings Series, 335-341.

Gelhar, L.W. and Wilson, J.L., 1974. Ground water quality modeling. Ground Water, 12: 399-408.

Hilgard, E.W. and Loughridge, R.H., 1895. The distribution of the salts in alkali soils. Calif. Agric. Exp. Sta. Bull., 108, 14 pp.

Hilgard, E.W. and Loughridge, R.H., 1906. Nature, value, and utilization of

alkali lands, and tolerance of alkali by cultures. Revised reprints of Calif. Agric. Exp. Sta. Bulls. 128 and 133, 73 pp.
Jacobson, T. and Adams, R.M., 1958. Salt and silt in ancient Mesopotamian Agriculture. Science, 128: 1251-1258.
Jury, W.A., Fluhler, H. and Stolzy, L.H., 1977. Influence of soil properties, leaching fraction, and plant water uptake on solute concentration distribution. Water Resour. Res., 13: 645-650.
Jury, A., Frenkel, H., Fluhler, H., Devitt, D. and Stolzy, L.H., 1978a. Use of saline irrigation waters and minimal leaching for crop production. Hilgardia, 46: 169-192.
Jury, W.A., Frenkel, H. and Stolzy, L.H., 1978b. Transient changes in the soil-water system from irrigation with saline water: 1. Theory. Soil Sci. Soc. Am. J., 42: 579-585.
Jury, W.A., Frenkel, H., Devitt, D. and Stolzy, L.H., 1978c. Transient changes in the soil-water system from irrigation with saline water: II. Analysis of experimental data. Soil Sci. Soc. Am. J., 42: 585-590.
McNeal, B.L., Oster, J.D. and Hatcher, J.T., 1970. Calculation of electrical conductivity from solution composition data as an aid to in situ estimation of soil salinity. Soil Sci., 110: 405-414.
Nelson, R.W., 1978. Evaluating the environmental consequences of groundwater contamination: 1, 2, 3, and 4. Water Resour. Res., 14: 409-450.
Nir, A. and Lewis, S., 1975. On tracer theory in geophysical systems in the steady and nonsteady state. Part I. Tellus, 27: 372-383.
Oster, J.D. and Rhoades, J.D., 1975. Calculated drainage water compositions and salt burdens resulting from irrigation with river waters in the Western United States. J. Environ. Qual., 4: 73-79.
Oster, J.D. and Wood, J.D., 1977. Hydro-salinity models: sensitivity to input variables. In: J.P. Law and G.V. Skogerboe (Eds). Proc. Nat. Conf. Irrig. Return Flow Quality Management. U.S. Environmental Protection Agency and Colorado State University, p. 253-259.
Oster, J.D., Raats, P.A.C. and Dirksen, C., 1974. Calculation of water uptake distribution from observed steady salinity profiles. Agron. Abstr., p. 14.
Parlange, J.Y., 1980. Water transport in soils. Ann. Rev. Fluid Mech., 12: 77-102.
Peck, A.J., 1973. Analysis of multidimensional leaching. Soil Sci. Soc. Am. Proc., 37: 320.
Peck, A.J., 1975. Effect of land use on salt distribution in the soil. In: A. Poljakoff Mayber and J. Gale (Eds) Ecological Studies. Analysis and Synthesis. Vol. 15 Plants in Saline Environments. Springer, Berlin.
Peck, A.J., 1977. Development and reclamation of secondary salinity. In: J.R. Russell and E.L. Greacen (Eds) Soil factors in crop production in a semi-arid environment. Univ. of Queensland Press in association with the Australian Society of Soil Science Inc. St. Lucia, Queensland, 327 pp.
Peck, A.J. and Hurle, D.H., 1973. Chloride balance of some farmed and forested catchments in south-western Australia. Water Resour. Res., 9: 648-657.
Peck, A.J., Johnston, C.D. and Williamson, D.R., 1981. Analysis of solute distributions in deeply weathered soils. Agric. Water Manage., 4: 83-102
Raats, P.A.C., 1974a. Movement of water and salts under high frequency irrigation. Proc. 2nd Int. Drip Irrig. Congr. San Diego, Calif, July 7-14, 1974.
Raats, P.A.C., 1974b. Steady flow of water and salt in uniform soil profiles with plant roots. Soil Sci. Soc. Am. Proc., 38: 717-722.
Raats, P.A.C., 1975. Distribution of salts in the root zone. J. Hydrol., 27: 237-248.
Raats, P.A.C., 1977. Convective transport of solutes in and below the root zone. Proc. Int. Conf. on "Managing Saline Water for Irrigation: Planning for the future". Lubbock, Texas: 290-298.
Raats, P.A.C., 1978a. Convective transport of solutes by steady flows. I. General theory. Agric. Water Manage., 1: 201-218.
Raats, P.A.C., 1978b. Convective transport of solutes. II. Specific flow problems. Agric. Water Manage., 1: 219-232.

Raats, P.A.C., 1980a. Multidimensional transport of solutes in saturated and in unsaturated soils. Neth. J. Agric. Sci., 28: 7-15.
Raats, P.A.C., 1980b. The distribution of uptake of water by plants: inference from hydraulic and salinity data. AGRIMED seminar on the movement of water and salts as a function of the properties of the soil under localized irrigation, 6-9 November 1979, Bologna (It.). (In press).
Raats, P.A.C., 1980c. The supply of water and nutrients in soilless culture. Proc. 4th Int. Congr. on Soilless Culture.
Rawlins, S.L., 1973. Principles of managing high frequency irrigation. Soil Sci. Soc. Am. Proc.,37: 626-629.
Rice, R.C. and Raats, P.A.C., 1980. Underground travel of renovated wastewater. J. Environmental Engineering Division of the American Society of Civil Engineers. 106: 1079 - 1098.
Van Schilfgaarde, J., 1977. Minimizing salt in return flow by improving irrigation efficiency. In: J.P. Law and G.V. Skogerboe (Eds). Proc. Nat. Conf. Irrigation Return Flow Quality Management, U.S. Environmental Protection Agency and Colorado State University.
Wierenga, P.J. and Sisson, J.B., 1977. Effects of irrigation on soil salinity and return flow quality. In: J.P. Law and G.V. Skogerboe (Eds). Proc. Nat. Conf. Irrig. Return Flow Quality Management. U.S. Environmental Protection Agency and Colorado State University. p. 115-121.

ANALYSES OF SOLUTE DISTRIBUTIONS IN DEEPLY WEATHERED SOILS

A.J. PECK, C.D. JOHNSTON AND D.R. WILLIAMSON
C.S.I.R.O. Division of Land Resources Management,
Wembley, 6014, Western Australia.

ABSTRACT

Peck, A.J., Johnston, C.D. and Williamson, D.R., 1981. Analyses of solute distributions in deeply weathered soils. Agric. Water Manage., 1981.

Methods of analysis of salt distributions in soil profiles are developed and applied to interpret Cl^- concentrations observed at 4 forested sites in south-western Australia. Both transient diffusion and steady-state diffusion-convection models are examined. The latter is favoured. The role of diffusion in Cl^- transport at 3 of these sites is significant, and in a section of one profile it may dominate convection. Water extraction by roots appears to be restricted to depths of less than 6 m at 3 of the sites with little water moving downwards through the soil at this depth; only 0.06% of average rainfall at the driest site. Maxima in the observed Cl^- profiles may result from water movement to greater depths through root channels, or essentially no convection in the lower profile, or a small upward flux of water in the lower region.

1 INTRODUCTION

The history of development of secondary salinity in Western Australia, and the impact on both agriculture and the water resources of this region have been discussed by Hillman (1981). In this paper we focus attention on the storage and distribution of solutes in lateritic soils and deeply weathered materials which blanket the Archean shield of granitic and gneissic rocks in this region.

Salinas (salt lakes) were a feature of the inland landscape of south-western Australia when European settlement began about 150 years ago. Such areas of obvious salt accumulation in an essentially undisturbed landscape can still be seen to the east of the present agricultural zone. As the native woodland was cleared and dryland crops and pastures were introduced, it became evident that soluble salts were widely distributed in the soils and underlying weathered material. By comparing samples of similar soils in adjacent cleared and uncleared areas, Teakle and Burvill (1938) argued that at most sites removal of the native vegetation resulted in a decrease in the salinity of surface soils. Similar observations have since been reported from the United States (Ferguson,

1976) and elsewhere in Australia (Hartley, 1980; Kilpatrick, 1980).

In Western Australia, recent soil salinity studies have centred on the storage and distribution of soluble salts in areas of relatively high rainfall (600 to 1400 mm yr^{-1}), particularly native Eucalypt forests which may be cleared in the future for agriculture, bauxite mining, forest products or other reasons (Dimmock et al., 1974; Batini et al., 1976; Herbert et al., 1978; Johnston, 1980; Johnston et al., 1980; Johnston and McArthur, 1981). These data from about 330 cored bore-holes provide a very substantial data base on the quantities and distributions of dissolved salts in the landscape.

The objectives of this paper are to develop methods for the analysis of solute distributions in soil profiles, and to apply these methods in an examination of data from boreholes representative of those in the data base referred to above. An understanding of processes and mechanisms contributing to the observed solute distributions provides a foundation for the identification of areas where certain land uses may cause problems of secondary salinity.

2 ANALYSIS OF SALINITY PROFILE DATA

2.1 Solute Transport Theory

Bresler (1973, 1981) has reviewed the theory of solute transport by convection, dispersion and diffusion during transient flow in porous materials such as soils. Water and solute flow in a one-dimensional, non-hysteretic, vertical system are described by:

$$J_w/\rho = q_w = -D_w \partial\theta/\partial z + K \quad (1)$$

and:

$$J_s = -D_s \partial c/\partial z + c\, q_w \quad (2)$$

respectively. In these equations J_w [dimensions $ML^{-2}\,T^{-1}$] is the mass flux density of the soil solutions; ρ [ML^{-3}] is the density of the soil solution; q_w [LT^{-1}] is the volume flux density of the soil water; D_w [L^2T^{-1}] is the water-flow diffusion coefficient; θ [L^3L^{-3}] is volumetric water content; z [L] is the space coordinate (positive down); K [LT^{-1}] is the hydraulic conductivity; J_s [$ML^{-2}T^{-1}$] is the mass flux density of solute; D_s [L^2T^{-1}] is the effective solute diffusion coefficient in the soil; and c [ML^{-3}] is solute concentration in the water.

Neglecting solute adsorption or exchange on soil particle surfaces (which is believed to be negligible in our examples) conservation of mass of water and solute respectively requires that:

$$\partial(\rho\theta)/\partial t = -\partial J_w/\partial z + W_* \quad (3)$$

$$\partial(\theta c)/\partial t = -\partial J_s/\partial z + S \quad (4)$$

where t [dimension T] is time; W_* [$ML^{-3}T^{-1}$] is a water source-strength; and S [$ML^{-3}T^{-1}$] is a solute source-strength.

Equations (1) through (4) lead to general partial differential equations for water and solute transport. Assuming constant and uniform ρ:

$$\partial\theta/\partial t = \partial(D_w \partial\theta/\partial z)/\partial z - \partial K/\partial z + W \quad (5)$$

and:

$$\theta\partial c/\partial t = \partial(D_s \partial c/\partial z)/\partial z - q_w \partial c/\partial z + S - cW \quad (6)$$

where $W = W_*/\rho$ is a volumetric water source-strength.

When soil characteristics D_w and K (functions of θ) and the effective diffusion coefficient D_s (a function of θ and q_w) are known, and source functions W (z,t), S (z,t) are specified, solutions of equations (5) and (6) of the form $\theta=\theta$ (z,t) and c=c (z,t) may be found which satisfy the initial and boundary conditions of a particular problem. Some solutions in this class are presented by Bresler (1980). We approach the task of identifying characteristics of the accumulation and transport of solutes in deeply weathered soils by examining simple models, and calculating system characteristics by inverse methods.

2.2 Steady-State Models

In this class of model it is assumed that although inputs and outputs of water and solute may vary seasonally, in the region considered a balance exists between their rates averaged over a period of several years. Moreover, this balance is assumed to have been maintained for such a long time that a dynamic equilibrium or true steady-state has been reached. In this condition both θ and c at any depth may vary about their time-averaged values $\bar{\theta}$ and $\bar{c}$, but $\partial\bar{\theta}/\partial t$ and $\partial\bar{c}/\partial t$ are negligible. Since seasonal effects are greatest at the ground surface or near a fluctuating water table, and these effects are attenuated rapidly with distance, observed values of θ and c will normally be closest to $\bar{\theta}$ and $\bar{c}$ in intermediate sections of the profile.

2.2.1 Effect of water uptake by roots

Gardner (1967) considered the salt profile beneath a crop irrigated with saline water. Neglecting solute diffusion or dispersion, and any extraction of salts by plant roots, he showed how the steady-state solute profile could be related to water uptake (W<0) by the roots. Equivalently, from equations (2) and (6) we find that :

$$W(z) = -(J_s/c^2)dc/dz \qquad (7)$$

where J_s is the spatially uniform solute flux density. When diffusion and hydrodynamic dispersion contribute to the solute flux density, and solute uptake may occur, W is related to the solute concentration profile c(z) by:

$$W(z) = \frac{1}{c}\left[\frac{d}{dz}\left(D_s \frac{dc}{dz}\right) - \frac{D_s}{c}\left(\frac{dc}{dz}\right)^2 - \frac{1}{c}\left(J_{so} + \int_0^z S\,dz\right)\frac{dc}{dz} + S\right] \qquad (8)$$

where J_{so} is the solute flux density at z=0.

2.2.2 Diffusion opposing convection

Gardner (1965) also examined the steady-state solute concentration distribution which may develop when diffusion or dispersion opposes convection, and there are no sources or sinks of water or solute within the soil. He assumed that the net flux density of solute was zero. A more general expression may be derived from equation (6) for the concentration distribution, in which J_s, q_w and D_s are spatially uniform and non-zero:

$$(c - J_s/q_w)/(c_* - J_s/q_w) = \exp[q_w(z-z_*)/D_s] \qquad (9)$$

where $c=c_*$ at a chosen depth $z=z_*$ in the region.

2.3 Transient Diffusion Models

We consider the solute concentration distribution in a region where q_w, W and S may be neglected, and D_s/θ is essentially uniform. Under these conditions equation (6) reduces to:

$$\partial c/\partial t = (D_s/\theta)\partial^2 c/\partial z^2 \qquad (10)$$

Solutions of this equation have been derived for many different initial and boundary conditions in the mathematical analysis of heat conduction (Carslaw and Jaeger, 1959).

2.3.1 Concentration boundary condition

When there is a spatially uniform concentration c_i of diffusant at $t=t_i$ and a constant concentration c_* is maintained at $z=z_*$ for all $t>t_i$, the appropriate solution to equation (10) is (Carslaw and Jaeger, 1959, p.59) :

$$(c-c_i)/(c_*-c_i) = \mathrm{erfc}[(z-z_*)\{\theta/4D_s(t-t_i)\}^{\frac{1}{2}}] \qquad (11)$$

where erfc is the complementary error function.

2.3.2 Flux boundary condition

Another model we examine assumes a spatially uniform concentration c_i of diffusant at $t=t_i$ and a constant flux F_* maintained at $z=z_*$ for all $t>t_i$. The solution to equation (10) for these conditions is (Carslaw and Jaeger, 1959, p.75):

$$F(z)/F_* = \mathrm{erfc}[(z-z_*)\{\theta/4D_s(t-t_i)\}^{\frac{1}{2}}]. \qquad (12)$$

3 DATA EXAMINED

Solute distributions determined from 4 cored boreholes in south-western Australia were chosen for detailed analysis. Since the solute distributions may be related to characteristics of local soils, vegetation and climate, this section of our paper includes brief descriptions of the environment of this region.

3.1 Environment of South-Western Australia

The region of interest consists of deeply weathered granitic and gneissic rocks of the Great Plateau of Western Australia. Elevations range between 260 and 400 m with local relief often about 75 m but increasing to the west where a major fault line (the Darling Scarp) is associated with more deeply incised valleys (Mulcahy et al., 1972). The dominant soil type is lateritic gravel in a yellow sandy matrix which is often about 1 m deep, but occasionally extends to 5 m. This material has a hydraulic conductivity at saturation of about 3 m d^{-1} (Sharma, M.L., personal communication, 1980). Towards the divides, and in some areas well down into valleys the gravel stones are cemented to a porous or massive lateritic ironstone. Beneath this surface material there is an acid, mottled or pallid kaolinitic clay which extends to basement rocks at depths which are often about 25 m and occasionally as much as 50 m. The hydraulic conductivity of the pallid zone materials is highly variable within an area of order 1 km^2, but the geometric mean of K in such an area (about 2×10^{-3} m d^{-1}) varies little between areas separated by as much as 200 km (Peck et al., 1980).

Sharma et al. (1980) have examined the ionic composition of solutes at various depths in these soil profiles and the underlying groundwaters. They conclude that the ratios of Na^+, K^+, Ca^{2+}, Mg^{2+} and SO_4^{2-} to Cl^- are generally very similar to those in sea water. Small differences are believed to reflect contributions from current weathering processes close to bedrock, and differential uptake of ions by plant roots closer to the ground surface. In this study Na^+ amounted to from 7 to 59% of the exchangeable metal ions which ranged from 3.2×10^{-6} eq. g^{-1} (ESP 44%) to 48.3×10^{-6} eq.g^{-1} (ESP 9%).

This region experiences a typical mediterranean climate with warm (25°C) dry summers and cool (10°C) moist winters. Fifty per cent of the rainfall, which declines from a maximum of about 1300 mm yr^{-1} near the Darling Scarp to about 600 mm yr^{-1} only 65 km further inland, falls in the winter months of June, July and August. Class A pan evaporation ranges from 1400 to 2000 mm yr^{-1} with daily rates of 8 to 10 mm in January and less than 2 mm in July.

3.2 Details of Borehole Sites

Detailed descriptions of small catchment areas in which 3 of the boreholes are located have been reported by Bettenay et al. (1980), and no unique features of the fourth site have been noted other than the slightly higher rainfall. Some characteristics of the sites are given in Table 1.

TABLE 1
Borehole site data

Borehole Number	Rainfall (mm yr^{-1})	Cl^- Accession J_{so}(g m^{-2} d^{-1})	Slope Position
2A02	1300	2.2×10^{-2}	lower
1351	1150	2.0×10^{-2}	mid
8251	800	8.3×10^{-3}	lower
1551	1150	2.0×10^{-2}	upper

At the time of sampling, all of the borehole sites were located in areas of native Eucalypt forest which had been only selectively cut for lumber. According to Carbon et al. (1980) who examined soil samples from 3 of these boreholes and 22 others in the region, average root length per unit soil volume decreases by about two orders of magnitude from the soil surface to a depth of 2 to 4 m. At greater depths the decrease of root density is less rapid. Some roots were found at the 18 m depth of unsaturated soil profiles.

Given that there are continuing solute imputs in rainfall, and there had been minimal disturbance of vegetation at the time of sampling at each site, it is reasonable to hypothesize that the solute concentration profiles were close to equilibrium. The greatest variation from equilibrium would be expected close to the ground surface due to seasonal wetting.

3.3 Methods of Sampling

Methods of subsampling soil cores, and extraction of solutes were generally the same as those described by Johnston et. al. (1980). For the present purposes we chose to examine Cl^- concentration distributions. Chloride was estimated from measured electrical conductivity κ(mS m^{-1}) of soil:water extracts. The concentration of chloride c(mg L^{-1}) in the extract was calculated using our correlations of :

$c = 1.98\,\kappa - 0.8$ (n=134, r=0.885)

for $\kappa < 4.1$, and :

$c = 2.84\,\kappa - 4.3$ (n=387, r=0.998)

for $4.1 < \kappa < 200$.

An examination of sources of experimental error in the estimation of Cl^- concentration in the soil solution suggests that total errors are unlikely to exceed $\pm$ 15%.

3.4 Data Smoothing

Since it was necessary to compute both the first and second derivatives of c(z) to determine W(z), we first fitted a cubic spline function to the original data points, and differentiated this function. In this procedure the weighting factor for each data point was empirically adjusted until the spline fit was subjectively judged to be satisfactory. Note that our derived data are computed from the cubic spline, and not directly from original data points.

3.5 Effective Diffusion Coefficient

As there are no measurements of $D_s(\theta)$ in the soils of this region we have used the relationship :

$$D_s = 1.6 \times 10^{-4}\,(1.08\theta^2 - 0.16\theta)\ m^2\,d^{-1} \qquad (13)$$

for $\theta > 0.24$ (fitted to data for Cl^- diffusion in clay soil from Porter et al., 1960) and :

$$D_s = 1.6 \times 10^{-4}\,(2.94\theta^3 - 0.332\theta^2 + 0.00925\theta)\ m^2\,d^{-1} \qquad (14)$$

for $\theta < 0.24$. Note that these expressions neglect any effect of solution velocity on D_s. According to Olsen and Kemper (1968) this contribution is negligible for $(q_w/\theta) < 10^{-3}$ m d^{-1}.

4 RESULTS AND DISCUSSION

4.1 Steady State Analyses

4.1.1 Borehole 2A02

The Cl^{-1} concentration profile, c(z) and moisture profile θ(z) at this site are shown together with a summary of profile morphology in Fig. 1. This profile is typical of the monotonic form at high rainfall sites as discussed by Johnston et al. (1980). Monotonic profiles are more commonly found in upper landscape positions except in very high rainfall sites such as this. To some indefinite depth, θ may be less, and c more than their time-averaged values at this site because the profile was sampled in the dry, summer season.

Assuming a steady-state profile, and uniform Cl^- flux density equal to the rate of accession ($S=0$; $J_s=J_{so}$), the liquid flux density q_w and source strength W were computed using equations (2) and (8). These results are shown on the right of Fig. 1. The largest solution velocity ($=q_w/\theta$) in this profile is of order 10^{-3}m d^{-1}, but the diffusive term is less by about a factor 100 than the convective term at this point. Thus only gross errors in D_s, which appear unlikely, will affect q_w. Since the diffusive term is negligible, Gardner's (1967) expression and equation (7) could be used to calculate W.

We note that, at this site with average rainfall about 1300 mm yr^{-1}, the flux density of water downwards through the soil is estimated to be only 240 mm yr^{-1} at z=1 m, and 110 mm yr^{-1} at z=9 m. Thus more than 80% of the water loss appears to take place above the 1 m depth in the soil, but about 8% of rainfall reaches the unconfined water table at the 9 m depth.

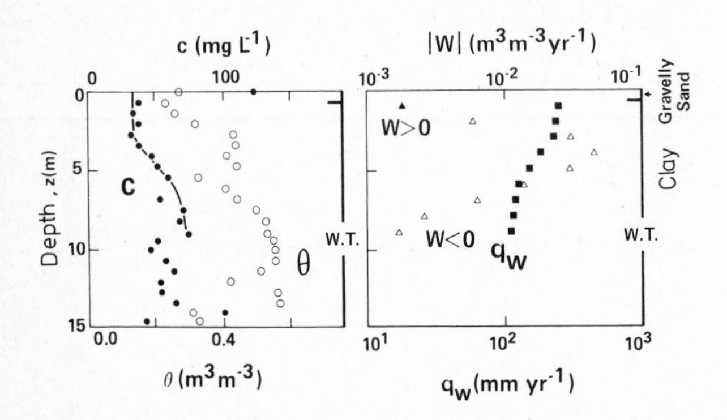

Fig. 1. Profiles of observed Cl^- concentration c and water content θ, and the cubic spline (solid curve) fitted to the Cl^- data points for borehole 2A02. The water flux density q_w and source-strength W shown were computed using equations (2) and (8) and assuming $J_S=J_{SO}$. Positive values of W are indicated by solid triangles. Negative values of W (open triangles) correspond to extraction of water from the soil. A summary of the soil profile morphology is shown to the right of the figure, where the depth of the permanent water table is indicated by "WT".

Since W<0 at all depths in this profile, we conclude that there is extraction of water by roots throughout the unsaturated zone. The maximum rate of extraction appears to occur close to the top of the finer-textured pallid zone, and within this zone ln(-W) decreases roughly linearly with depth. Similar variation of W with z has been observed in studies of agricultural plants (Feddes et al., 1974).

4.1.2 Borehole 1351

Fig. 2 presents data from this borehole in the same format as Fig. 1. The Cl^- profile at this site is typical of the bulge form discussed by Johnston et al. (1980). Solute concentration increases rapidly with depth in the interval z=2 to 5 m which corresponds to the uppermost pallid zone material at this site. Since the profile was sampled in July, water contents near the ground surface are probably higher, and Cl^- concentrations lower than their year-average values. Thus estimated values of q_w near the ground surface may be greater than their true averages.

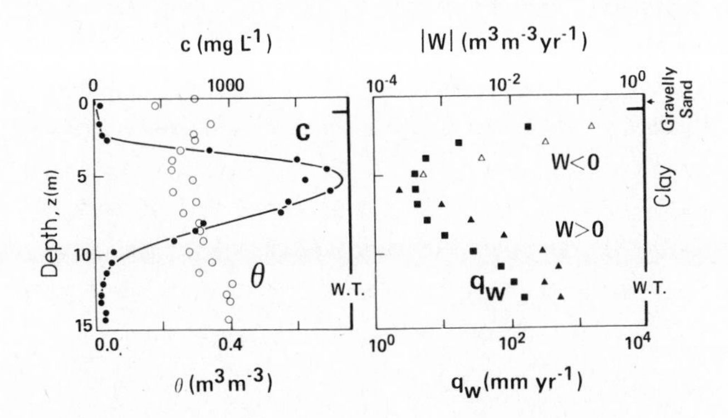

Fig. 2. As Fig. 1, but for borehole 1351.

At this site too, the magnitude of q_w suggests that the effect of solution velocity on D_s may be neglected, and the diffusive contribution to J_s is always less than that of convection. However, because dc/dz is much larger in parts of this profile than it was in the previous example, the diffusive term accounts for as much as 30% of J_s.

According to equation (2) with uniform J_s (=J_{so}), at this site only 17% of rainfall moves downwards through the soil at z=2 m, and at z=6 m this flux is less than 0.5% of rain. But in the region z>6m, q_w increases with depth reaching 153 mm yr^{-1} at z=13 m which is about the depth of the unconfined groundwater. As in the previous example, W is negative and ln(-W) decreases roughly linearly with depth in the interval z=2 to 5 m reflecting water extraction by roots. At greater depths, however, W must be positive to provide the calculated increase of q_w. Fig. 2 shows that W increases with depth to about z=11 m where it becomes relatively uniform.

We suggest that a mechanism for positive values of W in the lower profile is flow in preferred pathways which bypass the soil matrix. This mechanism has been demonstrated at one site in this region using dye as a tracer (Hurle, D.H., personal communication, 1979). According to this hypothesis, the total injection between z=6 and 13 m, an amount of 149 mm yr^{-1}, would be part of the loss of water from near-surface layers, probably associated with an ephemeral perched water table in the coarser-textured surface soil during the rainy season.

Clearly our assumption that all of the Cl^- input moves through the soil matrix fails when the bypass mechanism is operating. The total solute flux density should be divided between soil matrix and bypass channels, and a solute source term associated with the water injection. We have examined solutions to this problem, but they are not reported here.

In the following section we discuss a method for estimating J_s and q_w from c(z) if these flux densities are uniform over some range of depth. Application of this method to the site 1351 data yields results which are less definitive than those presented later. However, there is some support for estimated uniform flux densities q_w = 8.4×10^{-7} m d^{-1} and J_s = 4.8×10^{-3} g m^{-2} d^{-1} over the depths z=6 to 10 m. Note that this zone terminates at about the upper limit of observed water table elevations at this site. Moreover, if only 24% of the Cl^- input moves through the soil matrix, there would appear to be no evidence for water injection into the unsaturated zone at this site.

4.1.3 Borehole 8251

Fig. 3 presents primary and derived information for this borehole. Again the Cl^- profile takes the bulge form with high concentrations typical of lower rainfall areas in this region. There is an increase in c from the surface to a maximum at about z=6 m which is in the pallid zone clays. Below the maximum of c (z) there is an almost linear decrease with depth. At this site the interface between surface soil and the deeper clays is quite sharp at z=0.8 m. The profile was sampled in mid-summer so that estimates of q_w from c(z) near the ground surface may be less than their true averages.

The scatter of both c(z) and $\theta(z)$ data points shown in Fig. 3 is somewhat greater than that at the other 3 sites. Some of the deviations are larger than expected from experimental error. Perhaps they reflect samples taken from preferred pathways as discussed in the previous section. We note that the deviations which are outside the estimated error band are all points of anomalously low concentration.

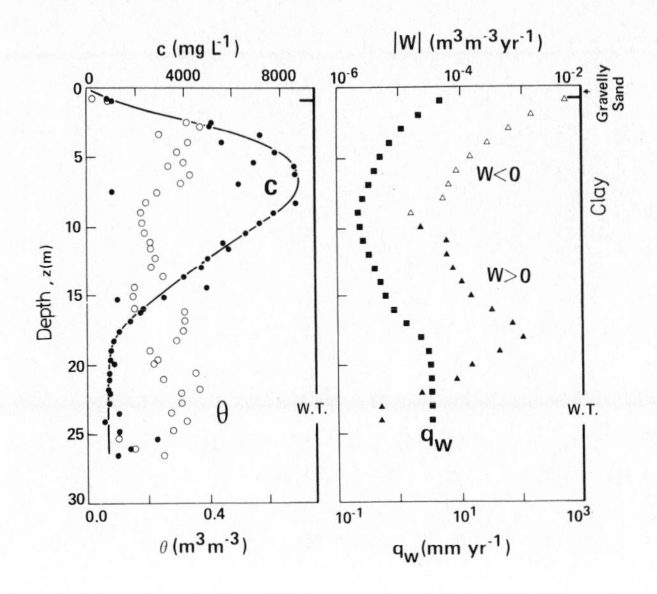

Fig. 3. As Fig. 1, but for borehole 8251.

Throughout the profile at this site, q_w is estimated to be less than 10^{-5} m d^{-1}. Thus solution velocity effects on D_s may be neglected, but in many parts of this profile the diffusive transport of Cl^- is comparable to that by convection. These components are almost equal at z=10 m. At z=1 m in this profile, q_w is estimated to be only 2.5 mm yr^{-1} or 0.3% of rainfall. The liquid flux density decreases even further to a minimum of 0.04% of rainfall at z=7 m, and then apparently increases with depth as in the case of borehole 1351. This variation of q_w is reflected by first negative and then positive values of W, as before.

An alternative method was developed to examine the Cl^- concentration distribution at this site. Given the concentration c, its depth gradient dc/dz and the effective diffusion coefficient D_s at any depth in the profile, equation (2) represents a straight line in the J_s (q_w) plane. Each point on this line represents a possible combination of J_s and q_w for the particular depth. Fig. 4 shows lines drawn for a number of depths in the profile for borehole 8251. The value of this method is that points of intersection between lines for different depths represent the possibility of the same J_s and q_w at more than one depth.

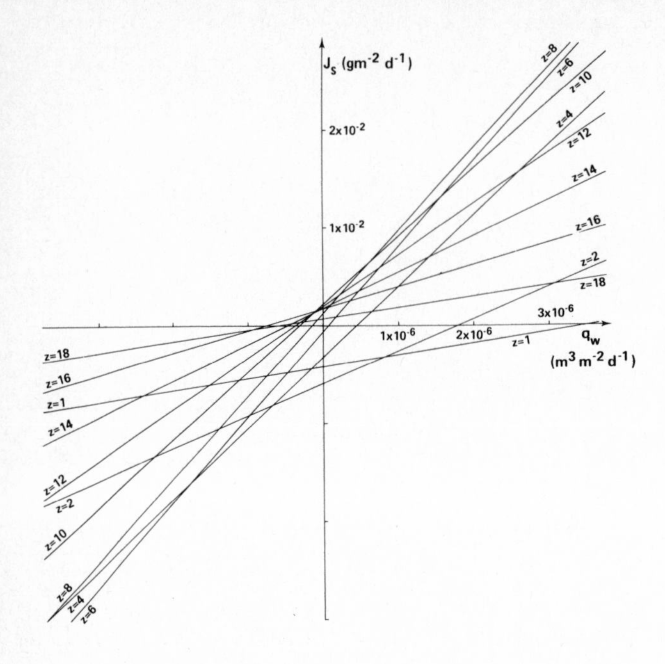

Fig. 4. Lines representing the possible combinations of Cl^- flux density J_s and water flux density q_W at given depths computed from equation (2) and fitted c(z) and θ(z) data for borehole 8251. Numbers next to the line indicate the depth z in metres.

Fig. 5 shows points of intersection of the lines representing data from the Cl^- concentration profile at neighbouring 1 m intervals of depth for site 8251. There is a cluster of intersection points for depths in the interval z=11 to 17 m which suggests a region of uniform flux densities. By averaging the coordinates of points in this cluster, we conclude that $J_s = 1.5\text{x}10^{-3}$ g m^{-2} d^{-1} and $q_W = -9.8\text{x}10^{-6}$ m d^{-1} in this interval.

According to equation (9), in an interval of spatially uniform D_s, J_s and q_W in a steady-state profile there will be a linear relationship between $\ln(c-J_s/q_W)$ and depth. Fig. 6 shows the fitted c(z) data for borehole 8251 plotted in this form using the values of J_s and q_W quoted above. Clearly the data are very closely fitted by a straight line over the depth interval z=9 to 17 m.

The above analysis indicates that the Cl^- flux density at about z=10 m is only 18% of the input in rainfall at this site. The difference may be interpreted as current accumulation of solute in the profile above the 10 m depth, or loss of solute by some mechanism above this depth, or downward movement of 82% of the Cl^- input in preferred pathways, or some combination of these processes. A mechanism for solute loss from the profile would be lateral flow of perched groundwater which develops in the surface soil at this site during the wet season.

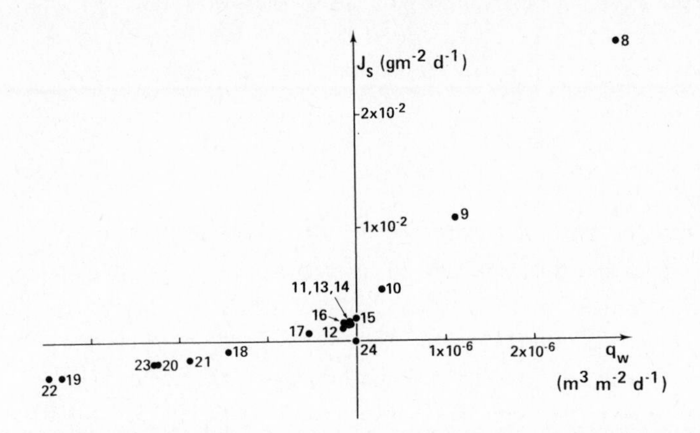

Fig. 5. Points of intersection of the $J_S(q_W)$ lines for successive 1 m depth intervals in 7<z<24 m for borehole 8251. The greater depth of the pair used to compute each point is shown next to the point.

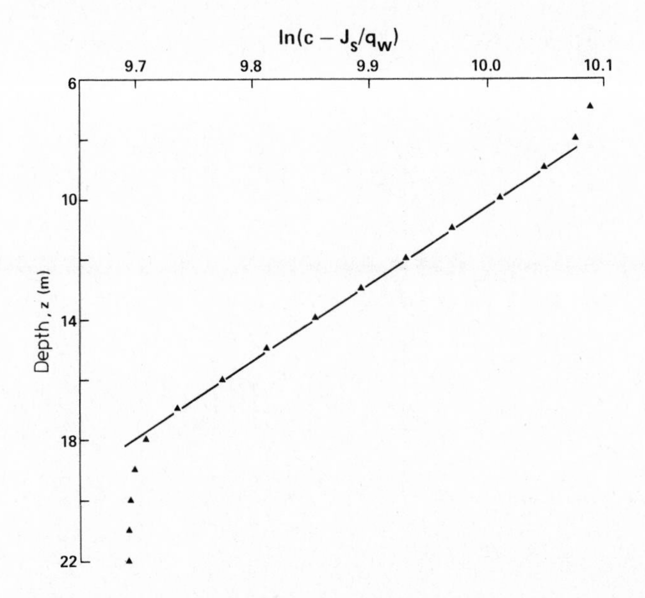

Fig. 6. The fitted Cl^- concentration data of borehole 8251 plotted in the form $\ln(c-J_S/q_W)$ as a function of z. Values of J_S and q_W were estimated from the cluster of points in Fig. 5. The data indicate a zone of uniform q_W and J_S in the interval z=9 to 17 m.

The most significant features of the estimated q_w in the z=9 to 17 m interval are its upward direction and very small magnitude, about 0.04 mm yr^{-1}. The magnitude of q_w explains the almost linear c(z) in this region as the result of steady diffusion between regions of different, but constant concentration.

We recall that when J_s was assumed equal to J_{so} throughout the profile it was concluded that q_w increased from 0.22 to 1.3 mm yr^{-1} over the depth interval z=9 to 17 m. Quite clearly, a reduction of J_s to only 18% of J_{so} has a large relative effect on the estimated q_w although the change is small in absolute terms as a proportion of local average rainfall.

4.1.4 Borehole 1551

This borehole is in a gentle saddle close to the divide of the same small catchment as 1351. Fig 7 presents the observed Cl^- concentration distribution and computed $q_w(z)$ and W(z) as before. The bulge profile form at this site is often observed in areas of some concavity in the land surface, even in upper-slope locations. Soil water contents to the 2 m depth suggest the possibility of recent wetting which is consistent with the winter sampling date. Thus estimates of q_w from c(z) near the ground surface may be greater than their true time-averaged values.

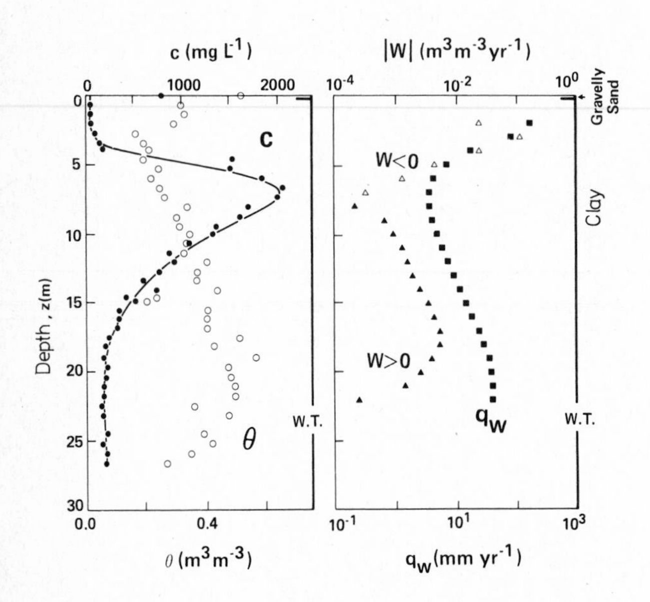

Fig. 7. As Fig. 1, but for borehole 1551.

At z=2 m the estimated q_w/θ is about $1.4x10^{-3}$ m d^{-1} so that solution flow may affect the magnitude of D_s, but since the diffusive term is more than a factor 10^3 less than that representing convection at this depth the error is quite negligible. At greater depths q_w is much less, so diffusion is relatively larger and at some depths its contribution to J_s is about 10% of that of convection.

The transition from surface soil to relatively fine-textured clays occurs at a depth of only 0.2 m at this site. Only 14% of the rainfall is estimated to move downwards through the soil at z=2 m, and this flux density decreases to a minimum of about 3% of rainfall at z=6 or 7 m. As in the case of borehole 1351, there is an apparent increase of q_w with depth in the region z>8 m reaching about 14 mm yr^{-1} at the unconfined water table ($z \approx 15$ m).

Fig. 7 shows that W is negative and that log(-W) decreases roughly linearly with depth to z=7 m. At greater depths positive values of W are computed from the increasing q_w. W appears to increase from z=8 to 15 m and then remain relatively constant to z=19 m. It may be significant that W becomes relatively constant at the water table depth.

The procedure described in the previous section was applied to determine magnitude of q_w and J_s in any zone of constant flux densities at this site. In this case the points of intersection for depths z=10 to 17 m appeared to cluster about a mean of $J_s = 1.6 \times 10^{-3}$ g m^{-2} d^{-1} and $q_w = -1.3 \times 10^{-6}$ m d^{-1}.

As in the case of borehole 8251, the estimated J_s below z=10 m is much less than (about 8% of) the local Cl^- accession rate. Since rainfall is much greater at borehole 1551 than the previous site, we believe that current accumulation of solute is less likely to be the explanation for the differences between J_s at z=10 m and J_{so} at this site. Other possible mechanisms are discussed in the previous section.

According to this estimate of q_w, the water flux density is directed upwards in the z=10 to 17 m interval at a rate of about 0.5 mm yr^{-1}. Such a conclusion is not unreasonable given the site location in a saddle. This upward flux results in the curvature of c(z) below the maximum of c at this site, which may be contrasted with the almost linear c(z) in the lower profile of borehole 8251 (see Fig. 3). Again we may note that assuming $J_s = J_{so}$, throughout the profile at this site, estimates of q_w increase from 4.8 to 23 mm yr^{-1} in the z=10 to 17 m interval. This is to be contrasted with the alternate estimate of $q_w = -0.5$ mm yr^{-1} throughout this interval.

4.2 Transient Diffusion Analyses

Sections of bulge profiles which are non-linear below the maximum are similar in shape to concentration distributions observed during transient diffusion. We have used the Cl^- concentration data from borehole 1551 (see Fig. 7) to examine two simple diffusion models using the theory presented in section 2.3 of this paper.

4.2.1 Concentration boundary condition

To examine this model, it was assumed that $z_* = 8$ m, $c_* = 1878$ mg L^{-1} and $c_i = 184$ mg L^{-1} as indicated by the data. Fig. 8 shows the fitted Cl^- concentration data from borehole 1551 below z=8 m plotted in the form $\text{erfc}^{-1}(x)$ as a function of z where :

$$X = (c-c_i)/(c_* - c_i). \tag{15}$$

As predicted by this model, the data lie close to a straight line over the interval z=8 to 18 m, but at greater depths the fitted Cl^- concentrations are less than those predicted. From the slope of the linear section, the time since the Cl^- concentration at z_* was increased to its present level may be estimated. Taking D_s/θ from equation (13) for the average θ in the interval z=8 to 18 m, $t-t_i$ is calculated to be 840 yrs. Although the data fit this model quite well, we are not aware of any event in this region which could have initiated or substantially increased Cl^- accumulation in the soils about 1000 yr B.P.

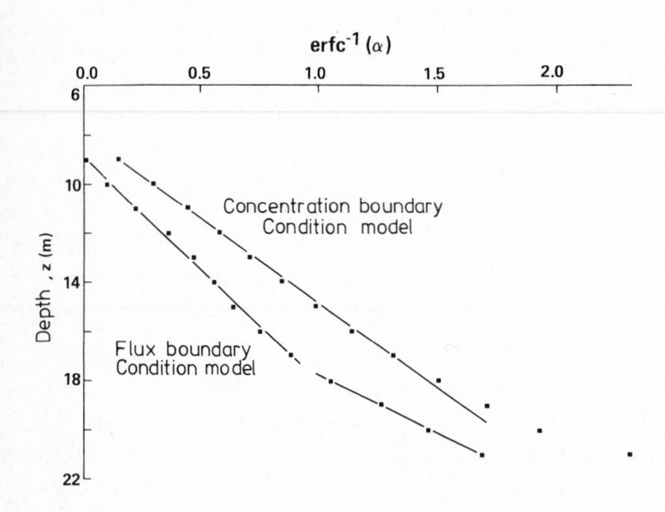

Fig. 8. The fitted Cl^- concentration data from borehole 1551 compared with two transient diffusion models: (i) Constant concentration boundary condition where c is related to X by equation (15) and erfc^{-1} (X) is plotted as a function of z. (ii) Constant flux boundary condition where the Cl^- flux density is related to Y by equation (16) and erfc^{-1} (Y) is plotted as a function of z. The variable α equals X in the first model, and Y in the second.

4.2.2 Flux boundary condition

This model may appear to be more plausible since a relatively constant Cl^- input is provided at present by accession in rainfall. Two cases were examined: In the first the Cl^- flux F_* at $z_* = 9$ m was computed from $D_s dc/dz$ at that depth, and in the second F_* was assumed equal to the independently measured Cl^- accession rate at the site of borehole 1551. Fig. 8 shows the data for the first case only ($F_* = 4.4 \times 10^{-3}$ g m^{-2} d^{-1}) plotted in the form $\mathrm{erfc}^{-1}(Y)$ as a function of z where :

$$Y = F/F_* . \tag{16}$$

Over the interval z=9 to 17 m the data fall close to a straight line in accord with the model, but there is a distinctly different slope to the line fitting data over z=18 to 21 m. A similar result, with greater change of slope was found for the second case. From the slope of the line fitting data in the interval z=9 to 17 m, and the value of D_s/θ for the average θ in this interval, $t-t_i$ was computed to be 1500 yr.

The basic agreement between times estimated by these two transient diffusion models is a consequence of the form of equation (10). As discussed before, no significance can be attached to the reasonable fit of the model for major accumulation of solute beginning about 1000 yr B.P. We note that the more linear section of c(z) beneath the maximum of c in the other bulge profiles is not well matched to either of the transient diffusion models.

5 SUMMARY AND CONCLUSIONS

Although parts of the Cl^- concentration profile in borehole 1551 are of the form expected in transient diffusion, the evidence supporting these models is not strong. Zones of uniform q_w and J_s in boreholes 8251, 1551 and possibly 1351 support the steady-state model, but it is readily shown from equations (3) and (4) that this is not sufficient evidence of steady conditions. However, we favour this model of salt distribution in deep profiles beneath essentially undisturbed vegetation.

The estimates of average water velocities (q_w/θ) in the profiles examined suggest that the velocity has little effect on D_s. Moreover the diffusive contribution to solute transport is so much less than the convective component in the region of maximum q_w that even an order of magnitude error in D_s has a negligible effect on J_s. At greater depths, however, there are zones in the bulge profiles where diffusion is at least comparable with convection in Cl^- transport. This situation is most evident at the lowest rainfall site. At the highest rainfall site, the role of diffusion is negligible, and Gardner's model would be adequate to relate c(z) to W(z).

Conclusions relating to the variation of q_w and W with depth in the profile are dependent on assumptions about the variation of J_s. If J_s is spatially uniform and equal to the average rate of Cl^- deposition in rainfall, then:

(i) There is extraction of water from the soil (presumably by roots of the native forest vegetation) to a depth of about 9 m at site 2A02, and about 6 m at the other sites.

(ii) There is an apparent injection of water into the soil below a depth of about 6 m at the site of all bulge profiles. This phenomenon could reflect the result of water movement down root channels from a perched water table. The total injection is about 13% of rainfall at site 1351, but less than 1% of rainfall at sites 1551 and 8251.

(iii) Since the calculated values of W are net average rates, it is possible that there is water uptake by roots at depths greater than 6 m, but at rates less than the local total injection.

(iv) At the 3 higher rainfall sites, between 14 and 18% of local rainfall moves downwards through the soil matrix at a depth of z=2 m; at z=6 m this flux density is still 9% of rainfall at the highest rainfall site, but less than 1% at sites 1351 and 1551.

(v) At the lowest rainfall site only 0.2% of rainfall passes through the soil at z=2 m, and 0.06% at z=6 m.

There is evidence to support Cl^- flux densities at depths between about z=11 and 17 m at sites 8251 and 1551 which are less by factors 6 and 12 respectively than the local average rates of Cl^- deposition in rainfall. In these regions it is concluded that there are spatially uniform water flux densities. This flux density is negligible (-0.04 mm yr^{-1}) at site 8251 so that diffusion is the dominant Cl^- transport mechanism in this part of the profile, and accounts for the almost linear section of c(z). At site 1551 the water flux density between z=10 and 18 m is estimated to be -0.5 mm yr^{-1} which results in the curvature of c(z) in this region of the profile. The small upward flux of water at this site reflects net discharge from the permanent groundwater which is consistent with the site location in a gentle saddle.

6 ACKNOWLEDGEMENTS

We wish to acknowledge the contributions made by Dr. J.D. Watson, in discussions during his visit with us, and the efforts of our drilling crew, particularly Mr. J. Moynihan, who often worked under difficult conditions to obtain soil core samples.

7 REFERENCES

Batini, F.E., Selkirk, A.B. and Hatch, A.B., 1976. Salt content of soil profiles in the Helena catchment, Western Australia. Res. Paper No. 23, Forests Dept. of W.A., Perth.

Bettenay, E., Russell, W.G.R., Hudson, D.R., Gilkes, R.J. and Edmiston, R., 1980. A description of experimental catchments in the Collie area, Western Australia. Tech. Paper No. 7, CSIRO Div. of Land Resour. Management, Perth.

Bresler, E., 1973. Simultaneous transport of solutes and water under transient unsaturated flow conditions. Water Resour. Res., 9: 975-86.

Bresler, E., 1981. Transport of salts in soils and subsoils.Ag.Wat.Man., 4: 35-62

Carbon, B.A., Bartle, G.A., Murray A. and Macpherson, D.K., 1980. The distribution of root length, and the limits to flow of soil water to roots in a dry sclerophyll forest. Forest sci., 26(4).

Carslaw, H.S. and Jaeger, J.C., 1959. Conduction of heat in solids. Oxford University Press, Oxford.

Dimmock, G.M., Bettenay, E. and Mulcahy, M.J., 1974. Salt content of lateritic profiles in the Darling Range, Western Australia. Aust. J. Soil Res., 12: 63-9.

Feddes, R.A., Bresler, E. and Neuman, S.P., 1974. Field test of a modified numerical model for water uptake by root systems. Water Resour. Res., 10: 1199-1206.

Ferguson, H., 1976. The salt status of saline seep area soils. In: Regional Saline Seep Control Symp. Proc. Bull No. 1132, Montana State University, Bozeman, Mont. p.86.

Gardner, W.R., 1965. Movement of nitrogen in soil. In: Soil Nitrogen. Am. Soc. Agron., Madison, Wisc. pp. 550-72.

Gardner, W.R., 1967. Water uptake and salt distribution patterns in saline soils. In: Isotope and radiation techniques in soil physics and irrigation studies. Int. Atomic Energy Agency, Vienna, pp. 335-40.

Hartley, R.E.R., 1980. Non-irrigated salinity on Kangaroo Island, South Australia Rep. S3/80, Soil Cons. Brnch., S.A. Dept. of Agric., Adelaide.

Herbert, E.J., Shea, S.R. and Hatch, A.B., 1978. Salt content of lateritic profiles in the Yarragil catchment, Western Australia. Res. Paper No. 32, Forests Dept. of W.A., Perth.

Hillman, R.M., 1981. Land and stream salinity in Western Australia. Agric. Water Manage.4: 11-18

Hingston, F.J. and Galaitis, V., 1976. The geographic variation of salt precipitated over Western Australia. Aust. J. Soil Res., 14: 319-35.

Johnston, C.D., 1980. Salt content of soil profiles in bauxite mining areas of the Darling Range, Western Australia. Tech. Paper No. 8, CSIRO Div. of Land Resour. Management, Perth.

Johnston, C.D., McArthur, W.M. and Peck, A.J., 1980. Distribution of soluble salts and water in soils of the Manjimup woodchip licence area, Western Australia. Tech. Paper No. 5, CSIRO Div. of Land Resour. Management, Perth.

Johnston, C.D. and McArthur, W.M., 1981. Sub-surface salinity in relation to weathering depth and landform in the eastern part of the Murray River catchment area, Western Australia. Tech. Paper No. 9, CSIRO Div. of Land Resour. Management, Perth.

Kilpatrick, J.R., 1980. Salt distribution in small catchments. M.Sc. Thesis, Latrobe Univ., Melbourne.

Mulcahy, M.J., Churchward, H.M. and Dimmock. G.M., 1972. Landforms and soil of an uplifted peneplain in the Darling Range, Western Australia. Aust. J. Soil Res., 10: 1-14.

Olsen, S.R. and Kemper, W.D., 1968. Movement of nutrients to plant roots. Adv. Agron., 20: 91-151.

Peck, A.J., Yendle, P.A. and Batini, F.E., 1980. Hydraulic conductivity of deeply weathered materials in the Darling Range, Western Australia. Aust. J. Soil Res., 18: 129-38.

Porter, L.K., Kemper, W.D., Jackson, R.D. and Stewart, B.A., 1960. Chloride diffusion in soils as influenced by moisture content. Soil Sci. Soc. Amer. Proc., 24: 460-3.

Sharma, M.L., Williamson, D.R. and Hingston, F.J., 1980. Water pollution as a consequence of land disturbance in south-west of Western Australia. In: Trudinger, P.A. and Walter, M.R. (Eds.), Biogeochemistry of Ancient and Modern Environments. Springer Verlag. 429-39.

Teakle, L.J.H. and Burville, G.H., 1938. The movement of soluble salts in soils under light rainfall conditions. J. Agric. West. Aust., 15: 218-45.

TRANSPORT OF SALTS IN CATCHMENTS AND SOILS

T. TALSMA
CSIRO Division of Forest Research,
Canberra, A.C.T., 2600.

ABSTRACT

Talsma, T. 1981. Transport of salts in catchments and soils. Agric. Water Manage., 1981.

Rainfall-runoff, and stream discharge-salinity relationships of forest catchments are shown to be consistent with mean sub-soil hydraulic conductivity and rain intensity-duration data. Similar analyses should be applicable in the study of dryland salinity problems. Stream water salinity did not vary with discharge rate in a catchment with highly weathered acid soils, that had long solute travel times and where depth dependent exchange reactions occurred during transport of anions. Such anion exchange behaviour is rare in irrigation soils, which usually differ in clay mineralogy, pH and base saturation. It is shown for Yandera loam that, under flow rates and solute concentration gradients which occur in irrigated conditions, the convective solute flux exceeds the diffusive flux by two orders of magnitude. An order of magnitude calculation shows that in most cases this is likely for the dryland situation too. Consideration of the concept of critical depth of water table for surface soil salinisation indicates that such depths may not differ as much from the irrigation case as is expected from isothermal flow theory.

1 INTRODUCTION

The study of salt transport through soils and across landscapes has largely been associated with salinity problems arising in irrigated agriculture and, to some extent, with nutrient dispersal below the root zone in intensively managed agricultural systems where, due to heavy use of fertilizer, eutrophication of drainage water may occur. The maintenance of good water quality for downstream (including urban) use from native and plantation forest catchments is also a current concern, particularly where more intensive management practices such as clear felling and burning are carried out.

The problem of dryland salinity, in Western Australia as well as elsewhere, has been apparent for a considerable time. However, compared with irrigation areas, such problems have emerged more slowly and have had less immediate economic consequences. Widespread clearing of trees and deep rooted scrub vegetation for dryland agriculture has generally disturbed the hydrologic balance less than irrigation development, so that the time to establish a new equilibrium in both the regional salt and water balance may be rather long. Also, the space scale of such disturbances is usually an order of magnitude larger than in irrigation development.

Mechanisms of vertical and lateral salt transport are similar for secondary salinisation in irrigation areas, catchment water quality changes under altered forest management and dryland salting after clearing, although transport rates may differ. Much of the detail of salt and water transport is already available (e.g. Gardner, 1965; Peck, 1971) and this is further discussed by Bresler (1981) and Raats (1981). Here, a physical basis will be explored to indicate which parts of a landscape are likely to yield shallow sub-surface and surface runoff rather than deep groundwater flow. This appears to be a largely unresolved question over much of the Western Australian wheatbelt (Peck, 1978a). Differences in chloride transfer behaviour between highly weathered forest soils and the irrigation soils of eastern Australia will also be discussed. Consideration is further given to some aspects of transport rates and mechanisms under irrigation and dryland conditions, and at what depth below the soil surface free groundwater is likely to cause surface salinisation under dryland conditions.

2 CATCHMENT STUDIES

2.1 Runoff Behaviour

In a recent study Talsma and Hallam (1980) developed a rapid assessment technique of measuring and evaluating catchment hydraulic conductivity at various soil depth intervals. It may be expected, a priori, that, where the mean hydraulic conductivity of surface or shallow sub-soil in a (wet) catchment is lower than frequently occurring, long duration rain intensities, substantial surface runoff should result. Alternatively, catchment water yield should be predominantly in the form of groundwater flow where the mean hydraulic conductivity exceeds such rain intensities. Soil hydraulic conductivities for two contrasting forest catchments are given in Table 1, and runoff hydrographs of these catchments following a series of rain storms are shown in Fig. 1. The rain intensities shown are for 4-hourly periods; note that $10^{-6} ms^{-1} = 3.6$ mm hr^{-1}.

The combined data of Table 1 and Fig. 1 indicate that the mean hydraulic conductivities provide a satisfactory basis for explaining the contrasting runoff hydrographs. For example the peak rain intensities during the first storm (8-9 Apr. 1974) were around $1.4 x 10^{-6}$ ms^{-1} for a 16-hour period, well above the mean hydraulic conductivities from 0.3 m soil depth downwards at the Greens catchment but well below those for Picadilly over the whole profile depth. Integration of the areas under the sharp peaks of the Greens hydrograph indicates that some 5%

of the total rain (200 mm) occurred as quick return flow. At Picadilly maximum groundwater runoff occurred some 4 days after the onset of heavy rain. Changes in stream water quality in relation to discharge rate for the same catchments are discussed below.

TABLE 1
Mean hydraulic conductivities of two ACT forest catchments at four depth intervals (in m).

Catchment	Soil	Hydraulic cond. ($10^{-6}ms^{-1}$)			
		0-0.1	0.3-0.6	0.7-1.0	1.1-1.4
Picadilly	krasnozem	>100	7.0	6.4	1.6
Greens	podzolics shallow earths	> 10	1.0	0.4	Rock

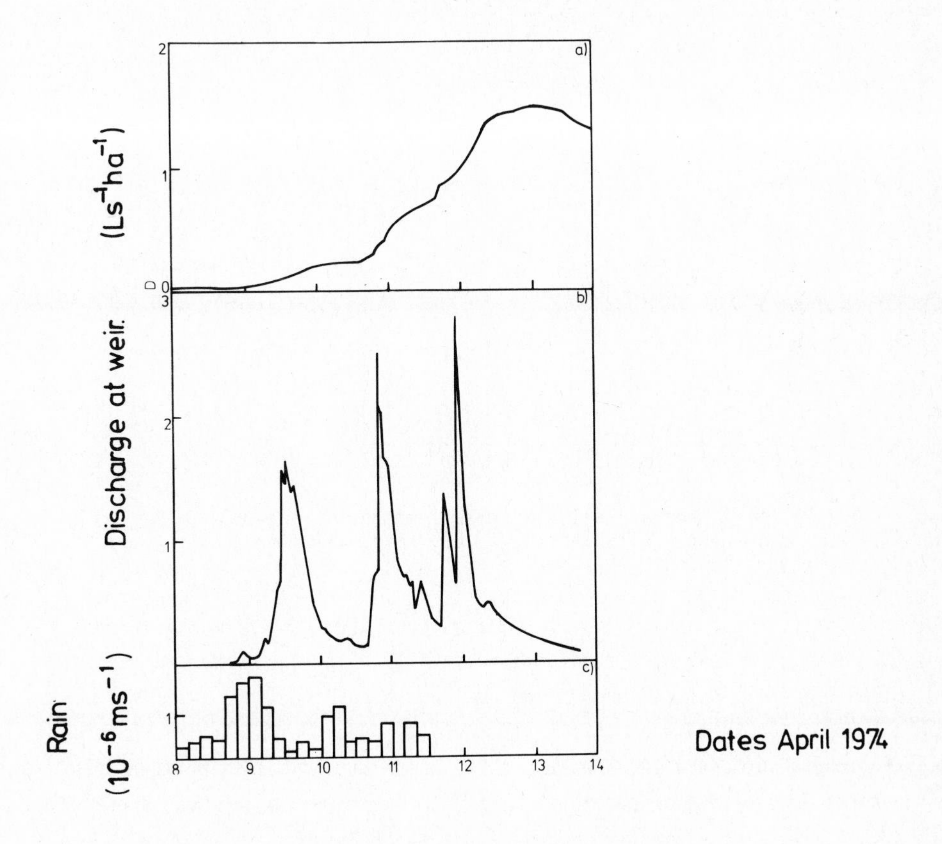

Fig. 1. Catchment hydrographs 8-14 April 1974 (a) Picadilly, (b) Greens, (c) Rain intensities over 4-hourly intervals.

2.2 Stream Water Quality in Relation to Runoff Behaviour

Catchment discharge rate - salinity relationships, during 1974/5, for the Picadilly catchment with deep and permeable soils, and for the Greens catchment with shallow and slowly permeable soils, are shown in Fig. 2. The electrical conductivity (EC) - stream discharge relationships are similar to those for Ca^{2+}, Mg^{2+}, Na^{+} and Cl^{-} (not shown in Fig. 2). Potassium, which is abundantly available and readily soluble from accumulated litter at the soil surface in forest catchments, is shown separately.

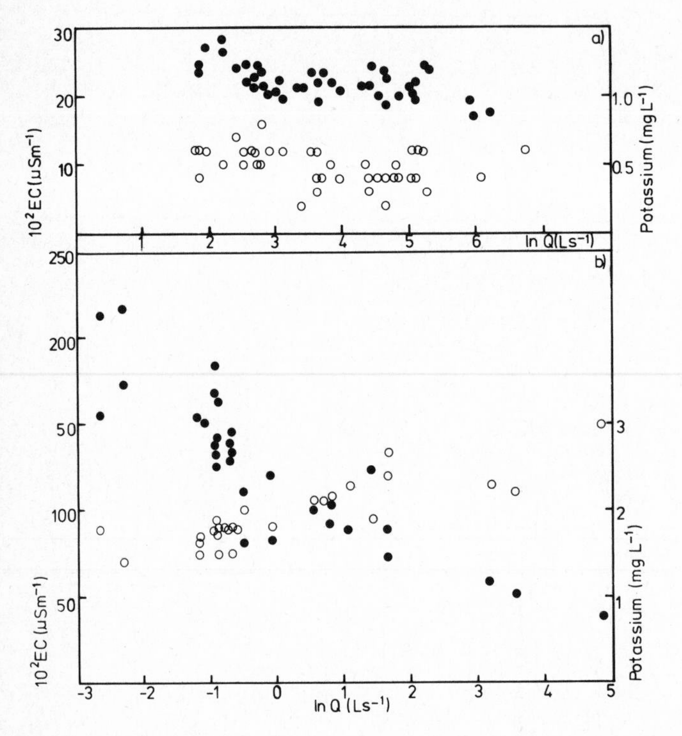

Fig. 2. Catchment discharge - salinity relationships during 1974/5 for (a) Picadilly and (b) Greens. Closed circles, EC, left scale; open circles, potassium, right scale.

Total salinity, expressed as EC, as well as potassium concentration varies very little with increasing discharge at Picadilly (Fig. 2a). Here all water and solute transport occurs through the soil and appears only as groundwater flow. It is noted in this context that solute transport times in this catchment are very long (e.g. Jury, 1975; Raats, 1981). The mean distance between discharge streams is of the order of 300 m, aquifer depth >20 m and mean annual discharge 0.4 m. The Picadilly catchment was control burnt in 1973, to reduce surface litter from 17 tonnes ha^{-1} to around 5 tonnes ha^{-1}. This temporarily released readily soluble cations and anions on the soil surface. However, no peak solute concentrations have been detected to date in the stream water.

Different relationships are apparent for the Greens catchment (Fig. 2b). Dilution of salt concentrations in groundwater (base flow) with substantial quick return and overland flow at high discharge rates yields a decreasing relationship between EC and $\ln Q$ (Q = discharge rate) with low overall electrolyte concentration at high discharge. However, for the readily soluble potassium, available at the soil surface, there is a tendency for an increase in concentration at increasing discharge. Such tendencies also exist for soluble iron and total phosphate, but not for other major cations and anions.

Comparison of soil profile hydraulic conductivities with prevalent rain intensities, in studies of wet forest catchments, has thus provided a qualitative basis for explaining differences in runoff hydrographs and streamflow-salinity relationships. Similar trends would also be expected in the dryland salinity situation, although the detail may differ in both cation and anion composition of stream water, and in initial and subsequent quality of runoff water from salinised surfaces (see e.g. Loh and Stokes, 1981).

Water which does not join the stream as surface runoff enters the soil profile, where it leaches accumulated salts by miscible displacement. Some aspects of this process are considered next, for both irrigated soils, and forest soils of eastern Australia.

3 SALT TRANSPORT DURING LEACHING

3.1 Non-reactive Solutes

Steady state one-dimensional transport of salts that do not react with the soil exchange complex, and are not excluded from part of the pore space (negative adsorption), is described by Gardner (1965) and Bresler (1981):

$$\partial C/\partial t = D\partial^2 C/\partial z^2 - v\,\partial C/\partial z \tag{1}$$

where C is solute concentration; t is time; D is the dispersion coefficient; z is distance and v is velocity of the solute. In most studies electrolytes such as Cl^- and NO_3^- are taken as non-reactive,and this has been confirmed in many laboratory and some field studies. Only a few appropriate field studies exist on leaching in Australian irrigated soils but their experimental data have been successfully interpreted by assuming that equation (1) is valid. Examples are a study on nitrate movement by Wetselaar (1962), see Gardner (1965), and two-dimensional chloride leaching data of Talsma (1967), which were satisfactorily explained by Jury (1975) using a flow model that did not involve exchange reactions or exclusion mechanisms. Most Australian irrigation soils are, apart from their immediate surface horizons, largely base-saturated, their pH is neutral to slightly alkaline (Groenewegen, 1961; Talsma, 1968) and clay minerals are predominantly illite and smectite. (Stace et al., 1968).

3.2 Reactive Solutes

Most forest soils in the eastern Australian highlands differ from irrigated soils on river flood plains. They may be strongly base-unsaturated, have low pH values and contain substantial amounts of aluminium and iron oxides. These are colloids with constant surface potential and a pH-dependent charge. At low pH, such soils are often significant anion adsorbers (Keng and Uehara, 1975; Black and Waring, 1979).

In a recent study Talsma et al. (1980) leached a 90 kg ha^{-1} surface applied KCl solution through such a soil near the Picadilly catchment. The resulting distribution of both soluble chloride and potassium are shown in Fig. 3 for various amounts of drainage water that passed beyond 1.5 m soil depth. Soluble chloride did not move beyond 1.2 m soil depth, and became increasingly adsorbed in the deeper part of the soil profile (Talsma et al., 1980).

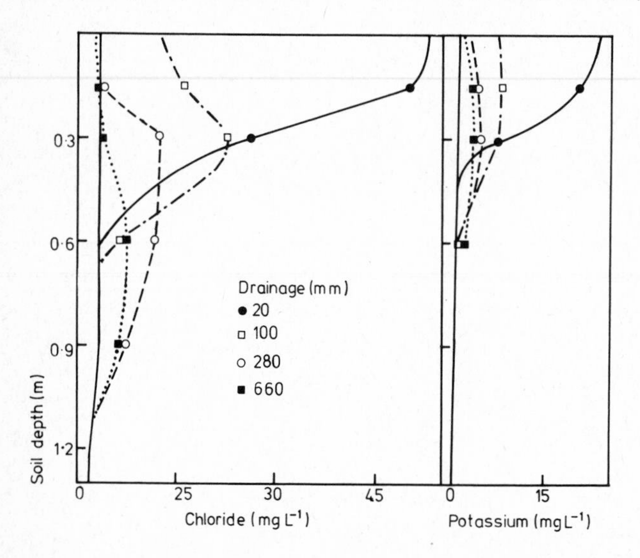

Fig. 3. Distribution of soluble chloride and soluble potassium in a highly weathered, acid soil profile after drainage of various quantities of leaching water. Nearly vertical solid lines: distributions of chloride and potassium before 1000 mg L^{-1} KCl application.

One-dimensional reactive solute transport may also be described by an equation formally identical to equation (1) by using a retarded dispersion-coefficient $D_0=D/R(C)$ and a retarded pore water velocity $v_0=v/R(C)$, where $R(C) = 1+(\rho/\theta)\,\partial S/\partial C$ is a retardation function due to reversible sorption. C and S are the solute concentration and adsorbed quantities of electrolyte, ρ is bulk density and θ is volumetric moisture content.

Thus, for instantaneous reversible sorption the slope of the equilibrium isotherm, $\partial S/\partial C$, determines the magnitude of solute dispersion and solute velocity. Talsma et al. (1980) found that the non-linear relation $S = 20C^{0.6}$ (S in mg kg^{-1}, C in mg L^{-1}) was an adequate representation of the potassium isotherm over the whole soil profile depth to 0.7 m, but that chloride adsorption became stronger at increasing soil depth. In the range of concentrations of interest in their study, chloride isotherms were linear and varied from $S = -40+1.1C$ near the soil surface to $S = -30+2.6C$ at 0.9 m depth. Thus the values of R(C) for potassium increased sharply at lower solute concentrations, whilst those for chloride at any particular depth interval remained constant with changing solute concentration but increased, from R(C)=4.5 at 0.15 m soil depth to R(C)=10.8 at 0.9 m depth. Black and Waring (1979) observed similar trends for anion adsorption in weathered soils in south-east Queensland.

It is seen therefore that modelling of chloride dispersion by using equation (1) with D_o and v_o would be difficult. The practical implication of these results is that, in addition to the long travel times discussed in the previous section, solute concentrations reaching stream water in catchments, with soils of similar chemical properties as at Picadilly, would be very much retarded and strongly attenuated. Such soils are also reported from Western Australia (e.g. see Stace et al.,1968, profile D, p355).

3.3 Transport by Convection and Diffusion

In many treatments of salt movement through soil it is assumed that convective transport dominates, although interactions between convective and diffusive transport is recognised as important in many cases (Gardner, 1965; Peck, 1971). An estimate of the separate contributions of convective solute flow, $v\bar{C}$, where v = pore water velocity, $\bar{C}$ is average chloride concentration, and diffusive flux, $D\partial C/\partial z$, where D is the diffusion coefficient for chloride in water, z is distance, may be readily made from available field data. Table 2 contains such data from a salinisation study on Yandera loam (Talsma, 1963). It is assumed that $D=10^{-10}m^2s^{-1}$ for a loam soil at volumetric moisture content, θ=0.25 (Gardner, 1965).

It is seen that even where there is a strong concentration gradient (Nov.-Dec. period near the soil surface during development of a salt crust) that convective flux dominates. Using similar arguments, Peck (1971) adopting $D=10^{-9}m^2s^{-1}$ for the diffusion coefficient in bulk solution rather than $D=10^{-10}m^2s^{-1}$ for moist soil concludes that, at a mean pore velocity of $10^{-9}ms^{-1}$, molecular diffusion would move more salt than would convection. Approximate calculations of Hurle and Johnston (1979) indicated that some 100 mm yr^{-1} may be available in Western Australia for leaching after clearing. In the steady state this would represent a pore water velocity ($\Theta \simeq 0.20$) of about $10^{-8} ms^{-1}$. Thus transport by convection would still be the major process. However, for fluxes much smaller than 100 mm yr^{-1}, diffusive transport may become important (see e.g. Peck et al., 1981). It is finally of interest to note that Smiles and Philip (1978) observed essentially piston flow displacement, in horizontal potassium chloride displacement experiments, over a large range of flow velocities. Their inferred dispersion coefficient, which was approximately equal to the product of moisture content and the molecular diffusion coefficient of potassium chloride in water, was virtually independent of the soil moisture content.

TABLE 2

Estimates of convective and diffusive solute flux of chloride in Yandera loam ($\Theta \simeq 0.25$, $D \simeq 10^{-10} m^2 s^{-1}$, $\Delta z \simeq 0.2$ m)

Period	Soil Depth (m)	v (ms^{-1})	$\bar{C}$ ($kg\ m^{-3}$)	ΔC ($kg\ m^{-3}$)	$v\bar{C}$ ($kg\ m^{-2} s^{-1}$)	$D \partial C/\partial z$ ($kg\ m^{-2} s^{-1}$)
Oct.	0.05-0.25	2.10^{-7}	15	14	3.10^{-6}	7.10^{-9}
	0.35-0.55	2.10^{-7}	7.5	4	$1.5.10^{-6}$	2.10^{-9}
Nov.-Dec.	0.05-0.25	6.10^{-8}	15	40	9.10^{-7}	2.10^{-8}
	0.35-0.55	6.10^{-8}	6.5	2	4.10^{-7}	1.10^{-9}

4 SALINISATION OF SURFACE SOILS

Salt accumulation of the soil surface usually results from capillary movement, sustained by evaporation, of saline groundwater at rather shallow depth. A detailed analysis of the factors involved in this process has been given by de Vries (1957). For one-dimensional, vertical water movement the total moisture flux, q, in both liquid and vapour phases, is given by:

$$q = -D_\theta(\partial\theta/\partial z) - D_T\ (\partial T/\partial z) - \Sigma D_{C_i}(\partial C_i/\partial z) - K \qquad (2)$$

where the terms on the right hand side of the equation describe water movement under the influence of gradients of moisture content, Θ; temperature, T; solute concentrations, C_i; and gravity respectively. Here, the D's are soil water diffusivities and K=hydraulic conductivity. Consideration of likely maximum gradients in temperature and solute concentration indicated (de Vries, 1957) that movement under these gradients would only be important at evaporation rates of the order of $10^{-9} ms^{-1}$ (32 mm yr^{-1}) or less. Hence, equation (2) may usually be reduced to two terms and, for steady state evaporation from a water table:

$$E = q = -D_\Theta(\partial\Theta/\partial z) - K \quad (3)$$

Integrating and solving in terms of z, equation (3), can be expressed in terms of the soil moisture suction, S, profile (Gardner, 1958):

$$z = \int dS/(1+q/K) \quad (4)$$

Gardner presented analytical solutions, while Talsma (1963) obtained numerical solutions for equation (4) for a number of field soils. In both cases only liquid transport was assumed; the contribution of vapour transport, obtained as a correction, was quite small for the evaporation rates considered.

Talsma (1963) concluded, from the shape of the resulting soil depth - steady capillary flux relationships, that the critical depth for satisfactory salinity control occurred at soil evaporation rates of $10^{-8}ms^{-1}$. This provided a physical basis for earlier experimental observations. The critical depth was originally defined by Polynov (1930) as the 'maximum height above the water table, to which salts contained in groundwater can rise under natural conditions both by capillary rise and diffusion'. Recently Peck (1978b), considering the likely solute concentrations (~ 10 kg m^{-3}) in groundwater, together with rainless periods (150 days) during which uninterrupted capillary transport would operate, suggested that a more plausible criterion for dryland salinity development would be a steady flux of $10^{-9}ms^{-1}$. Water tables associated with such a flux would be much deeper. For example, for Yolo light clay, a steady flux of $10^{-8}ms^{-1}$ can be maintained from 1 m depth, but $10^{-9}ms^{-1}$ can be maintained from 3.1 m soil depth.

Peck (1978b) cautioned against indiscriminate application of this criterion. Some of the consequences of this choice of steady flux of evaporation rate can be stated by reconsidering the fluxes (equation(2)) due to movement under thermal and solute concentration gradients, which at rates of $10^{-9}ms^{-1}$ should not be neglected (de Vries, 1957). As an example, Philip (1957) showed for deep water tables (small evaporation rates), that a downward heat flux inhibited evaporation more than an upward one increased it. The decrease of evaporation as a function of net shortwave radiation is shown in Fig. 4 for Yolo light clay with water table depths at 1 and 3 m. Net shortwave radiation in the summer for the Western Australian wheatbelt would be around 100-150 cal m^2s^{-1}, so evaporation suppression (Fig. 4) is quite significant for the lower water table.

Finally, Talsma (1963) found very reasonable agreement between field data and the theory of steady state isothermal flux in the liquid phase when no limit was set on evaporation by the moisture conduction properties of the soil (generally from early autumn to late spring). However in the summer period evaporation from the soil was less than that predicted. These observations were made on field plots that received no irrigation water and were thus subjected to quite long periods of upward capillary transport. Thus it would appear, from the preceding arguments, that the criterion suggested by Peck (1978b) would result in critical water table depths that are generally overestimated.

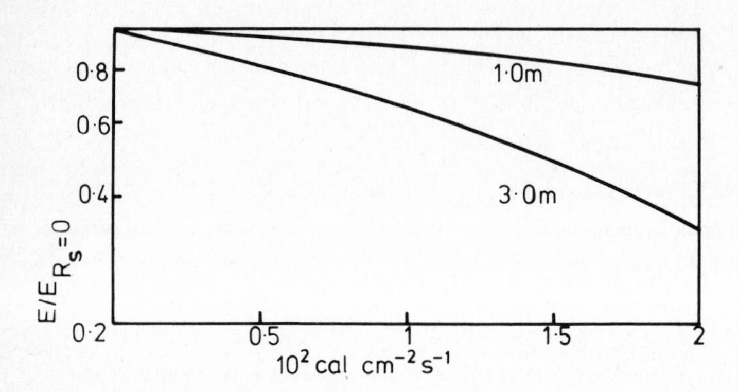

Fig. 4. Decrease of steady evaporation, E, with net incoming radiation, R_s, for Yolo light clay with a water table at 1 and 3 m depth (after Philip, 1957).

5 SUMMARY AND CONCLUSION

In this review of salt transport through catchments and soils, measured soil properties have been used to investigate:

(a) the partition of overland and groundwater flow in catchments;
(b) relationships between catchment discharge and stream water quality;
(c) anion adsorption in weathered acid soils;
(d) the relative importance of convective and diffusive salt fluxes;
(e) critical water table depths for surface soil salinisation under irrigated and dryland conditions.

The methodology used to investigate these aspects of salt transport has been taken from experimental studies carried out in eastern Australian catchments and soils. Such techniques should be equally useful in studies of other areas, such as the dryland salinity problem in Western Australia.

6 REFERENCES

Black, A.S. and Waring, S.A., 1979. Adsorption of nitrate, chloride and sulfate by some highly weathered soils from south-east Queensland. Aust. J. Soil. Res., 17: 271-282.
Bresler, E., 1981. Transport of salts in soils and subsoils. Agric. Water Manage., 4: 35-62
Gardner, W.R., 1958. Some steady state solutions of the unsaturated moisture flow equation with application to evaporation from a water table. Soil Sci., 85: 228-232.

Gardner, W.R., 1965. Movement of nitrogen in soil. In: Bartholomew, W.V. and Clark, F.E. (Editors), Soil Nitrogen. Am. Soc. Agron. Inc., Madison, Wisc., USA. Agronomy 10: 550-572.
Groenewegen, H., 1961. Composition of the soluble and exchangeable ions of the salty soils of the Mirrool Irrigation Area (N.S.W.) J. Soil Sci., 12: 129-141.
Hurle, D.H. and Johnston, C.D., 1979. On the physical basis of salinity in the Darling Range of South Western Australia. Proc. Hydrology and Water Resources Symp., Perth, 164-165. The Institution of Engineers, Canberra, Australia.
Jury, W.A., 1975. Solute travel time estimates for tile drained fields. II. Applications to experimental studies. Soil Sci. Soc. Am. Proc., 39: 1024-1028.
Keng, J.C.W., and Uehara, G., 1974. Chemistry, mineralogy and taxonomy of Oxisols and Ultisols. Soil Crop Sci. Soc. Fl. Proc., 33: 119-126.
Loh, I.C. and Stokes, R.A., 1981. Predicting stream salinity changes in south-western Australia. Agric. Water Manage., 4: 227-254
Peck, A.J., 1971. Transport of salts in unsaturated and saturated soils. In: Talsma, T. and Philip, J.R. (Editors), Salinity and Water Use. Macmillan, London, pp 109-123.
Peck, A.J., 1978a. Salinisation of non-irrigated soils and associated streams: A review. Aust. J. Soil Res., 16: 157-168.
Peck, A.J., 1978b. Note on the role of a shallow aquifer in dry land salinity. Aust. J. Soil Res., 16: 237-240.
Peck, A.J., Johnston, C.D. and Williamson, D.K., 1981. Analyses of solute distributions in deeply weathered soils. Agric. Water Manage., 4: 83-102
Philip, J.R., 1957. Evaporation, and moisture and heat fields in the soil. J. Met., 14: 354-366.
Polynov, B.B., 1930. Determination of critical depth of occurrence of the groundwater level salinising soils. Izv. Sector Hydrotechnics and Hydrotechnical Constructions. No. 22. Leningrad.
Raats, P.A.C., 1981. Residence times of water and solutes within and below the rootzone. Agric. Water Manage., 4: 63-82
Smiles, D.E. and Philip, J.R., 1978. Solute transport during absorption of water by soil: Laboratory studies and their practical implications. Soil Sci. Soc. Am. J., 42: 537-544.
Stace, H.C.T., Hubble, G.D., Brewer, R., Northcote, K.H., Sleeman, J.R., Mulcahy, M.J. and Hallsworth, E.C., 1968. A Handbook of Australian Soils, Rellim Tech. Publ., Glenside, S.A.
Talsma, T., 1963.The control of saline groundwater. Meded. Landbouwhogesch., Wageningen, 63: 1-68.
Talsma, T., 1967. Leaching of tile-drained, saline soils. Aust. J. Soil Res., 5: 37-46.
Talsma, T., 1968. Environmental studies of the Coleambally Irrigation Area and surrounding districts, III: Soil Salinity. Bull. (Land Use Series) Water Conserv. Irrig. Commission, N.S.W.., 2: 35-48.
Talsma, T. and Hallam, P.M., 1980. Hydraulic conductivity measurement of forest catchments. Aust. J. Soil Res., 18: 139-148.
Talsma, T., Mansell, R.S. and Hallam, P.M., 1980. Potassium and chloride movement in a forest soil under simulated rainfall. Aust. J. Soil Res., 18: 333-342.
Vries, D.A. de, 1957. Soil water movement and evaporation from bare soil. Proc. 2nd Aust. Conf. Soil Sci., Melbourne , 1957. Vol. I, Part II, pp 48, 1-12. CSIRO, Melbourne.
Wetselaar, R., 1962. Nitrate distribution in tropical soils: III. Downward movement and accumulation of nitrate in the subsoil. Plant and Soil, 16: 19-31.

SALINE SEEP DEVELOPMENT AND CONTROL IN THE NORTH AMERICAN GREAT PLAINS - HYDROGEOLOGICAL ASPECTS

M.R. Miller
Montana Bureau of Mines and Geology, U.S.A.

P.L. BROWN
U.S.D.A., Fort Benton, U.S.A.

J.J. DONOVAN, R.N. BERGATINO, J.L. SONDEREGGER AND F.A. SCHMIDT
Montana Bureau of Mines and Geology, U.S.A.

ABSTRACT

Miller, M.R., Brown, P.L., Donovan, J.J., Bergatino, R.N., Sonderegger, J.L. and Schmidt, F.A., 1981. Saline seep development and control in the North American Great Plains - hydrogeological aspects. Agric. Water Manage., 1981.

The widespread occurrence and rapid growth of saline seeps on or adjacent to cultivated drylands has become one of the most serious conservation problems in the Great Plains region of North America. Dryland salinity, hardly recognized 35 years ago, has now taken approximately 0.8 million ha out of crop production. Equally serious as the loss of arable land is the local and potential regional deterioration of surface and shallow groundwater resources which, in many areas, are the primary sources of potable water. Significant concentrations of trace metals, particularly selenium, as well as high nutrient levels, have been found in many ground and surface water samples. A number of livestock, wildlife, and fish kills have been noted and are believed to be directly related to the saline seep problem.

The best solution to the problem is to utilize precipitation where it falls, before it moves beneath the root zone. A number of conservation practices have been identified, but three of the most successful control practices are: a) growing deep-rooted perennial crops; b) switching to flexible, intensive cropping systems; and c) draining selected upland, freshwater potholes. On one research site where these practices were applied during the past 10 years, significant results include: lowering the water table an average of 2.5 m; a decrease in salinity of groundwater by approximately 25%; a 75% reduction in soil salinity in the seep area, from the surface to 0.6 m; and a decrease in the salt-affected area from 12 ha to less than 0.4 ha.

1 INTRODUCTION

Saline seeps are recently developed saline soils in nonirrigated areas that are wet some or all of the time, often with white salt crusts, where crop or grass production is reduced or eliminated. Once developed, they often grow at an average rate of about 10% a year, soon making it difficult to farm across and very time consuming to farm around.

During the last 35 years, large outbreaks of saline seep have appeared in the semiarid Northern Great Plains, which encompass large portions of the prairie provinces of Canada - Alberta, Saskatchewan, and Manitoba - and the northern plains states of the United States - Montana, and North and South Dakota (Fig. 1). This is the major grain-growing region in Canada and one of the major grain-growing areas of the United States. The alarming growth of the dryland salinity problem, particularly during the last 10 years, has brought about an active diversified research program involving numerous investigators from state, provincial, and federal organizations. Research has centered around three general areas:

a. Engineering - new machinery design and application, surface and subsurface drainage.

b. Agriculture - new and flexible cropping systems; soil and water management; plant-water use; rooting depths, and salt tolerance of adaptable crops; weed disease, and residue problems.

c. Hydrogeology - formation, development, causes and controls, salt loads, water quality, groundwater flow systems, local and regional impacts.

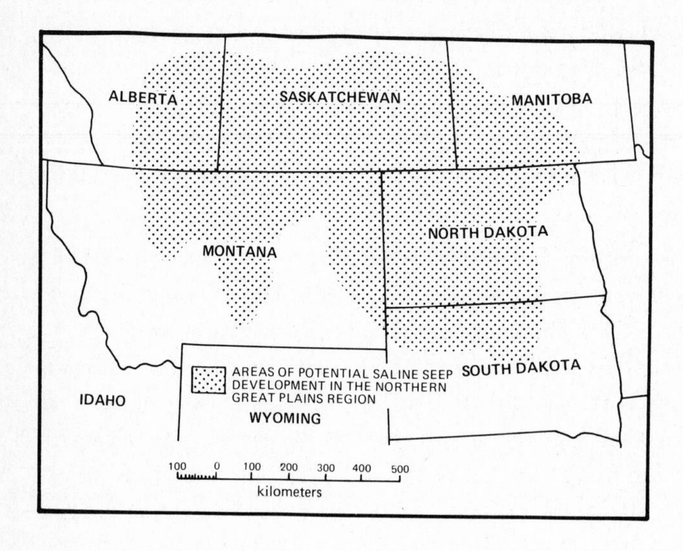

Fig. 1. Area of potential saline seep development on the Northern Great Plains.

The objective of this paper is to summarize the hydrogeological aspects of the problem emphasizing formation and development, regional extent and potential impacts, effective control practices and detailed evaluation of one specific research site over a 10-year period.

2 SALINE SEEP FORMATION AND DEVELOPMENT

The origin and development of dryland salinity has been characterized and documented at a number of research sites across the Northern Great Plains including 25 areas in Montana (Luken, 1962; Ballantyne, 1963; Greenlee et al., 1968; Miller, 1971; Ferguson et al., 1972; Bahls and Miller, 1973; McCracken, 1973; Halvorson and Black, 1974; Miller et al., 1975; Worcester et al., 1975; Custer, 1976; Doering and Sandoval, 1976; Halvorson and Reule, 1976; Doering and Sandoval, 1978; Holm, 1978; Lewis et al., 1979).

A schematic diagram illustrating the formation of a typical saline seep is shown in Fig. 2. Research results show that dryland salinity is caused by a combination of cultural, climatic, and hydrogeological conditions. The importance of any one of these factors may vary significantly from area to area; but the formation of saline seep follows the same general process throughout the region. This process starts with excess water percolating downward beneath the root zone, picking up soluble salts, accumulating on shallow, less-permeable layers (typically shale) and forming a local groundwater flow system. The flow system moves saline water from the recharge to the discharge area (seep) where it evaporates, depositing the salts on the surface.

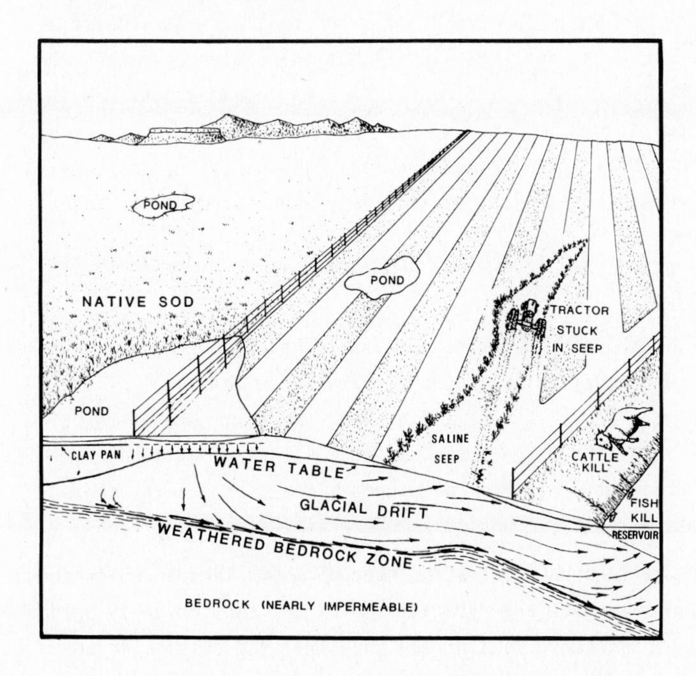

Fig. 2. Schematic diagram illustrating the formation of a saline seep.

2.1 Cultural Factors

Any land use practice which allows excess moisture to migrate downward through the soil profile beneath the root zone, can contribute to the formation and growth of dryland salinity. A number of land use practices have been noted as contributing to saline seep, but by far the most important change in the northern plains is the widespread use of the alternate crop-fallow (summer-fallow) farming system. Most soils store only 100 to 200 mm of water in the root zone during a fallow period. Once recharged by precipitation, any additional water entering the soil moves to the water table and may resurface downslope as a saline seep. Sandy soils which have very limited holding capacity and allow water to infiltrate rapidly, can readily recharge the local groundwater flow system. Because of the widespread use of the summer-fallow system (5.5 million ha in Montana alone), relative to other land use practices, it becomes apparent why it is the dominant cause.

Practices which cause excess snow accumulation, such as shelter belts, single-row windbreaks, railroad and highway cuts, snow fences, and snowbanks left by snowplows, can locally contribute to saline seep growth. In Alberta, Sommerfeldt (1976) has documented the influence of large snowbanks on the growth of nearby seeps. On one research site in Montana, the melting of a 4 to 5 m snowbank left by snowplows raised the water table adjacent to the roadway more than 1.3 m. Constructed ponds, dugouts and small farm reservoirs can be important locally if they are located on recharge areas and leak significant quantities of water. On one research site in central Montana, a 1.5 ha saline area appeared just down-gradient from a small leaky stockwater reservoir. The upper portion of the seep was located on native sod. Dryland salinity has been occasionally noticed up-gradient from roadbeds crossing drainways, particularly when culverts are absent, too small or plugged.

2.2 Climatic Factors

The climate of the Northern Great Plains region east of the Rocky Mountains is arid to semiarid and is characterized by moderately low but widely variant precipitation (250 mm to 450 mm a year); low humidity; warm summers; cold winters; and many days of sunshine and wind. The region is in the westerlies wind belt throughout the year, so frequent large frontal storms migrate across the region, which commonly experiences a rapidly changing sequence of weather conditions.

Because of numerous dry air masses originating over northern Canada and sweeping southward across the plains, winters are typically cold, but snow cover is seldom more than 0.1 m. Strong winds remove most of the snow from the cultivated upland areas and deposit it in drifts along the north and north-east facing slopes.

When the drifts melt in the spring, the underlying soil is readily recharged with meltwater, enhancing seep development (commonly called "north slope alkali" in the early days) in these areas. Mild weather occasionally occurs a few times each winter because of moderate to strong winds from the west or south-west, locally referred to as "chinooks". The western portion of the region lies in the "chinook" belt. About half of the total annual precipitation falls during April, May and June when evaporation is low and crop water-use is limited due to recent sowing and crop emergence. Crop yields are directly related to the rainfall during these months. As a general rule, but with numerous exceptions, the largest outbreaks of dryland salinity occur in areas with more than 350 mm of annual precipitation. Close correspondence between local precipitation and water table fluctuations indicates that water is moving through the soil profile beneath the root zone and accumulating on the underlying bedrock. Beneath fallow areas, a water table rise of 50 mm to 3 m can be expected during wet years. These water table highs gradually decline during the rest of the year, but normally do not reach the previous year's low, indicating a continued buildup of water over the years. As a result, each succeeding wet cycle intensifies the dryland salinity problem.

2.3 Hydrogeological Factors

In addition to being directly associated with the alternate crop-fallow farming system, saline seep occurrence is closely related to the surficial and bedrock geology. A detailed evaluation of the geology and hydrology of the Northern Plains area is beyond the scope of this paper, but because of the widespread distribution and recent expansion of saline seeps and the potential deterioration of the surface water and shallow groundwater resources throughout the region, a brief description of the hydrogeological setting is in order.

2.3.1 Bedrock geology

The geological history of the North American Great Plains includes long periods of sedimentation, occasional emplacement of volcanic and plutonic igneous rocks, regional uplift, erosion, and glaciation. Throughout most of Paleozoic (225 to 600 million years ago) and Mesozoic (70 to 225 million years ago) time, thousands of meters of predominantly marine sediments were deposited in this region. At the end of the Mesozoic Era (Late Cretaceous time) and extending into Early Tertiary time (50 to 70 million years ago), the entire area was uplifted, faulted, and folded, producing the Rocky Mountains to the west and south, tilting the sedimentary rocks underlying the region gently to the north-east (0.5 to 3 degrees), and subjecting the area to erosion. Continued erosion during the Tertiary period stripped away the uppermost Cretaceous sediments, exposing vast areas of predominantly dark, salt-laden, marine shales of Cretaceous age. The erosion products were subsequently redeposited further east as nonmarine shale, silt, sand and coal of Tertiary age. As a result there remains a bedrock landscape essentially composed of four major units: a) Colorado Shale (oldest), b) Judith River-Claggett-Eagle Formations, c) Bearpaw Shale, and d) Tertiary Fort Union Formation (youngest). Equivalent units are present in the Dakotas and the Canadian provinces but have different geological nomenclature. Fig. 3. shows the areal extent of each unit in the Northern Plains regions.

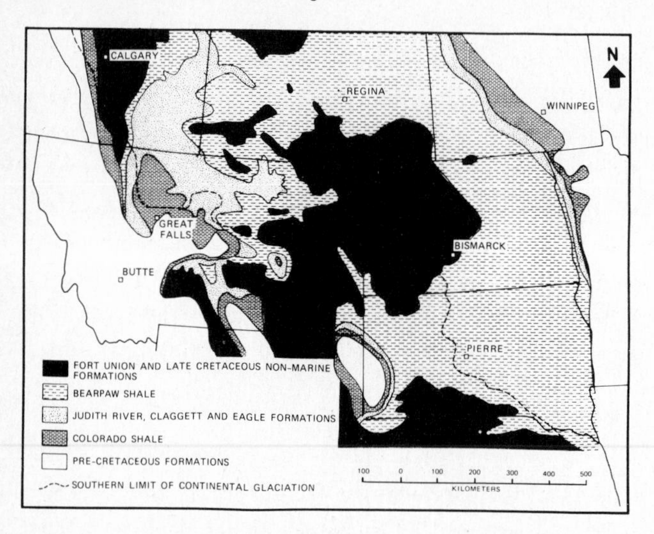

Fig. 3. Generalized geological map of the Northern Great Plains Region.

The Colorado and Bearpaw Shale units underlying large areas of Montana, Saskatchewan, Manitoba and the Dakotas, are typically thick (up to 650 and 350 m, respectively), black marine shales with numerous concretions and thin bentonite beds. Both units are loaded with salt, predominantly secondary gypsum, some calcite, and an abundance of other elements. Lithologic information obtained from more than 600 test holes (primarily located in the Colorado Shale region) indicates that the weathered zone at the till-shale contact is 1 to 10 m thick. The underlying unweathered shale is dry, suggesting that it is virtually impermeable.

The Judith River-Claggett-Eagle unit, covering large areas of Montana and Alberta, includes a number of marine and nonmarine shale and sandstone formations. Relative to the Colorado and Bearpaw, it has more silt and sand, but also contains numerous beds of bentonite, shale, and mudstone which perch most of the shallow groundwater. Undoubtedly some of the perched water leaks through to the next dense layer, or possibly joins a deeper groundwater flow system. In general, the occurrence of dryland salinity in areas underlain by this unit is not as extensive at the surface, but areal groundwater contamination is more widespread, suggesting more permeable materials and the presence of a deeper flow system.

The nonmarine Fort Union unit underlying eastern Montana, southern Saskatchewan, and western North Dakota consists of interbedded siltstone, shale, sandstone, coal and underclay. Water that percolates beneath the root zone moves relatively fast down through the silt and sand until it encounters the dense underclay beneath a coal seam where it forms a perched water table. The perched water moves laterally along the coal seam until it breaks out at the surface (Halvorson and Black, 1974; Doering and Sandoval, 1976).

The overall soluble salt load contained in each geological unit is reflected in the concentration of total dissolved solids (TDS) contained in the groundwater at the discharge site. In general, concentrations of TDS vary from 20 000 to 55 000 mg L^{-1} in the Colorado and Bearpaw Shale units; 10 000 to 25 000 mg L^{-1} in the Judith River-Claggett-Eagle unit; and 3 000 to 15 000 mg L^{-1} in the Fort Union unit (Miller et al., 1978).

In summary, the bedrock geology of the region is important in the formation of saline seeps for several reasons:

a. The bedrock units influence the type of material and amount of salt that are incorporated in the overlying glacial drift and associated groundwater.

b. The fresh shale, thin bentonite beds, and underclay create a virtually impermeable layer, prohibiting significant fluid movement within the units.

c. The weathered shale zone and thin coal seams provide a laterally extensive permeable layer that allows water stored in the overlying glacial till to migrate slowly downgradient, to accumulate in low areas or to drain into nearby coulees.

d. Because the shale is easily eroded, numerous valleys were cut into it. The valleys were subsequently filled with till during glacial time and, when the glacier retreated, it left a maze of abandoned channels and isolated depressions in which saline water could accumulate.

2.3.2. Surficial geology

During Pleistocene time (20 000 to 2 million years ago), much of the plains region was covered one or more times by glacial ice (Fig. 3). The ice left a mantle of unconsolidated, poorly sorted deposits called glacial till or drift, which filled most of the pre-existing valleys and coulees, and left a relatively flat to gently rolling glacial plain (swell and swale topography). The ice blocked the preglacial river systems, forcing the streams to change course numerous times and in some places, to cut new channels. Since the retreat of the glacial ice 20 000 years ago, the area has again been subjected to erosion by running water and the present-day drainage pattern has evolved.

In Montana, drill-hole information and exposures along major streams indicate that the thickness of glacial till ranges from 1 to 25 m, however, till thicknesses exceeding 100 m are not uncommon in Canada. Except for the upper metre, the entire till profile contains abundant calcium and sodium salts. To determine the potentially soluble materials present in the soils, subsoils, glacial till, and bedrock, 20 drill-hole samples from 19 research sites across Montana were batch-leached at 55°C, with a pH $\geq$ 5.0 for 21 hours to simulate accelerated weathering conditions. This, plus the use of a carbonate bomb, x-ray diffraction, and computer calculations employing water well samples have resulted in the following observations (Sonderegger et al., 1978):

a. Sufficient sodium is available in the shallow aquifer systems (upper 6 m) to maintain existing saline seeps for the next 25 years to 100-plus years.

b. Dolomite is believed to be forming in many of the seep discharge areas.

c. Alkaline earth content of the leachate may be primarily controlled by carbonate and sulfate minerals.

d. Total leachable dissolved constitutents ranged from 2.2 to 23.5 weight % (averaged 8.06%) of the original sample, and the weight percent of the noncarbonate portion varied from 0.66 to 10.1% (averaged 3.71%).

The till is predominantly unsorted clay and silt with well-rounded pebbles scattered throughout. X-ray analysis of the clay fraction of selected till samples indicated 80% montmorillonite, 15% illite, and 5% kaolinite or chlorite (Miller, 1971). Scattered sand and gravel lenses were encountered in some test holes; however, they are normally thin, discontinuous, and small in areal extent. Numerous vertical joints have been observed in the till both in surface exposures and in cores collected from test holes. These vertical joints are believed to accelerate the vertical movement of water through the till.

The glacial landscape of the Northern Plains is dotted with literally thousands of potholes of all sizes which hold water for varying periods of time. Frequently these ponds occupy part or all of the recharge area; contain large quantities of water during the wet years; and are actively cultivated during the dry years, disturbing the shallow claypan that normally develops beneath them. Three of these potholes were carefully monitored during the summer of 1975 to determine the amount of water added to the shallow groundwater flow system. Results shown in Table 1 clearly indicate that ponds can contribute significant quantities of water to the profile, seriously aggravating the seep problem downslope (Brown and Miller, 1978). The size and rate of growth of the seep are related directly to the amount of groundwater in storage (recharge area) upgradient from the saline area.

TABLE 1

Estimated water infiltration from three potholes.

Pothole No.	Original Size Hectares	Measurement Period Days	Est. Water Infiltration Hectare - m
Ry1	16.0	79	1.91
Hz1	1.9	70	0.28
Hz2	3.3	64	0.37

In summary, glacial drift has several important effects on saline seep development:

a. The poor drainage associated with the swell-and-swale topography typical of glaciated terrains allows water to pond for extended periods of time, especially during the spring months.

b. The extensive vertical joints and fractures of the glacial till allow water to percolate vertically through the drift.

c. Because the till and weathered shale have low horizontal permeability, lateral drainage of excess water in them is slow.

d. The mantle of glacial deposits provides an excellent soil-moisture storage reservoir ideally suited for intensive dryland farming. Unfortunately, it also provides an effective groundwater reservoir in which water can slowly accumulate until it appears at the surface as a seep.

e. The drift contains an extremely abundant supply of water-soluble salts throughout the region.

When considering the entire region, most of the saline seeps (95%) occur in the glaciated region where the clay till provides an excellent soil moisture reservoir ideal for dryland farming. In Montana, however, several large outbreaks of dryland salinity have appeared in the unglaciated central, south-central, and eastern portions of the state. These unglaciated areas are characterized by shallow soils derived from underlying Colorado, Bearpaw, and Fort Union units, on which the alternate crop-fallow farming system is extensively used. Because of the shallow and somewhat more permeable soils, seep formation is normally much more rapid because the excess water moves from the recharge to discharge area during the same year. Seeps in these shallow soils expand 50 to 200% a year during wet cycles.

3 REGIONAL EXTENT

As noted earlier, and illustrated in Fig. 1, hydrogeological conditions conducive to the formation of saline seep exist over more than 590 000 km^2 in the North American Great Plains (Miller, 1971; Miller et al., 1976). The dominant cropping system over this entire region is the alternate crop-fallow farming system. Accurate estimates of seep development are difficult to obtain, particularly for the earlier days (early 1950's to mid 1970's) when the problem was much smaller and citizens were unaware of its existence. Original estimates were mostly guesswork based on telephone and/or letter surveys of local, county, state and federal resource personnel. Once the problem became widespread and easily recognized, revised estimates were tabulated, generally being somewhat higher than previous estimates. The best estimate for the entire region was reported by Vander Pluym (1978), in which he estimated that about 0.8 million ha were severely affected by dryland salinity and possible another 2.7 million ha were affected to some extent.

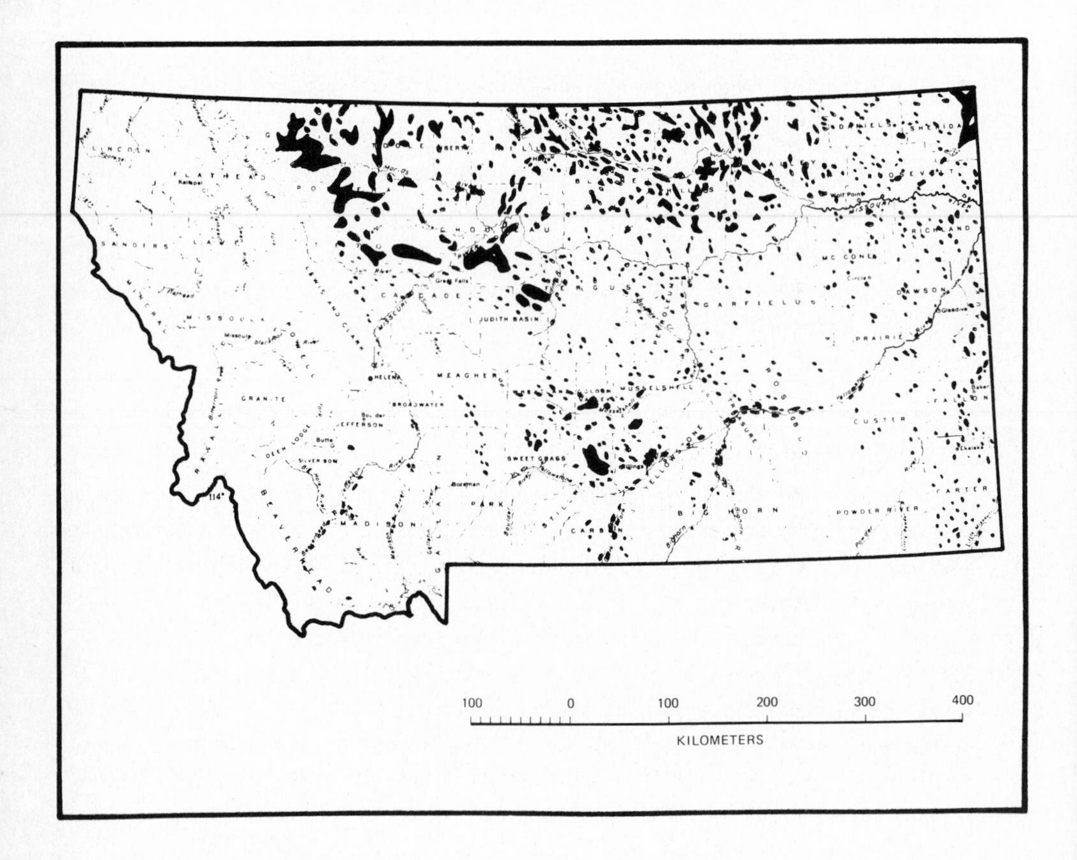

Fig. 4. Distribution of saline seeps in Montana - 1978.

During the last 10 years, estimates of seep-affected areas in Montana have risen from 32 000 ha in 1971 (survey from soil scientists of the Soil Conservation Service), to 57 000 ha in 1974 (survey of U.S. Department of Agriculture county committees), to 69 000 ha in 1976 (Montana Department of State Lands assessment), to an estimated 81 000 ha in 1978.

The general distribution of saline seeps in Montana is shown in Fig. 4. The map and the estimate of 81 000 ha are based on an aerial and field reconnaissance survey conducted by the Montana Bureau of Mines and Geology in 1978. The survey was made primarily to obtain a regional assessment of the saline seep problem and a water-quality inventory of the Montana plains (Miller et al., 1978). Dryland salinity areas were delineated on photo index sheets and on county highway maps, then transferred to base maps. The aerial survey allowed field crews to concentrate on critical areas, and provided the first uniformly documented distribution of saline areas in Montana.

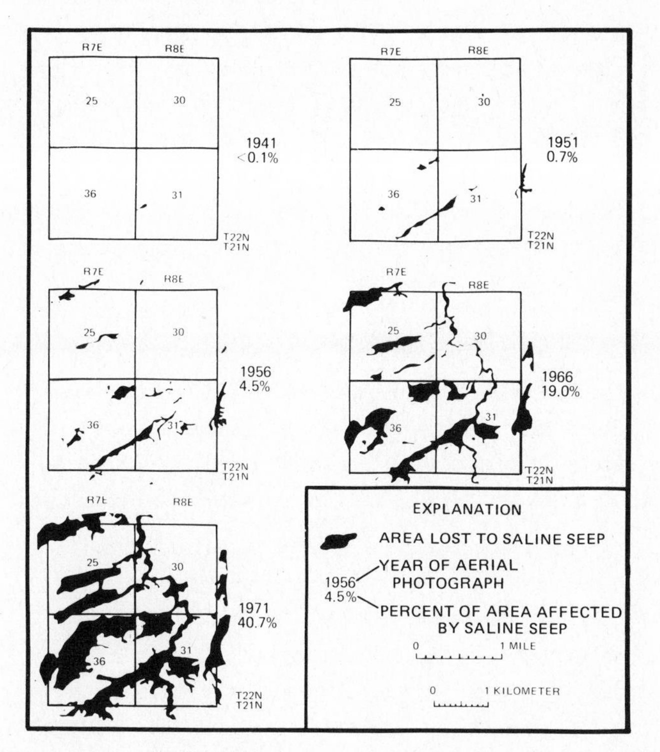

Fig. 5. Rate of growth of a saline seep in a 10 km^2 area of Montana (1941 to 1971).

On a regionwide basis, the area of saline seep appears to be expanding at an average rate of about 10% a year. The rate varies substantially from year to year, depending upon climate, but the general trend is toward significant increase. After each wet cycle (years with average to above average spring precipitation), expansion of seep areas by 20 to 200% is not uncommon. On the other hand, very little or no expansion may occur during dry cycles. To document the rate of growth, a 50 km^2 watershed near Fort Benton, Montana, was selected, and areas of dryland salinity were delineated on aerial photographs taken in the years 1941, 1951, 1956, 1966 and 1971. Over the 30-year period, the seep-affected area for the entire watershed increased from less than 0.1% to 19.4% (Miller et al., 1974; Bahls and Miller, 1973). In one 10 km^2 area, illustrated in Fig. 5, over 40% was taken out of production during the same period.

4 LOCAL AND REGIONAL IMPACTS

Equally serious as the loss of arable land is the local and potential regional deterioration of surface and shallow groundwater resources which, in many areas, are the primary sources of potable water. During a regional water-quality survey of wells, springs, streams, and reservoirs in 42 counties of Montana, encompassing roughly 305 000 km^2, more than 2 800 sites were evaluated and 452 samples were collected of which 247 were analyzed for trace elements (Miller et al., 1978). Of more than 2 800 sites, 14% had TDS less than 500 mg L^{-1}; 16% between 500 and 1 000 mg L^{-1}; 64% between 1 000 and 10 000 mg L^{-1}; and 6% had more than 10 000 mg L^{-1}. Because precipitation was above average during the 2-year period of the project, the reported TDS values obtained from all the surface water sites were probably low. As shown in Table 2, the concentrations of most constituents are considerably higher in wells penetrating the glaciated Colorado Shale area than in the glaciated Judith River-Claggett-Eagle region. Groundwater sampling of the Bearpaw Shale and Fort Union region was quite limited. In general, groundwater from the Bearpaw Shale area resembles the Colorado Shale, in chemical composition and concentration of dissolved solids. Fort Union groundwaters were similar to Judith River-type waters in composition but somewhat lower in concentration.

Significant concentrations of trace elements, particularly selenium, were found in many of the groundwater samples. Of the 161 samples analyzed for selenium, 40% had concentrations greater than the 0.01 mg L^{-1} limit for potable water set by the U.S. Public Health Service; some values were as high as 2.0 mg L^{-1}. The Se values were considerably higher in the Colorado Shale region, having a mean value of 0.308 mg L^{-1} with 59% of samples having more than 0.01 mg L^{-1} and 38% over 0.1 mg L^{-1} compared to the Judith River region, which had a mean concentration of 0.028 mg L^{-1} with 24% of samples having more than 0.01 mg L^{-1} and 7% having more than 0.1 mg L^{-1} (Miller et al., 1978; Donovan et al., 1979).

TABLE 2

Mean concentration of groundwater in Montana Plains.

	Colorado Shale Unit	Judith River-Claggett Eagle Unit
	Mean Concentrations (mg L^{-1})	
Ca	275	120
Mg	866	71
Na	1317	670
K	17.7	6.1
Fe	0.3	0.3
SiO_2	12	12
HCO_3	534	722
Cl	141	177
SO_4	6041	1143
NO_3	57	6
Se	0.308	0.028
Sr	4.6	2.0
TDS	9262	2928
pH	7.62	7.89

Numerous fish and livestock kills have been reported throughout the region, particularly in the Colorado Shale area. It is not uncommon for reservoirs in seep-affected areas to increase in TDS at a rate of 500 mg L^{-1} a year (Miller, 1971). Most of the fish kills occur when TDS increase to about 7 000 mg L^{-1}. Nitrate values in the reservoirs commonly exceed 100 mg L^{-1}. As far as is known, autopsies have not been performed to determine exact cause of death. Symptoms vary considerably; some animals (usually ones brought in from other areas) are killed almost immediately, others slowly go blind (possibly by nitrate poisoning), while others appear to dehydrate (possibly high TDS, particularly $MgSO_4$).

The lack of a detailed historical data base makes it difficult to quantify the effects of saline seep development on the surface and shallow groundwater resources of the region; however, data suggest that significant water-quality deterioration has occurred in the glaciated portion of Montana where dryland farming has been a way of life for many years. A number of areas in the unglaciated region in central and eastern Montana also appear to have serious contamination. Probably the most convincing statements come from the landowners. Some of the more frequent comments are:

a. "Over the last 5 (to 20) years we have had to drill 2 (to 4) wells, each one deeper than the last, to get good water."

b. "All the wells in the area have gone bad. That is why we hooked up to the rural water distribution system."

c. "During the last 5 (to 15) years, salty springs have appeared in my coulees and now the banks are sliding into the draw."

d. "I don't have any fresh water left on my place, so I guess I'll sell all my cattle and plow up the rest of my pasture."

e. "Last year my shelter belt started to die."

All of these statements, and many more, imply that the local groundwater flow system is out of equilibrium, causing the salts to be flushed out of the profile, thereby rapidly contaminating the water resources of the area.

5 DRYLAND SALINITY CONTROL PRACTICES

As previously mentioned, saline seeps are caused by cultural, climatic, and hydrogeological factors. Because the climatic and geological factors are natural and essentially unchangeable, solutions to the problem have to be centered primarily around the cultural aspects, with some mechanical modifications of the hydrologic system. State and federal agencies who are concerned about the problem are currently developing a list of conservation and crop production practices to help control saline seeps. These practices place the major emphasis on utilizing precipitation before it moves below the root zone. Some of the practices, including both cropping and mechanical procedures are:

a. Use of deep-rooted perennial crops such as alfalfa or alfalfa-grass mixture to dry out deep subsoils quickly. Alfalfa will root to 6 m in 4 to 5 years and use more than 760 mm of water (Brown et al., 1976; Brown and Cleary, 1978). This creates a significant storage reservoir for excess water.

b. Use of flexible cropping systems (Krall and Jackson, 1978). Flexible cropping means sowing when stored soil water and rainfall probabilities are favourable for a satisfactory yield, and fallowing only when crop prospects are unfavorable. Flexible cropping involves careful management and planning. Several of the most important methods to ensure the success of the system include:

1) Use soil moisture penetrometer regularly to determine available water.

2) Test soil regularly and fertilize accordingly. Adequate fertilization increases rooting depths from 0.3 to 0.6 m and increases water use from 25 to 75 mm.

3) If necessary, rotate safflower or sunflower every three or four years to use deep soil water not used by grains.

4) Delineate gravelly and sandy areas with drill rig, soil probe, and/or detailed soil maps. These areas are often important recharge areas for seeps. They should be farmed intensively and separately from the surrounding farmland. Deep-rooted perennials may also be sown on these areas.

5) Usually leave stubble standing over winter to trap snow and help prevent blowing and drifting conditions.

c. Sow alfalfa-grass mixture during late fall in and around saline seep areas. Use a variety of salt-tolerant grasses and repeat sowing in subsequent years to move stand into seep centre. Mow, remove or burn all vegetation each fall.

d. Consider returning areas of residual saline soils back to salt-tolerant grasses, to be used for hay or pasture.

e. Observe excess snow accumulations. If practical, remove or alter barriers that trap excess snow, leave standing stubble to prevent drifting snow, and construct roadways which will normally blow clean. Areas where snow accumulation cannot be avoided should be sown to deep-rooted perennials.

f. Establish grass waterways in drainageways or sluiceways to remove runoff water efficiently. Check and clean all culverts beneath road beds to avoid ponding of surface water, especially if the ponded water is in the recharge area. Install additional culverts as needed.

g. Drain upland recharge areas (Class I and II wetlands) that contribute water to saline seep, particularly those which are frequently cultivated.

1) Land shape any small area that collects and holds water for over 24 hours.

2) Consider irrigation of adjacent cropland from upland freshwater ponds that cannot be readily drained (Oosterveld, 1978).

3) Avoid cultivating large potholes that normally hold water for extended periods of time (Class III or higher wetlands). Once the underlying claypan is disrupted, seepage is commonly increased significantly.

4) Completely drain all man-made ponds and dugouts that contribute to dryland salinity, especially those which are no longer used.

h. Install observation wells 3 m or more in depth at strategic locations on each farm. Ideally, in order to effectively monitor change in local groundwater flow systems, the wells should be located in both the recharge and discharge areas, and be deep enough to reach the underlying impermeable layer. A rising water table, particularly in the recharge area, means that the crops and cropping practices should be changed quickly. A stable or falling water table shows that current cropping practices are using most or all of the precipitation as it falls.

It should be emphasized that cause and control of dryland salinity can vary significantly from farm to farm, and especially from area to area. As a general rule, however, cropping practices alone will control or reclaim about 60 to 70% of the saline seeps, while mechanical practices or a combination of cropping and mechanical practices will be necessary for control of the other 30 to 40%. Each farm needs to be carefully evaluated to determine all sources of excess water and to implement the appropriate control practices.

During the past year, a special saline seep extension team was funded by the State of Montana to assist farmers in the Triangle area (nine counties in north-central Montana). The purposes of the team are to evaluate saline seep problems on a farm-by-farm basis; to provide information and assistance to landowners and conservation districts; and to develop cooperative control and management plans. It is too soon to tell how effective the program will be, but it appears to be succeeding because more than 180 farmers are currently signed up to receive assistance.

6 RESEARCH RESULTS FROM ONE TYPICAL SITE-HANFORD-BRAMLETTE TEST AREA

The Hanford-Bramlette test site, about 56 km north-east of Great Falls, Montana, has an area of approximately 65 ha (Fig. 6). The site is characterized by a poorly drained upland recharge area (25 ha) sloping gently to the north and east into a lowland discharge area (20 ha), which has been extensively salinized. The site is dissected by two small coulees on the north-west and east sides. Past cropping history for the test area is as follows: field A, 1959 to 1969 crested wheat grass (soil bank program), 1970 barley, 1971 to 1974 alfalfa, 1975 to 1979 annual cropping; field B, 1959 to 1970 alternate crop-fallow, 1971 to 1979 alfalfa on western part, annual cropping on the remainder; field C, 1959 to 1970 alternate crop-fallow, 1971 to 1979 annual cropping; fields D and E, 1959 to 1979 alternate crop-fallow.

Since June 1969, 68 test holes have been drilled, logged, and cased on the site to determine depth to bedrock, amount of groundwater in storage, lithologic character of the glacial till and shale, and the distribution of salt load throughout the profile. In addition, water level changes have been monitored, water-quality samples have been collected and analyzed, aquifer and infiltration tests have been completed, three small drain systems have been installed and minitored, and water use by three cropping rotations has been evaluated (Miller et al., 1974;

Brown and Miller, 1978).

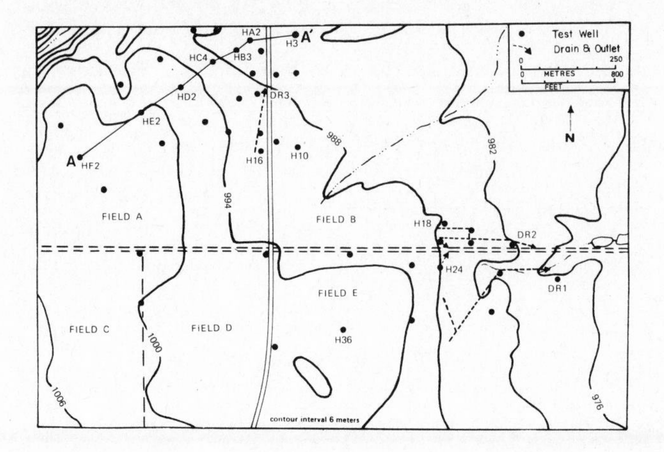

Fig. 6. Index and topographic map of the Hanford-Bramlette test area, Choteau county, Montana.

The area is mantled by glacial till (about 10 m thick) and is underlain by the Colorado Shale (more than 550 m thick). During drilling, two tills, representing two separate ice advances, were encountered - the upper till is normally thicker and is buff-to-tan, whereas the lower till is generally thin (less than 1 m), somewhat more dense, and light gray in color. The till is predominantly unsorted clay and silt with well-rounded pebbles throughout. A few discontinuous sandy lenses were found in some of the test holes. Numerous vertical joints, oxidized and filled with salt crystals (primarily selenite), were observed in cores and in exposures along coulee edges (Miller, 1971). Except for the upper metre, the entire till profile and the 1 to 3 m thick weathered zone of the shale contains about 6% total soluble salts by weight and more than 3% noncarbonate soluble salts by weight (Sonderegger et al., 1978). The configuration of the bedrock surface (Fig. 7) generally conforms to the surface topography. Two bedrock ridges sloping gently to the north-east are present, indicating preglacial drainage patterns and create small discharge areas for groundwater to accumulate.

A water-bearing zone averaging about 6 m thick was encountered in every test hole. Typically, water readily filled the test holes when the weathered shale zone was reached or when a thin saturated meltwater layer was encountered. The weathered shale horizon is the only slightly permeable layer with any lateral extent, indicating that most of the groundwater movement downslope takes place along this layer (Fig. 8). As expected, aquifer test and hydrograph data indicated very low transmissivity values, averaging 0.52 m^2 day^{-1} and a seepage rate of about 1.6 m^3 day^{-1} ha^{-1} (Miller et al., 1974). Water contained in the weathered zone is commonly under artesian pressure in the discharge areas. This pressure caused a number of test holes to flow small quantities of water (up to 0.06 L s^{-1}) for many days after they were drilled - one test hole (H16) located near the center of the discharge area flowed approximately 0.05 L s^{-1} for more than 2 years (1970 to 1972). Confined groundwater conditions are further supported by the high barometric efficiency (20 to 85%) exhibited by the test wells.

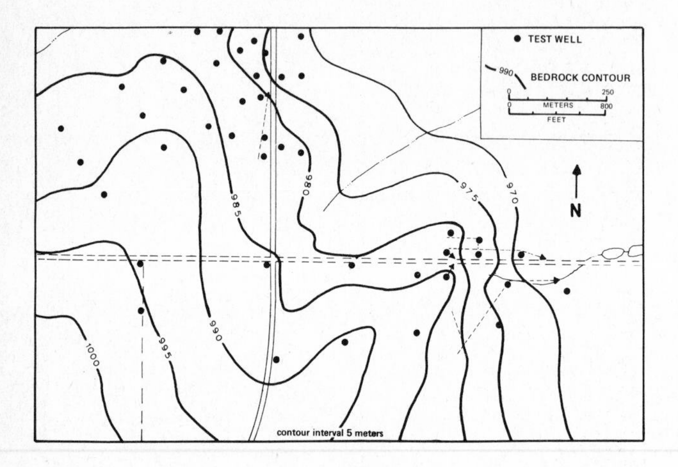

Fig. 7. Configuration of eroded surface of the Colorado Shale underlying the Hanford-Bramlette test area.

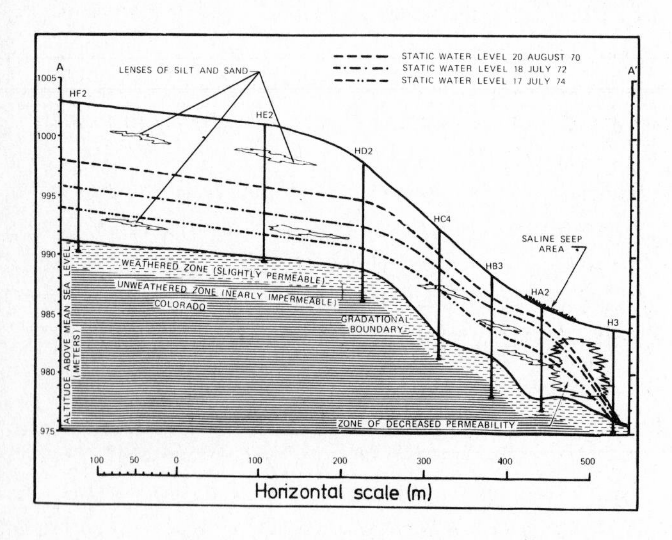

Fig. 8. Geological section A-A' across Hanford-Bramlette test area showing change in static water level from 1970 to 1974.

Infiltration rates from tests conducted in the area varied from 50 to 125 mm day^{-1}. One test, which consisted of adding 42 m^3 of water to a 81 m^2 area during a 60-hr. period (infiltration rate of about 85 mm day^{-1}), raised the water level in well HF2 (located 7 m from edge of plot) from 6.5 m to 5.8 m, 8 days after the test (Miller et al., 1974). It is estimated that 1 to 4% of the annual precipitation percolates through the till profile to join the shallow groundwater flow system under native sod conditions in the glaciated portion of the Great Plains. Based on long-term hydrograph records, it appears that the alternate crop-fallow farming system allows approximately 7 to 15% of the annual precipitation to pass through to the groundwater system. The higher percentage occurs during years with average to above-average spring precipitation.

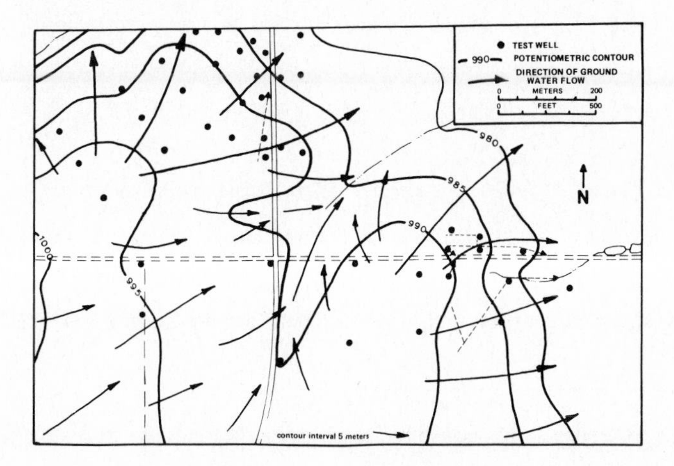

Fig. 9. Configuration of the potentiometric surface and direction of groundwater flow on the Hanford-Bramlette test area.

The shape of the potentiometric surface (Fig. 9) corresponds closely to the surface and bedrock configuration. The direction of groundwater flow, indicated by the arrows, delineates the local recharge (source) area responsible for gradual seepage into the coulee to the north-west and for the large discharge (saline seep) area occurring downgradient at the first pronounced slope break along the southern and eastern edge of the area (Fig. 12). The seep area first appeared in the early 1950's as a small spot along the east side, and spread rapidly along the slope break as well as upslope and downslope, salinizing approximately 20 ha (Fig. 12).

The occurrence of saline seeps at the slope break is believed to be caused primarily by (a) a pronounced decline in the hydraulic gradient, causing an immediate velocity reduction and resultant groundwater buildup, and (b) the presence of a groundwater barrier (zone of decreased permeability) a short distance downslope from the seep area, indicated by an abrupt disappearance of the perched groundwater (Fig. 8). Test hole H-3 was drilled into this barrier and encouraged a very tight, dense clay till, with a notable absence of thin meltwater layers. Only the weathered zone of the shale was saturated. Because most of the excess water stored in the till is derived from upland recharge areas, seep occurrence is usually most pronounced at the first break in slope and, depending on local recharge and leakage from upslope, other seeps may occur at the next lower slope break or groundwater barrier.

Research data indicate that the formation of saline seeps is primarily a result of local rather than intermediate or regional groundwater flow systems. Several observations supporting this conclusion are:

a. Potentiometric surfaces conform to local topography rather than the regional gradient.

b. Concentration of TDS changes rapidly within a small area, reflecting local recharge-discharge.

c. Seeps generally first occur at mid-slope rather than along coulee bottoms.

d. Local changes in cropping systems rapidly affect local groundwater conditions and size of nearby saline seeps.

e. Drill-hole data indicate perched watertable conditions overlying dry bedrock.

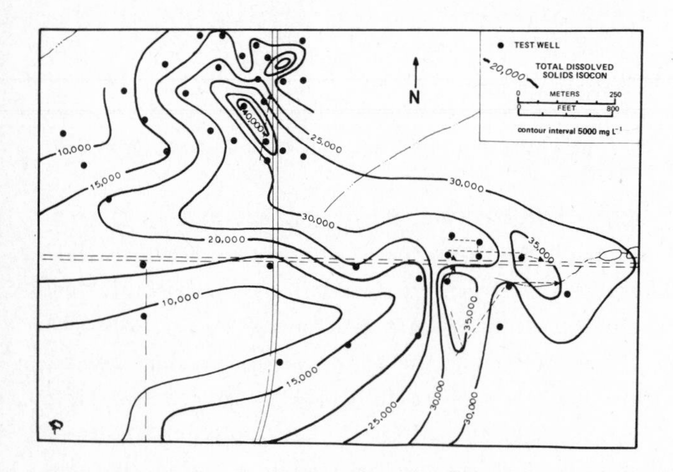

Fig. 10. Isocons of total dissolved solids (TDS) on the Hanford-Bramlette test area.

Data summarized from over 75 water-quality analyses show that TDS (Fig. 10) ranged from 3 200 mg L^{-1} in recharge areas to 40 900 mg L^{-1} in discharge areas, and averaged 22 500 mg L^{-1}. The concentrations of Na, Mg, and SO_4 in the water (averaging 3 300, 2 150 and 15 000 mg L^{-1}, respectively) are dominant as a result of dissolution of gypsum and ion exchange processes in the aquifer. Nitrate is typically very high (averaging 670 mg L^{-1} as NO_3), followed by HCO_3 (649 mg L^{-1}), Ca (373 mg L^{-1}) and Cl (245 mg L^{-1}). Concentrations of trace elements are usually high, with selenium being potentially the most toxic, varying from 0.06 to 2.0 mg L^{-1} (Donovan and Miller, 1980). In general, water picks up about 1 500 mg L^{-1} m^{-1} TDS as it migrates down through the soil, and roughly 150 mg L^{-1} m^{-1} TDS as it moves toward the discharge area.

Since 1970 three interceptor-and lateral-type drain tile systems have been installed in discharge areas at depths to 2 m below the surface; they are shown on Fig. 6. The drains were designed to accelerate the movement of shallow, unconfined groundwater and to lower the water table in seep areas. The 1 200 m of 100 mm diameter perforated tile was wrapped with a fine nylon mesh to minimize clogging by fines. The 180 m drain (DR3) installed in field A intercepted water from flowing well H16. This water provided an estimated 80 to 90% of the discharge of DR3; a total of 2 000 m^3 of water which averaged 33 700 mg L^{-1} TDS (71 000 kg of salt) during the 1970 to 1974 period (Miller and Hanson, 1979). During the first year of performance (1978/1979) drains DR1 and DR2 have flowed intermittently from 0 to 0.13 L s^{-1}, with flow ceasing in late fall and beginning in early spring. Estimated total flow of water from drains DR1 and DR2 for the year was 2 920 m^3; the water averaged 37 500 mg L^{-1} TDS (110 000 kg of salt). The drain lines, buried in dense clay till, had little or no effect on observation wells located 1 to 5 m away (Donovan and Miller, 1980). In fact, 3 flowing observation wells, H16, H8, H24 (Fig. 6), drilled to the weathered zone of the shale, had a combined discharge exceeding the flow of all three drain lines.

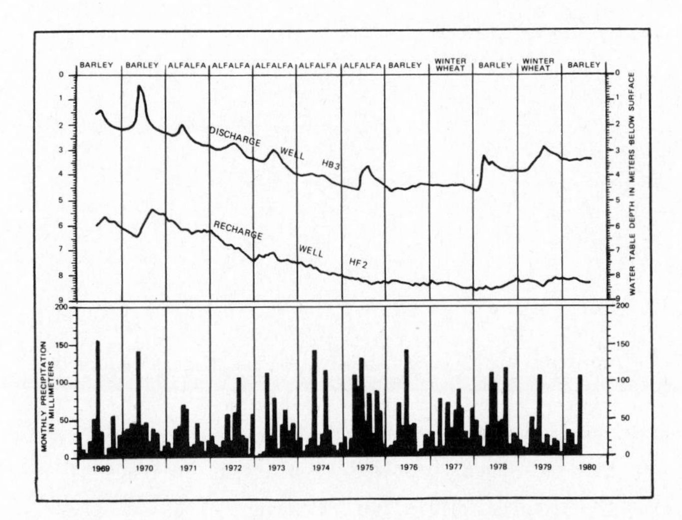

Fig. 11. Hydrographs of two wells on the Hanford-Bramlette test area to compare monthly precipitation and cropping history.

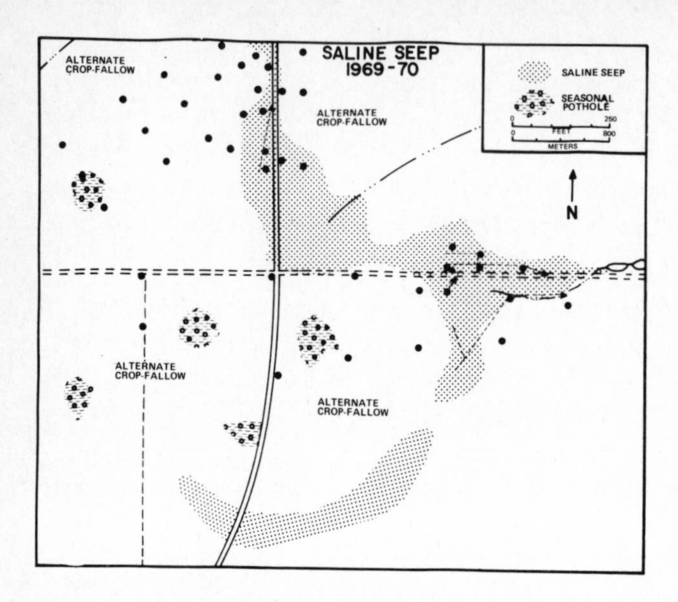

Fig. 12. Distribution of saline seep on the Hanford-Bramlette test area 1969 to 1970.

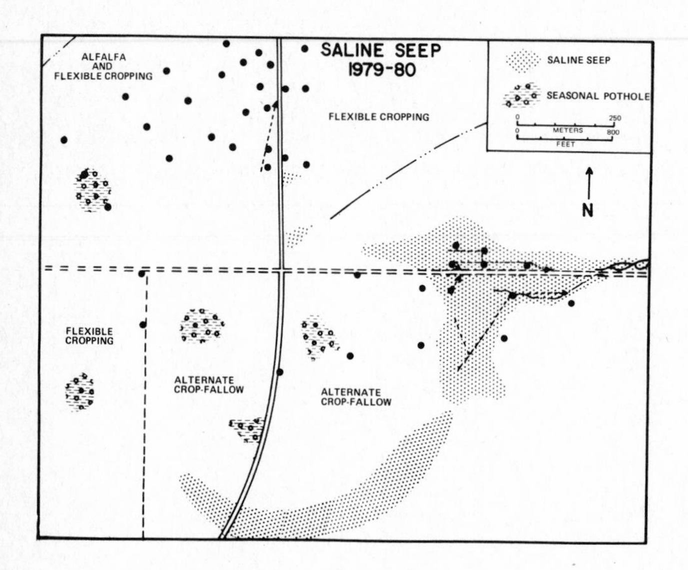

Fig. 13. Distribution of saline seep on the Hanford-Bramlette test area 1979 to 1980.

Because of low permeability, excessive till thickness, high content of montmorillonite clay, and serious waste-disposal problems, subsurface drainage of saline seep discharge areas appears to be unfeasible. However, by using other controls previously mentioned - flexible cropping, deep-rooted perennial crops, and surface draining of upland recharge areas - the salinity problem on the

Hanford portion (fields A and B) of the test site has been greatly reduced.

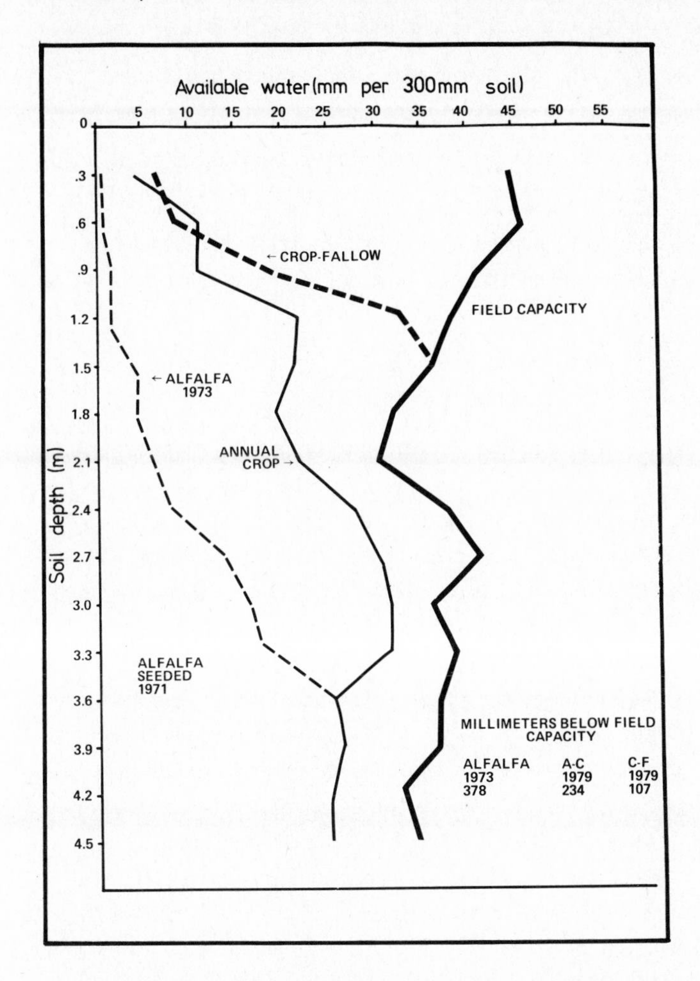

Fig. 14. Plant available water versus soil depth for several cropping systems.

In 1971, 32 ha of Ladak 65 alfalfa was seeded on the northern half of the site (all of field A and western part of field B). The alfalfa was left in for 5 years (1971 to 1975), and since then the fields were cropped annually, rotating barley and winter wheat. The changes brought about during the next 4 to 9 years are illustrated in Figs. 8, 11, 12, 13 and 14, and can be summarized as follows (Brown and Miller, 1978):

a. During the period 1971 to 1975, water levels had dropped an average of 2.9 m in the discharge areas and 2.0 m in the recharge area, indicating that the alfalfa used all current precipitation as it fell and dried out the deep subsoil as well (Figs. 8, 11, 14).

b. In 1976, soil samples showed that alfalfa roots had penetrated to 4.6 m and had extracted 480 mm of water from the soil (Brown and Miller, 1978).

c. Once water levels dropped in the discharge area, the surface salts were leached back down, reducing the soil salinity in the upper 0.6 m of soil by 70%.

d. A 5 to 20% decrease in TDS was observed in groundwater samples collected in seep areas.

e. The size of the saline seep was reduced from 12 ha to less than 0.4 ha. During the same period, the seep area on the south half (fields D and E) increased in size (Figs. 12, 13).

f. In 1976 and 1977, barley and winter wheat seeded into the discharge area yielded 70% of normal, and by 1979 yielded 100% of normal.

To investigate the length of time required for various cropping practices to refill the soil moisture reservoir, a strip of alfalfa in the recharge area was plowed up in 1974 and three rotations were established. The rotations were a) annual cropping, b) crop-fallow, and c) crop-crop-fallow. The results are summarized in Fig. 14, and show that after 6 years (1974 to 1979) the crop-fallow rotation had essentially refilled the soil moisture reservoir, whereas annual cropping was still 234 mm below field capacity. The crop-crop-fallow rotation (not shown) plotted about midway between the other rotations. This experiment indicates that the alternate crop-fallow farming system quickly recharged the deep subsoil (90 mm of water per each two-year crop-fallow period) following alfalfa; whereas, the annual cropping system maintained a relatively dry layer beneath the root zone. This suggests that intensive or flexible cropping, periodically rotating to deep-rooted crops, may be necessary to control dryland salinity problems in the Northern Great Plains region.

7 CONCLUSIONS

a. Geological conditions favorable for saline seep formation and development exist throughout the Northern Great Plains region. Areas underlain by the Colorado and Bearpaw Shale are the most susceptible to the problem.

b. Sufficient soluble salts are available in the upper 6 m of the soil profile to continue existing saline seeps for the next 25 to 100 years.

c. Regional water-quality inventories strongly suggest that shallow groundwater resources are being contaminated by the saline seep process. High concentrations of TDS, NO_3, and Se are the most troublesome.

d. The dominant land-use practice contributing to dryland salinity is the alternate crop-fallow farming system which allows significant quantities of water to move beneath the root zone and activate the local groundwater flow system.

e. The causes and controls of dryland salinity vary significantly from farm to farm and especially from area to area. As a result, the best chance for success is to evaluate the saline seep situation on a farm by farm basis.

f. The most effective methods used to control dryland salinity are: deep-rooted perennial crops, flexible cropping system, and surface drainage of cultivated upland recharge areas.

g. Solution of the problem will require the following:

1) A comprehensive demonstration and extension program to work directly with farmers.

2) An active research program to i) develop a variety of crops for incorporation into the flexible cropping system - the use of these crops will help control the problem throughout the region as well as bring an economic return to farmers; ii) investigate and solve such crop-related problems as: weed control, plant diseases, straw management, no-till systems, etc.; iii) investigate the distribution, behavior and potential for toxicity of selected elements in ground- and surface water of the Northern Great Plains - immediate attention should be given to selenium and nitrate; and iv) investigate in detail reported livestock and fish kills.

3) A regional observation well and water-quality sampling network to i) forecast trends in shallow groundwater conditions for planning cropping-system strategy in problem areas; ii) quantify long-and short-term changes in water quality; and iii) evaluate the effectiveness of various crops and cropping systems in controlling saline seep formation.

4) Continued cooperation and coordination among research, extension and demonstration programs to get the saline seep problem under control in the shortest amount of time.

8 REFERENCES

Bahls, L.L. and Miller, M.R., 1973. Saline seeps in Montana. In: Second Annual Report, Montana Environmental Quality Council, pp. 35-44.

Ballantyne, H.K. 1963. Recent accumulation of salts in the soil of southeastern Saskatchewan. Can. J. Soil Sci., 48: 43-48.

Brown P.L. and Cleary, E.C., 1978. Water use and rooting depths of crops for saline-seep control. In: Dryland-saline-seep control, Proceedings Meeting of Subcommission on Salt-Affected Soils at 11th. Int. Soil Sci. Soc. Congr., Edmonton, Alberta, Canada, pp. 7-1 to 7-7.

Brown, P.L., Cleary, E.C. and Miller, M.R., 1976. Water use and rooting depths of crops for saline-seep control. In: Proceedings on Regional saline-seep control Symposium, Bozeman, Montana. Montana Cooperative Extension Service, Bull. 1132: 125-136.

Brown, P.L. and Miller, M.R., 1978. Soils and crop management practices to control saline seeps in the U.S. northern plains. In: Dryland-saline-seep control, Proceedings Meeting of Subcommission on Salt-Affected Soils at 11th. Int. Soil Sci. Soc. Congr., Edmonton, Alberta, Canada, pp. 7-9 to 7-15.

Custer, S.G., 1976. Shallow groundwater salinization in dryland farm areas of Montana. Montana University Joint Water Resources Research Center, Report No. 79, 214 pp.

Doering, E.J. and Sandoval, F.M., 1976. Saline-seep development on upland sites in the Northern Great Plains. U.S. Dept. of Agr., ARSNC-32.

Doering, E.J. and Sandoval, F.M., 1978. Chemistry of seep drainage in southwestern North Dakota. In: Dryland-saline-seep-control, Proceedings Meeting of Subcommission in Salt-Affected Soils at 11th. Int. Soil Sci. Congr., Edmonton, Canada, pp. 4-1 to 4-14.

Donovan, J.J., Sondereggar, J.L., Miller, M.R. and Schmidt, F.A., 1979. Saline-seep progress report - investigation of soluble salt loads, controlling mineralogy and some factors affecting the rates and amounts of leached salts. Montana Bureau of Mines and Geology, Open-File Report No. 36, 32 pp.

Donovan, J.J. and Miller, M.R., 1980. Saline-seep drainage systems-semi-annual report. Montana Bureau of Mines and Geology, Open-File Report No. 54, 25 pp.

Ferguson, H., Brown, P.L. and Miller, M.R., 1972. Saline seeps on nonirrigated lands of the northern plains. In: Proceedings on Control of agriculture-related pollution in the Great Plains, Lincoln, Nebraska. Great Plains Agricultural Council, Publ. 60: 169-191.

Greenlee, G.M., Pauluk, S. and Bowser, W.E., 1968. Occurrences of soil salinity in drylands of southwestern Alberta. Can. J. Soil Sci. 48: 65-75.

Halvorson, A.D. and Black, A.L., 1974. Saline-seep development in dryland soils of northeastern Montana. J. Soil and Water Cons. 29(2): 77-81.

Halvorson, A.D. and Reule, C.A., 1976. Controlling saline seeps by intensive cropping of recharge areas. In: Proceedings on Regional saline-seep-control symposium, Bozeman, Montana. Montana Cooperative Extension Service, Bull. 1132: 115-124.

Holm, H.M., 1978. Saskatchewan soil salinity program - soil salinity crop tolerance testing. In: Dryland-saline-seep control, Proceedings Meeting of the Subcommission on Salt-Affected Soils at the 11th. Int. Soil Sci. Soc. Congr., Edmonton, Canada, pp. 5-11 to 5-30.

Krall, J.L. and Jackson, G.D., 1978. The flexible method of recropping. In: Dryland-saline-seep control, Proceedings Meeting of the Subcommission on Salt-Affected Soils at the 11th. Int. Soil Sci. Soc. Congr., Edmonton, Canada, pp. 7-23.

Lewis, B.D., Custer, S.G. and Miller, M.R., 1979. Saline-seep development in the Rapelje-Hailstone area - northern Stillwater County, Montana. U.S. Geological Survey, Water Resour. Investigations 79-107, 28 pp.

Luken, H., 1962. Saline soils under dryland agriculture in southeastern Saskatchewan (Canada) and possibilities for their improvement. Part 1. Distribution and composition of water-soluble salts in soils in relation to physiographic features and plant growth. Plant and Soil, 17: 1-25.

McCracken, L.J., 1973. The extent of the problem. In: Proceedings Alberta Dryland Salinity Workshop, Lethbridge.

Miller, M.R., 1971. Hydrogeology of saline-seep spots in dryland farm areas - a preliminary evaluation. In: Proceedings of saline-seep-fallow Workshop, Great Falls, Montana, 9 pp.

Miller, M.R., Schmidt, F.A., Smith, D.J., Shaw, R.L. and Sullivan, P.P., 1974. Saline-seep development, hydrologic response, and groundwater quality on the Highwood Bench, Montana. Montana Bureau of Mines and Geology, Open-File Report HY-74-1, 415 pp.

Miller, M.R., Sonderegger, J.L., Smith, D.J., Bergantino, R.N., Schmidt, F.A. and Bermel, W.M., 1975. Origin, development and regional extent of salinity problems in Montana - hydrogeological aspects. Saline-seep-report (1974) to Montana Department of State Lands, 57 pp.

Miller, M.R., Vander Pluym, H., Holm, H.M., Vasey, E.H., Adams, E.P. and Bahls, L.L., 1976. An overview of saline-seep programs in the state and provinces of the great plains. In: Proceedings on Regional saline-seep control Symposium, Bozeman, Montana. Montana Cooperative Extension Service, Bull. 1132: 4-17.
Miller, M.R., Bergantino, R.N., Bermel, W.M., Schmidt, F.A. and Botz, M.K., 1978. Regional assessment of the saline-seep problem and a water quality inventory of the Montana plains. Montana Bureau of Mines and Geology, Open-File Report 42, 417 pp.
Miller, M.R. and Hanson, T.L., 1979. Drainage of saline seeps, north-central Montana (Abs.). Am. Soc. Agric. Engnrs., 1978 winter meeting, Chicago, Illinois.
Oosterveld, M., 1978. Disposal of saline drainwater by crop irrigation. In: Dryland-saline-seep control, Proceedings Meeting of Subcommission on Salt-Affected Soils at 11th. Int. Soil Sci. Soc. Congr., Edmonton, Alberta, Canada, pp. 4-24 to 4-28.
Sommerfeldt, T.G., 1976. Snow trapping by windbreaks. In: Proceedings on Regional saline-seep-control symposium, Bozeman, Montana. Montana Cooperative Extension Service, Bull. 1132: 87.
Sonderegger, J.L., Donovan, J.J., Miller, M.R. and Schmidt, F.A., 1978. Investigations of soluble salt loads, controlling mineralogy and some factors affecting the rates and amounts of leached salts- saline-seep progress report. Montana Bureau of Mines and Geology Open-File Report 30, 31 pp.
Vander Pluym, H.S.A., 1978. Extent, causes and control of dryland saline seepage in the Northern Great Plains region of North America. In: Dryland saline-seep control, Proceedings Meeting of the Subcommission on Salt-Affected Soils, 11th. Int. Soil Sci. Soc. Congr., Edmonton, Alberta, Canada, pp. 1-4 to 1-58.
Worcester, B.K., Brun, L.J. and Doering, E.J., 1975. Classification and management of saline seeps in western north Dakota. North Dakota Farm Res. 33(1): 3-7.

TERRAIN, GROUNDWATER AND SECONDARY SALINITY IN VICTORIA, AUSTRALIA

J.J. JENKIN
Soil Conservation Authority of Victoria,
Kew, 3101, Vic.

ABSTRACT

Jenkin, J.J., 1981. Terrain, groundwater and secondary salinity in Victoria, Australia. Agric. Water Manage., 1981.

In Victoria secondary salinity occurs under diverse conditions and the individual characteristics of particular examples are intimately related to rock structure, lithology and landform as well as exhibiting the superimposed effects of present and past climate and varied land use. Ameliorative strategies, to be effective, must take these factors into account on both local and regional scales, as well as being compatible with the prevailing socio-economic conditions. It is suggested that procedures devised for soil and water salinisation control should and need not be mutually exclusive, and that strategies can be developed which not only produce a permanent solution to these problems, but also result in increased farm productivity.

1 INTRODUCTION

Secondary salinity is an increasing problem in Victoria. On a conservative estimate, at least 90 000 ha of the non-irrigated (dryland) cleared country is severely affected. In addition there is a similar or even larger area of incipiently salted land. As shown by many observations in recent years, these areas are not static, both expansions and new occurrences being common. Also many streams carry appreciable salt loads and reports of dam and bore waters becoming too salty for stock are legion.

The problem of descriptive salinity terminology has been reviewed by Peck (1978) and it is agreed that rationalisation of the confused multiplicity of terms is desirable (Jenkin, 1979). In the present paper "seep" or "salt seep" is used to describe areas which exhibit signs of active salinisation and which have an obvious surface outlet. Seeps which occur in apparently closed depressions are described as "salt pans" but should not be confused with Rowen's (1971) use of the term to include all forms of salt seep. "Salt scald" is reserved for cases where inherently saline subsoils have been exposed by erosion.

The degree of salinisation varies in both area and severity depending on regional and local geological conditions, including land form, the present and past climate and its ecological effects, and the specific forms of land use applied. All these aspects must be considered when ameliorative measures are proposed.

2 TERRAIN AND SECONDARY SALINITY

The geological and geomorphic conditions in which secondary (man-induced) salinisation occurs are diverse (Fig. 1). Most of western Victoria is affected in some degree, certain landscapes with particular severity. The distribution of salting in Victoria is summarised in Table 1, which is followed by a brief description of conditions in each of the terrain types.

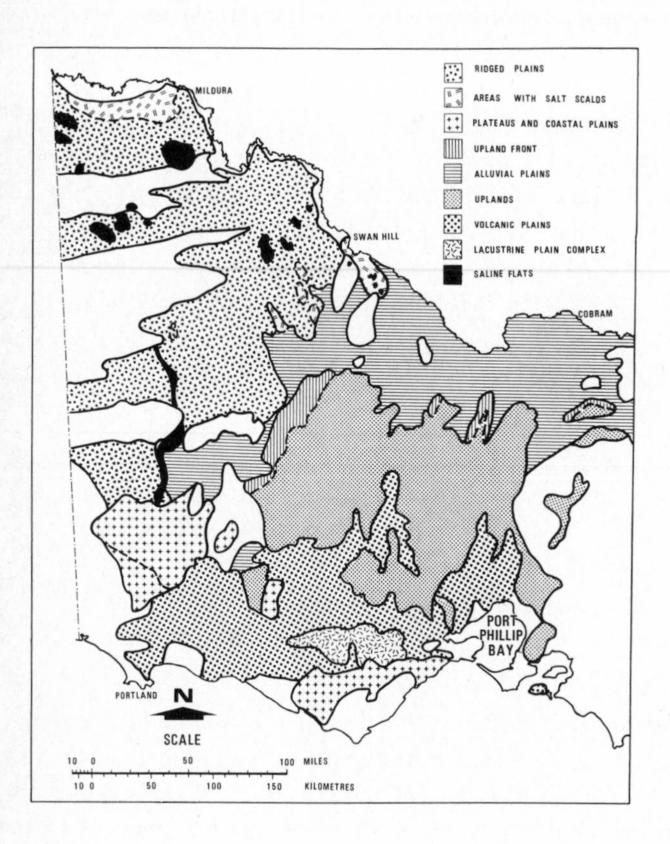

Fig. 1. Landscapes in which secondary salinity problems occur.

TABLE 1

Terrain type and dryland salting.

Terrain type		Dryland salting
(1) Ridged Plains		
(a) Dunes	-	widespread seeps and pans, including water channel seepage.
(b) Broad ridges and flats	-	seeps on flats, extensive in clearly defined areas; also water channel seepage.
(c) Clay plains	-	salt scalds, common in restricted areas.
(2) Plateaux and Coastal Plains		
(a) Laterite capped plateau)	-	hillside and valley seeps; saline stream flows.
(b) Laterite capped bench)		
(c) Dissected coastal plain)		
(3) Upland Front		
Hills and ridges adjoining riverine-plain flats		extensive seepage on flats near the break of slope; with or without saline stream flows.
(4) Alluvial Plains		
Flats with levee ridges and inset terraces	-	occasional seeps and pans, often accompanied by groundwater effluent flows.
(5) Uplands		
(a) Hilly country on folded sediments	-	valley side and valley floor seeps with saline stream flows.
(b) Prominent strike ridges	-	abundant seeps at break of slope, saline stream flow.
(c) Metamorphic aureole ridges	-	as above
(d) Undulating country on granite	-	occasional hill side seeps.
(e) Alluvial valleys	-	valley floor seeps; saline stream flows.
(6) Volcanic Plains		
Flat to undulating with numerous cones and maars	-	seeps and salt pans in limited areas.
(7) Lacustrine Plain Complex		
Flats with lakes, swamps, lunettes and low lava ridges	-	salt pans in depressions, seeps in peripheral shallow valleys and saline stream flows.
(8) Saline Flats		
Flats, often gypseous, with salt lakes and salinas	-	widespread salt pans and seeps.

2.1 The Ridged Plains

The ridged plains of north-western Victoria represent the many-stage retreat of the late Tertiary Murravian sea, combined with later extensive aeolian reworking and peripheral fluvial activity (Bowler and Magee, 1978; Jenkin, 1976, Lawrence, 1966, 1976). Hills (1939) drew attention to the NNW-trending sandstone ridges which Blackburn (1962) suggested were indicative of former shorelines. This relationship is now accepted (Bowler and Magee, 1978; Jenkin, 1976), and satellite imagery has shown that there are many more such shoreline traces than previously realised. Partly obscuring these ridges are extensive fields of E-W dunes which, towards the south-east, grade into slightly undulating sand sheets.

Seeps on the sides and at the base of dunes, as well as inter-dune salt pans are widespread, particularly in the northern half of the ridged area (Rowan, 1971). In the broad troughs between the NNW-trending ridges, large areas are salt affected. Frequently, the soils are gilgaied and produce a conspicuous cellular pattern following salinisation. This is the "broad plains" salting described by Rowan (Table 2). Since that time, considerable and rapid expansions have been observed (Speedie, personal communication, 1979).

TABLE 2
Salt affected land in north-western Victoria
(After J.N. Rowan, 1971)

	Number	Area (ha)
Salt pans and seeps at the base of E-W dunes	745	1300
Salt seeps beside channels	428	900
Salt pans and seeps on broad plains		
(i) In beds of former lakes	121	1900
(ii) Not in beds of former lakes (broad trough salting)	64	400
(iii) Within beds of large lakes or former lakes which also include non-saline surfaces	-	-
Plains with saline surfaces on eroded clay soils (salt scalds)	-	c.40 000

In these situations perched saline water tables are the principal factors in salinisation. The perching is of at least two dimensions. Firstly, there is a regional body of perched water lying on clays and marls of the Murray Group, and secondly, local perched bodies within the aeolian system. The regional perched water body is responsible for the extensive broad trough salting which is potentially the most serious problem in the north-west as any rise in the water table sufficient to cause surface salinisation has an immediate affect over large areas. Control is therefore more difficult than in the smaller dune seeps and pans with their limited local recharge.

In the case of dune-related salting, a distinction should be made between closed (salt pan) and open (salt seep) accumulation (Fig.2 a,b). In salt pans, the shape of the water table after rain suggests that there is appreciable recharge through the pan surface although there would also be a contribution from lateral throughflow.

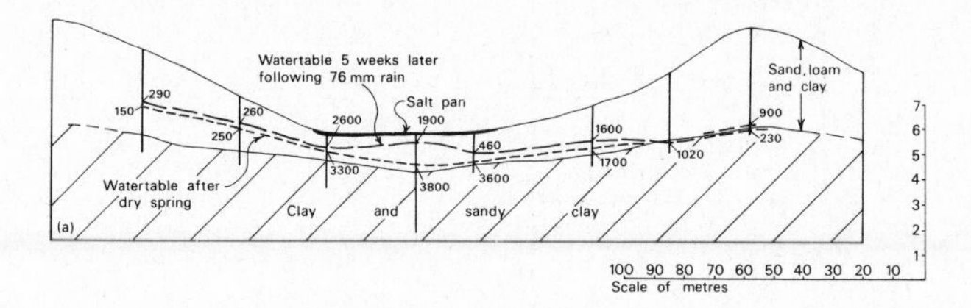

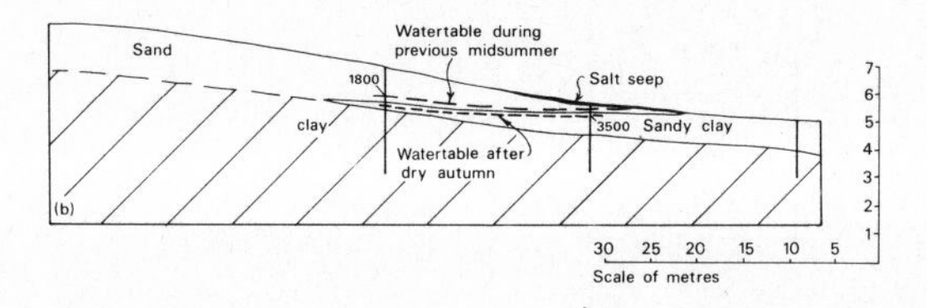

Fig. 2. Salt seeps in dune terrain west of Woomelang, north-western Victoria. Groundwater electrical conductivity as mSm^{-1}. (After Rowan, 1971).

Salinities of the soil profile, where there is some plant cover, normally increase from the surface downwards (from about 50 to 800 mSm^{-1}, 1:5 aqueous extract) indicating leaching. Accumulation then occurs in the groundwater zone just above the impervious clay substrate. This is also reflected in the groundwater salinity changes following recharge. Superimposed on the usual soil salinity profile in completely bare areas there is a concentration of salt on and in the top few centimetres (about 900 mSm^{-1}), below which the usual leaching profile occurs.

In the dune-base seep it appears that the perched water body builds up during the wet period and emerges at the break of slope. Groundwater salinities increase downslope and, in the bare areas, soil salinities increase only slightly (50 to 90 mSm^{-1}) compared with enclosed basins. This suggests that concentrated surface salts are washed to lower levels and widely distributed, although groundwater salinities still attain levels similar to those beneath the enclosed pans. Recharge is probably over the whole dune slope and the location of the seep governed by the topography and the sandy clay dune substrate. This is one of the many examples of seep emergence at a break of slope, a feature not necessarily associated with an impervious substrate.

Areas of salt scald, although they are not usually linked with secondary salting, are in geomorphic situations which suggest an underlying common factor in salinisation in this area (Fig. 3).

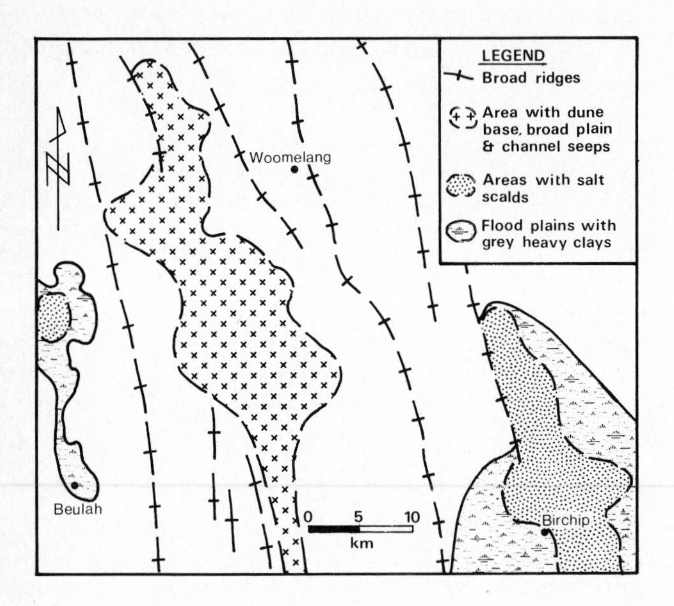

Fig. 3. Distribution of secondary salinisation in part of the southern Mallee, north-western Victoria.

2.2 Plateaux and Coastal Plains

Geomorphically, Victoria is noted for its step-like form (Jenkin, 1976). Of these level surfaces, the laterite-capped Dundas plateau in the west and the Western District coastal plain are particularly important as far as secondary salinisation is concerned.

On the Dundas plateau the laterite sheet lies on thin, Late Tertiary sediments and varied Palaeozoic sedimentary, igneous and metamorphic rocks. Dissection by the essentially radial streams has intersected the deep weathered zone to reach more or less fresh bedrock. Perched saline water bodies are general, lying on the kaolinitic mottled and pallid zones, and result in seep or spring lines, which are most conspicuous on the upper slopes of the steep valley sides. In many cases, these high-level seeps feed the streams and recent drilling has shown that the saline water in the valleys moves in a thin sandy alluvial horizon over a clay layer on relatively fresh bedrock, which lies at a shallow depth, and beneath the highly carbonaceous surface layer. The possibility that a general body of saline water exists, occupying fractures in the fresh bedrock and emerging in some valleys as springs, should not be discounted. Immediately south of the Dundas plateau, the lateritic sheet remains only as narrow caps on the interfluves, and high-level perched groundwater is less significant than it is further north. However, the Early Cretaceous parent rocks, feldspathic sandstones and mudstones, are likely to carry saline water in joints and fractures, although the rocks themselves are highly impermeable.

The coastal plain sediments in south-central Victoria also carry lateritic remnants ranging from extensive areas to elongate interfluve caps. Perched water table phenomena are similar to those in the Dundas Tableland but depositional clays sometimes take the place of the pedogenetic perching layers typical of the laterites.

2.3 The Upland Front

This is a complex zone which combines water and salt movement in deeply weathered and fresh bedrock with accumulation and concentration in contiguous, dominantly clayey sediments which lie at the upland edge of the Riverine Plain. Salting occurs at the break of slope along the sides of ridges projecting into the plain or in alluvial re-entrants flanked by low hills. Such occurrences are regarded as being particularly significant owing to their potential for rapid spread and have been the subject of detailed investigation over several years, outlines of which are given in later sections.

2.4 Alluvial Plains

On the Riverine Plain of northern Victoria salt seeps and pans appear in shallow depressions sometimes with associated groundwater discharge channels (Fig. 4).

Macumber's (1969; 1978) studies of the Loddon Valley have shown that the late Tertiary gravel valley infill (the deep lead, Fig. 5) is the main path of groundwater movement from the uplands to the plains. However, the hydrostatic pressure developed in this aquifer has caused it to discharge at weak points in the overlying aquitards, thus converting much of the Loddon Plain into a groundwater discharge zone (Fig. 6).

Discharge has fluctuated with climatic changes during the Quaternary, as the stratigraphy of the lower Loddon Plain indicates, and the present reactivation of this discharge is probably directly related to land use following settlement. Macumber (1978) has shown that this change, which has occurred within the last 75 years, represents a "fundamental shift in hydrologic equilibrium...in the regional groundwater flow systems" and suggests that similar discharge zones will also develop on the Campaspe (Fig. 6) and Goulburn Plains.

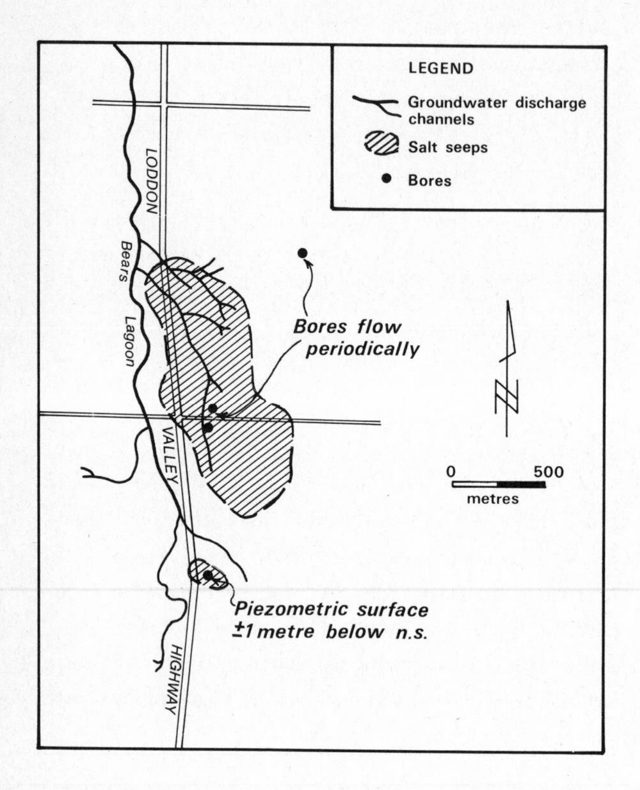

Fig. 4. Groundwater discharge on the Riverine Plain, northern Victoria. (After Macumber, 1978).

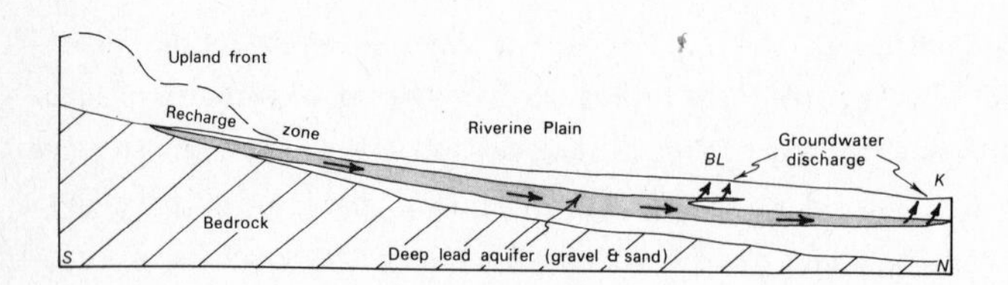

Fig. 5. Groundwater movement beneath the Riverine Plain, northern Victoria. (Based on Macumber, 1978).

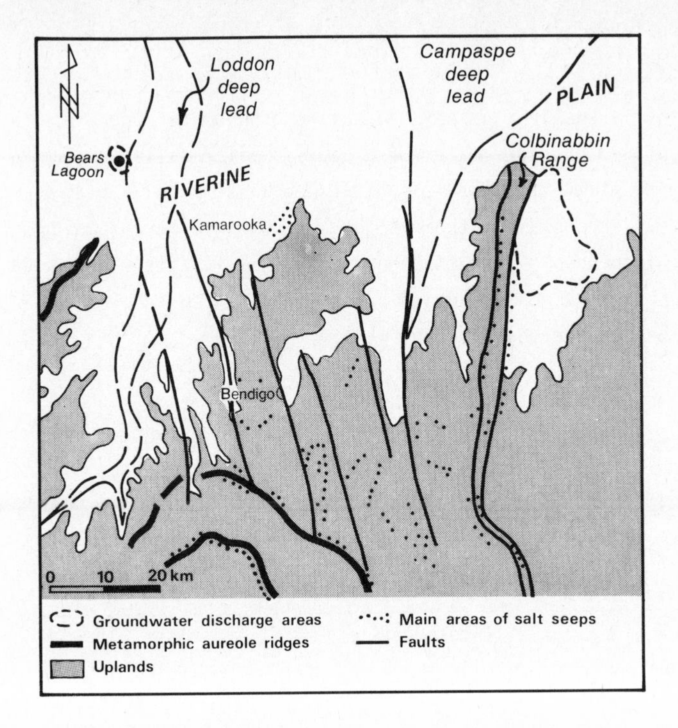

Fig. 6. Occurrence of salt seeps in north-central Victoria.

2.5 Uplands

Two main types of soil salinisation occur in the Uplands. These are hillside seeps, related to structurally controlled discharges of saline water, and valley floor seeps resulting from the general rise of the water table in the intermontane alluvial valleys, often combined with leakage of pressure water from one or more shallow confined aquifers.

Many of the narrow upland valleys in the Palaeozoic sedimentary terrain are deeply gullied and innumerable salt seeps emerge in their floors and along the gully sides to produce a small, but saline, permanent flow. The salinity is reduced only temporarily following rain (to about 30 mSm^{-1} after 50 mm), but soon returns to a level of approximately 700 mSm^{-1}. Salt seeps are either below the natural surface level or, because of the steepness of the valley sides, only small areas of flat or lower slope are affected. Down stream the valleys broaden, the seeps affecting progressively larger areas, usually to about 2 ha, and finally grade into continuous alluvial flats on which areas of many hectares may be affected. In one valley at Knowsley, east of Bendigo, for example, about 300 ha is obviously salinised, while in other valleys, seep areas up to 100 ha are common.

In the valley at Knowsley, a shallow, unconfined saline groundwater body is underlain by partially confined pressure water of broadly similar salinity (Table 3). Saline springs emerging above the level of unconfined water at site 5 suggest that the lower aquifers are leaking, their higher piezometric levels being consistent with this. It is probable that the pressure waters in this valley keep the shallow aquifer replenished, a feature which seems to be common in a number of physiographic regions, although the scale varies widely.

TABLE 3

Groundwater at Knowsley, near Bendigo

A. Salinity EC as mSm^{-1} (approximate $mg\ L^{-1}$ in brackets)

Site	Shallow	Intermediate	Deep
1	1292 (8400)	-	1486 (9660)
	1283 (8340)	-	1502 (9765)
2		-	2455 (15960)
	(in Permian glacial bedrock)		
		-	2580 (16770)
3	2206 (14340)	-	1902 (12360)
	1824 (11860)	-	1863 (13040)
4	1978 (12860)	-	1071 (6960)
	1848 (12010)	-	1013 (6587)
5	2383 (15490)	1978 (12860)	2005 (13030)
	2432 (15800)	1980 (12870)	2040 (13260)

B. Water levels (metres below natural surface)

Site	Shallow	Intermediate	Deep
1	2.41(1.8-3.7)*	-	2.37 (10.4)
	1.55	-	1.49
2	-	-	5.59 (4.9-9.8; 15.2)
	-	-	5.01
3	1.70(0.6-2.4)	-	1.52 (13.4)
	0.94	-	0.71
4	1.15(1.9)	-	2.36 (3.7-12.2)
	1.32	-	1.78
5	1.22(1.2-1.9)	0.11(5.5)	0.09 (6.7)
		+0.01	+0.01

* Depths water struck in brackets.

In the Axe Creek drainage basin, south-east of Bendigo, valleys in the middle section lie between gently rolling bedrock hills, and are extensively salinised. The alluvium is very thin and the seeps are emerging from fractures in the underlying sandstones and mudstones, leaving behind encrustations of magnesite. Further down the valley the alluvium is thicker and forms well developed terraces. Some of the lower lying soils are salinised, but the main feature is the emergence of saline flows at the fault-controlled breaks of slope at the valley sides. These flows join the main stream by way of the eroded channels of formerly shallow tributaries.

As with the upland front zone, the break of slope is a favoured site for the development of salt seeps. There are several situations in the uplands where this topographic factor is the principal determinant of the place of seep emergence.

These are:

(a) At the foot of a fault scarp with poorly developed alluvial fans.

(b) At the foot of a strike ridge of resistant rocks.

(c) Adjacent to the metamorphic aureole ridges which surround granitic intrusions.

(d) At the toes of alluvial fans or aprons formed along any of these ridges or scarps.

Salt seeps originating within the granite itself are usually rare, some of the very deeply weathered granite country providing the exception. In some areas however, the soils on the granites are affected by seepage from the metamorphosed sediments of the adjacent aureole. Seeps are also common on the opposite side of the ridge but do not appear to be in greater numbers than along other kinds of ridges or to be significantly distinctive chemically, as is sometimes suggested.

2.6 Volcanic Plains

Soil salinisation is sporadic on the volcanic plains, the main occurrences being localised in areas of old, deeply-weathered flows. In the Hamilton district of western Victoria the bauxitised and kaolinised Hamilton Basalt and the Dunkeld Basalt, with its solodic soils, are both affected, the salt seeps appearing at the base of slopes, in shallow watercourses and in depressions (Gibbons and Downes, 1964; Sibley, 1967). Some depressions form small ephemeral lakes and are often saline to some degree.

2.7 Lacustrine Plain Complex

This region is the site of a large former lake now reduced to strings of smaller lakes, swamps and flats with accompanying lunettes (Currey, 1964). Although the substrate is volcanic plains basalt, this is mainly blanketed by lacustrine sediments and plays little direct part in the widespread salinisation. The common salt pans and seeps are probably derived from relict salt remaining after the lake contracted. Peripheral drainage channels carry saline flows and the valley floors are extensively affected by seepage.

2.8 Saline Flats

The salinas and gypsum playas and flats of north-western Victoria are the result of groundwater discharge from extensive deep aquifers which have their intake in the piedmont zone to the south-east (Lawrence, 1966, 1975, 1976; Macumber, 1969, 1978). They have fluctuated with climatic variation in the recent geological past and are essentially natural features. Since settlement however, a significant secondary component has been added. Former salinas and gypsum flats are being reactivated and existing ones are expanding. Lunettes are ubiquitous in these areas and seeps often appear at the break of slope along the ridge base.

Lawrence (1966) has pointed out that the gypsum, deposited at the sites of ancestral lakes, is a precipitate from saline groundwater lying a metre or so beneath the surface. If the water table outcrops however, sodium chloride is the major precipitate, and seasonal fluctuation in water level causes the magnesium-rich bitterns remaining after the halite crystallisation to drain away (Cane, 1962). Reactivation, or the formation of new salinas, should repeat this sequence. Consequently it is expected that the gypseous flats will eventually be dominated at the surface by sodium chloride, and that local groundwaters will become enriched with magnesium. A progressive chemical deterioration would therefore accompany any expansion of salinisation in these areas.

3 AQUIFER RECHARGE AND DISCHARGE

The dryland salinity phenomena described in the previous section can only be satisfactorily explained in terms of groundwater movement. Soil throughflow, as postulated by Cope (1958), is clearly inadequate in all the Victorian physiographic situations in which persistent salting occurs.

The reasons for this are:

(a) The quantity of available soluble ions stored in the soil (A,B and B/C horizons) is inadequate to maintain continuing salt seeps.

(b) The excessively low horizontal hydraulic conductivities of the soil themselves compared with their substrates.

(c) The actual presence in every case investigated so far of saline groundwater, either unconfined or under pressure or frequently both.

These points will now be considered with reference to water and salt movement over the whole profile with emphasis on two critical areas, the Palaeozoic sedimentary uplands and the uplands front zone.

3.1 Water and Salt Movement

3.1.1 The sedimentary uplands

(i) The area under natural conditions. Closely folded and faulted Palaeozoic sandstones, mudstones and slates are characteristic of about 85% of the upland area (Fig. 1). Consequently, this sedimentary terrain has been selected for detailed study. The most intensive research has been conducted in a north-northwest-trending belt immediately to the east of Bendigo, extending from the metamorphic aureole ridge in the south to Kamarooka in the north, a distance of 60 km (Fig. 6). The belt lies along the strike of the bedrock sediments and there is no essential change in the lithologically repetitious sequence in that direction. However, from south to north changes occur in other parameters, these being:

(a) Average annual rainfall decrease from 525 to 425 mm.

(b) Average annual evaporation increase, P/E ranging from about 0.4 in the south to 0.3 in the north. It is only for three months in mid winter that precipitation exceeds evaporation, and then only by a maximum of about 20 mm.

(c) Topography, from steep hills and deep, narrow valleys, through hilly to undulating country with broader thinly alluviated valleys, to gently undulating country with generally poorly defined drainage channels.

(d) Soils, which in the steep country are dominated by shallow, stony gradational profiles, followed in the gentler country by red-brown duplex profiles, the clay B horizons of which become thicker and heavier towards the north. The soils on the alluvium and younger colluvium are mostly yellow duplex types with red-brown sodic duplex soils tending to replace them in the north.

(e) Weathering, the depth and intensity increasing towards the north, but not in phase with the present soils as the preservation of palaeosols also increases northwards.

(f) The quantity of soluble ions stored in the regolith (which includes the soil and weathered zone) increases appreciably towards the north.

(ii) The area following clearing for agriculture. On the removal of forest, as judged by comparison between forested and cleared catchment pairs, three phenomena occur:

(a) There is an appreciable increase in the amount of water in the regolith.

(b) The B and C horizons remain at field capacity for most of the year, while under forest they dry out rapidly and only retain this level of moisture for a very brief period following rain.

(c) Soluble ions are leached from the regolith in amounts generally ranging from about 20 to 70% of their original concentration.

The soil moisture averaged over the year is appreciably greater in the cleared country than it is under forest. This is so for all areas examined, of which the undulating Palaeozoic sedimentary country is typical (Fig. 7). Also, owing to their continuing dampness, the soils in cleared areas could always be penetrated to a greater depth by driven tube than those under forest. The depths at which the two curves diverge usually vary between 0.2 and 0.4 m and it is also above this level that wide fluctuations in soil moisture occur. Below that level, down to about 0.5 m, there are marked seasonal differences between the two areas. In the cleared areas the soil wets up at the break of season and remains at field capacity for at least half the year. Under forest, on the other hand, after water-repellent fungal mycelia have become saturated, the sub-soil wets up rapidly, but only for a very short period. The soil water content is soon reduced by the active transpiration of trees whose roots are concentrated below 0.4 m, and the normal water content of the sub-soils under forest is below wilting point. The surface soils however tend to be slightly damper under forest than in cleared areas. This is probably related to the relatively low density of actively transpiring, shallow-rooted plants in the forest, as well as reduced evaporation from tree-shaded ground.

The occurrence of free sub-surface water is also significant. Bores in forested areas invariably dried up before mid summer, while most of those in the equivalent cleared basins contained water throughout the year. The changes in soil moisture regime are accompanied by marked changes in soluble ion concentrations in the soils. The electrical conductivity peak which is so prominent under forest, coinciding with the top of the B/C soil horizon and the maximum concentration of tree roots, is lost following clearing. There is also an appreciable loss in total ions which is reflected in the electrical conductivity curves (Fig. 8). In the gentle terrain some of the ions concentrate at depth in the valley, but in most cases it is expected that the leached ions move out of the catchment in interflow (flow in the normally unsaturated zone) and deeper groundwater flow.

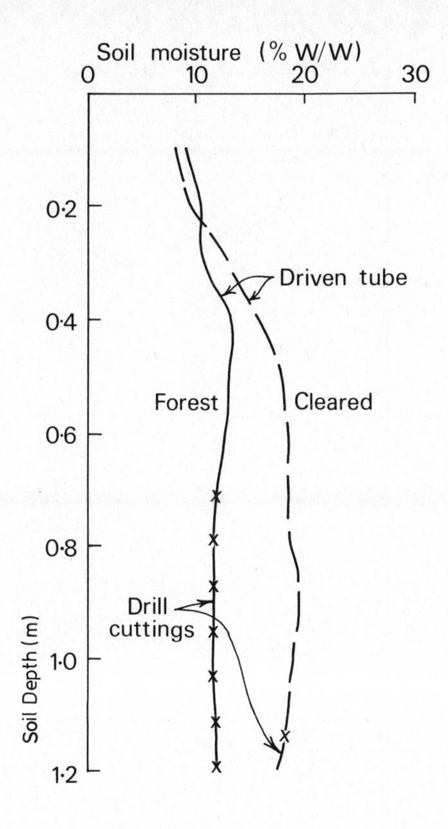

Fig. 7. Average soil moisture in undulating Palaeozoic sedimentary country, south-east of Bendigo, Victoria.

Similar patterns occur in the very gentle country in the north but the differences in the top few metres are not as marked as in the southern catchments, possibly because the lower rainfall results in less leaching and the much thicker, heavy clay subsoil and weathered zone retards percolation.

The soluble ions move at different rates during leaching resulting in ionic differentiation, specific ions tending to concentrate at particular levels in the landscape and the profile. The more mobile ions, principally sodium and chloride, tend to concentrate low in the valleys and deep in the profile. The differences between the ionic proportions in small forested and cleared catchments demonstrates some of the compositional changes which occur during leaching. In the change from forest to cleared country in the same terrain type, concentrations of both cations and anions in the A horizon become more varied, and the maxima are considerably reduced. The cation maxima also change towards a greater proportion of ($Na^+ + K^+$) at the expense of Mg^{2+} and, to a lesser extent, Ca^{2+}. The relative anion concentrations also change towards a higher HCO_3^- and lower SO_4^{2-} with the Cl^- proportion remaining relatively constant. Shifts also occur in the proportions of C horizon ions, the cations trending towards higher ($Na^+ + K^+$) than under forest. Thus, clearing produces a greater diversity of ionic proportions, particularly in the A horizon, but to some extent over the whole profile as well.

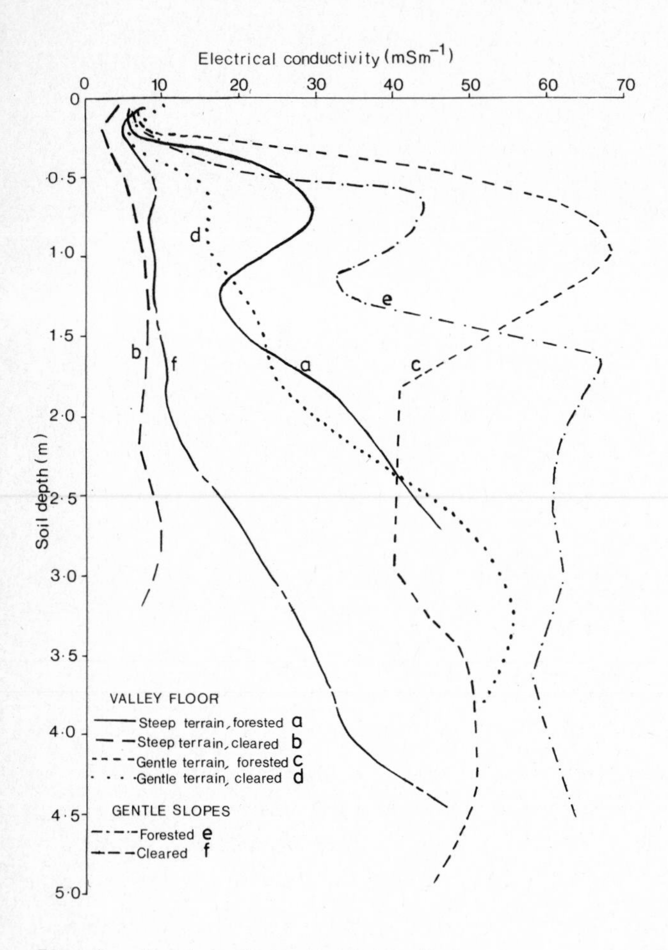

Fig. 8. Electrical conductivity (1:5 aqueous extracts) in soils on Palaeozoic sediments, south east of Bendigo, Victoria.

3.1.2 The upland front

The flats at the junction of the uplands and the Riverine Plain are particularly liable to dryland salting, as a small rise in the level of the highly saline groundwaters can affect a large area of land. Several of these occurrences have been documented and rapid, extensive increases noted, particularly after a series of wet years. Two examples with different geological characteristics will be described, viz.

(i) Kamarooka - flats without ordered drainage lying along the west side of a sedimentary bedrock ridge (Fig. 6).

(ii) The Colbinabbin Range - alluvial fans, along greenstone (basic volcanic) and shale ridges, which merge with alluvial flats to the east and west (Fig. 6).

(i) Kamarooka. Already about 900 ha have been severely affected by secondary salinisation, that is the soil is either quite bare or carries a variable cover of unproductive species such as sea barley grass (*Hordeum maritimum*). A similar area is incipiently affected, with sea barley grass starting to replace perennial rye grass, the clovers having previously disappeared, and occasional bare patches appearing. In addition, there are at least 10 000 ha with saline waters at 2 m or less below the surface. The salinity hazard in this area is therefore regarded as extreme.

The area studied in detail is shown in Figs. 9 and 10. Here the bedrock sediments have been very deeply and intensely weathered, and store considerable quantities of soluble ions, on average about six times the amount in similar rocks in the steep country only 45 km to the south. To the west, the ridge gives way abruptly to plains, while to the east, the transition is much more gradual.

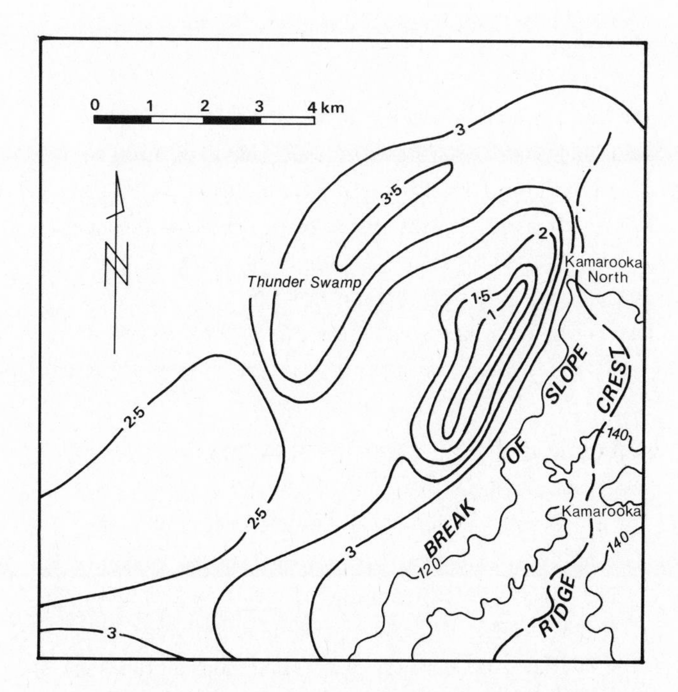

Fig. 9. Piezometric levels of the upper groundwaters, Kamarooka, Victoria. Contours are in metres below the present surface.

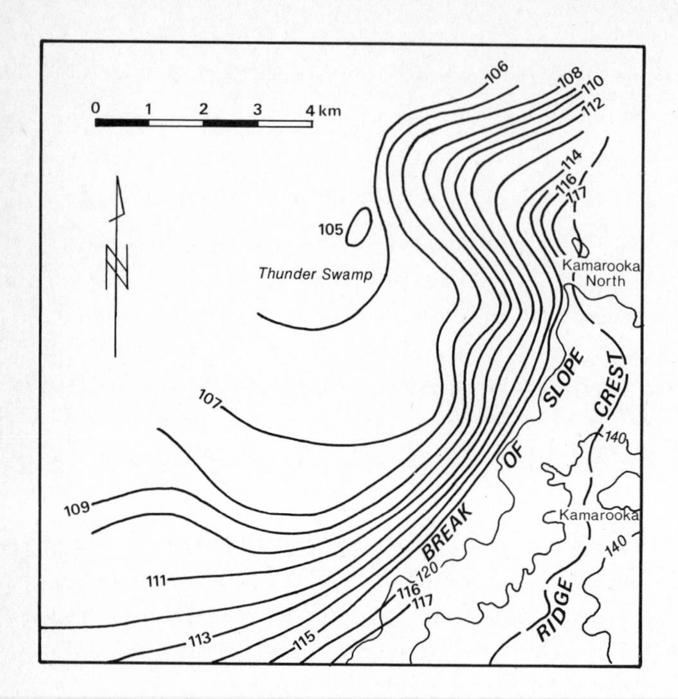

Fig. 10. Reduced levels of the upper groundwaters, Kamarooka, Victoria. Contours are in metres above sea level.

The salt affected area to the west, despite its flatness, is geomorphically complex, with a history of swamp and lake formation interspersed with surface forms and subsurface sediments indicating periodic external drainage. At present the drainage is not clearly integrated and the occasional floods which occur find their way to the main streams by anastomosing, ill-defined routes. The soils are dominantly heavy clay on both the bedrock and the plains sediments. Top soils, where they have been preserved, are sandy loams. On the lower slopes of the ridge, gravelly colluvial deposits give way to the clayey alluvial, and finally paludal sediments.

Drilling indicates that, down to 30 m there are two or three levels at which water enters the bores. In some places these waters seem to be interconnected, water levels being the same in all bores in the one piezometer nest, yet in others the waters seem to be acting independently, water levels being markedly different in the three bores. The piezometric level of the upper waters (Fig. 9) reflects subtle differences in the surface level, the obviously salted area, defined approximately by the 2 m contour, occurring in a slight depression less than 0.5 m below the general surface level.

The reduced level contours of the upper water surface show distinct gradients sloping from the ridge and subsurface bedrock highs, north-westwards across the plain (Fig. 10). This suggests that groundwater movement is from the ridge outwards into the plain and not along the side of the ridge or inwards towards the ridge from the pressure waters present in shoestring sands beneath the plain. A local source for the groundwaters is therefore postulated. Despite the rapid response of bore water levels to rain (several hours), the recharge site is by no means obvious. Two-dimensional modelling carried out by W. Trewhella and B. Smith of the State Rivers and Water Supply Commission suggests that water movement at a particular point on the salinised flat is vertical, yet the infiltration rates and hydraulic conductivities of the alluvial and pedogenetic clays are extremely low. It is therefore proposed that a conducting zone at depth is the controlling factor and that it is located in the bedrock. If the salt flat itself is excluded, the slopes or crest of the ridge are the only possible recharge areas. The top soils on the slopes are clay loams and absorb water readily, but the subsoils are similar to the clays on the flat and are highly impermeable (Table 4). It is possible that the slope top-soils store water and allow it to infiltrate slowly to the bedrock through preferred paths, such as cracks or old root holes, or that the infiltration is sufficient to produce a head which can affect the bores.

TABLE 4

Infiltration rates and hydraulic conductivity at Kamarooka.
B horizon of red duplex soil on ridge slope
(P. Dyson, personal communication, 1980).

Site	Infiltration ($mS^{-1} \times 10^{-6}$)				Site	Hydraulic conductivity ($mS^{-1} \times 10^{-6}$)			
						Fresh Water		Saline Water (5000 mg L^{-1} NaCl)	
1	0.04	0.06	0.03	0.06	A	0.9		0.4	1.5
2	0.06	0.09	0.08	0.08	B	0.6	0.5	0.5	0.2
3	0.04	0.06	0.06	0.06	C	1.0	1.0	0.3	0.4
4	0.07	0.06	0.08	0.05	D	0.9	0.5	0.7	0.6
					E	2.1		0.3	1.0
					F	0.4	1.4	0.4	0.5

The possibility that the few relatively high rates are the most significant should not be overlooked as these may represent preferred paths of water movement.

It should be pointed out that the actual quantities of water involved are quite small, as the permeability of all the materials is very low, and that water movement and pressure transmission will be along restricted paths which are not aquifers in the accepted sense of the term. The other possible recharge area is on the ridge crest itself where soils tend to be thin and stony, and the bed-rock actually outcrops in places.

TABLE 5

Downslope changes in groundwater chemistry.
(P. Dyson, personal communication, 1980).

A. Chemical data: transects from ridge outwards

	Site	EC mSm^{-1}	pH	K^+	Na^+	Ca^{2+}	Mg^{2+}	Cl^-	HCO_3^-	SO_4^{2-}
Transect 1	6	2630	7.7	1.7	198	22	85	248	3.9	55
	7	2650	7.8	1.0	182	24	68	232	2.8	40
	8	3460	7.7	1.3	235	23	95	317	2.6	34
Transect 2	9	3740	7.7	1.2	266	36	101	341	2.9	62
	10	3880	7.7	1.2	265	46	115	346	1.9	80
	11	2850	7.8	1.4	217	31	83	273	2.7	56

B. Changes relative to Cl^-

Site	K^+	Na^+	Ca^{2+}	Mg^{2+}	HCO_3^-	SO_4^{2-}	Concentration Factor*
6-7	-34.73	1.63	15.54	-13.55	-21.76	-20.82	-6.45
7-8	-41.37	-133.88	-25.27	-10.48	-65.54	-72.46	30.18
9-10	-1.47	-1.85	26.31	12.39	35.95	27.56	1.47
9-11	36.30	1.14	14.80	5.66	27.26	65.33	19.63

Ionic concentrations in milliequivalents L^{-1}

*Concentration factor = percentage Cl^- difference between sites.

Groundwater recharge downslope from the ridge is also suggested by chemical changes (Transect 1, Table 5). Depletion of Na^+, K^+, SO_4^{2-} and HCO_3^- occurs in the downslope direction and is attributed to precipitation of $CaCO_3$ and $CaSO_4$, and to cation exchange. In Transect 2, however, the enrichment in SO_4^{2-}, K^+, Ca^{2+} and Mg^{2+} may indicate solution of ions previously precipitated in a swamp environment (P. Dyson, personal communication, 1979). The salinities shown in Table 5 are typical of the groundwaters on the plain in and around the salinised area to at least 8 km towards the west and north-west.

(ii) The Colbinabbin Range. The Range is essentially a double ridge projecting into the Riverine Plain. Salt seeps and saline stream flows are widespread and numerous on both flanks of the Range over a distance of more than 30 km.

The main range consists principally of Cambrian greenstones (altered basic volcanic rocks) while cherts dominate in the secondary range (Fig. 11). The Range lies between two high-angle thrust faults which are thought to have been active periodically at least since Ordovician times, the last identified movements being in the Pleistocene as the alluvium of the plain is affected (Bowler and Harford, 1966) and the adjacent salt lakes are probably due to groundwater discharge in a recently downwarped basin (Fig. 6).

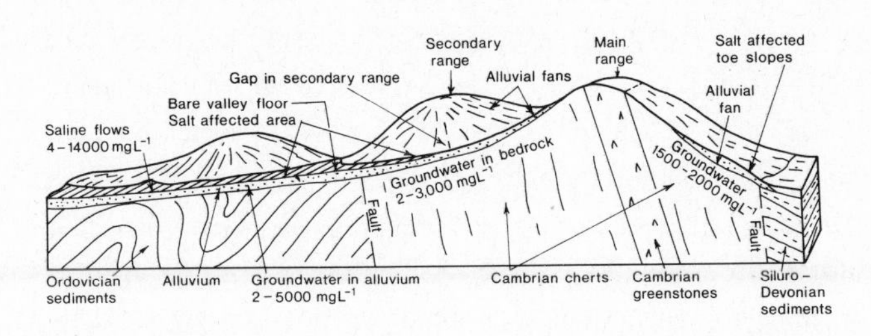

Fig. 11. Secondary salinisation along the Colbinabbin Range. (After Jenkin and Irwin, 1975).

Both ridges, which together make up the Range, are flanked by smoothly concave alluvial fans, consisting of red friable clays, which coalesce longitudinally to form continuous aprons. Between the ridges, the fans also coalesce downslope to produce a smoothly U-shaped median valley segmented by low saddles. Each segment drains through a gap in the secondary range then, via broad shallow valleys between low Ordovician bedrock hills, to the Riverine Plain. The fans on the east side of the main range often merge imperceptibly into the nearby floodplain, although there is sometimes a slight break of slope at the junction. On this side, salt seeps often appear at the toes of the fans but may also occur in slight depressions or at local breaks of slope higher up the fan. Salt springs coming directly from the bedrock are also known. On the west side of the Range, seeps often appear first in the gaps in the secondary range and then migrate both up and downstream. Here too, however, some seeps first appear higher up the valleys.

Two transects were drilled, one on either side of the Range, and the results, combined with surface observations, suggest that the initial seepage occurs at points controlled by geological structure. This includes lithological discontinuities, sub-fan bedrock topography, fracture zones related to faults, and jointing. The fan toe-slopes seem to be affected by saline water emerging at a higher level and moving downslope beneath the fan. Similarly, such seeps can emerge further upslope, being brought closer to the surface by a local bedrock high or a break in the smooth continuity of the fan surface. In the case of the gap salting in the secondary range it is possible that a blockage of subsurface drainage, caused by resistant beds or directly related to faulting, has resulted in the damming and near surface spill-over of saline groundwater. Groundwater salinities normally are not as high as at Kamarooka, possibly due to the generally steeper slopes and better drainage at Colbinabbin.

3.1.3 Factors controlling water and salt movement

The salt increment (Hutton and Leslie, 1958) at present is not sufficient to initiate and maintain salt seeps. In the cleared undulating country on Palaeozoic sedimentary bedrock it represents only 0.008% of the salts stored in the upper one metre of soil, while 4% of the amount of stored salts is discharged in run-off each year. At this rate, the soils should be entirely free of soluble salts in 25 years, provided there is no replacement from mineral breakdown. Considering that chloride is the dominant anion, and the rocks cannot yield the quantities required, this possibility can be ignored. Furthermore, as most of the catchments examined have been cleared for 80 to 100 years, and as these still retain an appreciable proportion of the soluble ions in the profile, it is suspected that a new equilibrium is established (cf. a similar suggested by Peck, 1978), possibly within about 20 years, following which the concentration of soluble ions will remain constant under any particular set of imposed conditions.

The level of salt storage in the soil at equilibrium, judging from catchments in a variety of Victorian landscapes and climates, seems to depend mainly upon:

(a) The local storage of soluble ions which is, to a large extent, a function of weathering intensity and is therefore related to past climatic history on a geological time scale.

(b) The hydrological properties, permeability for example, and mineralogical characters of the regolith such as chemical reactivity, ion exchange capacity and physical response to wetting.

(c) Present climate, particularly the relationship between evaporation and precipitation.

(d) The geomorphic environment, especially slope, relative elevation and other drainage characteristics.

(e) Land use. The form of land use as well as the time since clearing is important. For example, there is appreciably more leaching from an area under a cropping rotation than under semi-improved native pasture.

Specific paths of water and salt movement vary considerably in magnitude and should be considered on both regional and local scales. On a regional scale, they are directly controlled by geological structure, in the broad sense, and influenced by climate. On a local scale the controls are varied, and range from the specific nature of the rocks and soils (texture) and their superimposed micro-structures (joints and fractures) to biologically induced channels such as root holes, burrows, or the disruption of otherwise impervious layers by deep ripping, and the increase of surface detention by contour banking and ploughing.

Regionally, the general controls depend on the stratigraphic disposition of aquifers and aquitards, the relative efficiency of each, and their inter-connections. Water in fractured bedrock is often regarded as being unconfined, but intense weathering and soil formation may produce an effective confining layer and result in the development of pressures in the groundwater. In the upland country of west-central Victoria, the evidence cited points to the bedrock as the main groundwater transmitting medium.

This also applies where bedrock lies at shallow depth beneath fluvial sediments along the upland front. Two levels seem to be particularly important in this regard. These are the weathered, often rubbly top of the bedrock and the base of the weathered zone where it rests on fresh, less permeable bedrock. However, mining records and the occurrence of springs show that fresh bedrock can also transmit groundwater, although this is through fractures as the rocks themselves are effectively impermeable, and structural details control the points at which the water emerges. Groundwaters in fresh bedrock sometimes develop artesian pressures, and several of the abandoned deep mines at Bendigo have been flowing for fifty years or more. These waters are characterised by the presence of abundant hydrogen sulphide derived from pyrite, testifying to their passage through the fresh bedrock. However, fracture systems do not behave in the same way as the pores in granular media (Legrand, 1979) and should be viewed as preferred paths. There is no doubt that at least some of the recharge to superimposed sediments is achieved through the laterally outcropping and underlying bedrock.

3.2 Discharge and Salt Concentration

In the uplands there are commonly two forms of discharge. If an erosion gully is present it acts as a groundwater drain, sometimes crossed by a series of impermeable rock bars producing a succession of groundwater dams which eventually spill-over as gully-floor springs, resulting in a permanent saline flow. If the erosion gully is absent, the alluvium eventually becomes saturated with saline water and salt seeps appear on the valley floor. Discharges such as these, by flowing over the surface or through shallow permeable beds, may affect valley floors far downstream from the actual site of emergence. Examples are common in the Dundas Tablelands where streams are fed from perched seeps, and on the inner side of aureole ridges where saline water seeps from the fractured metamorphic rocks, affecting soils on the granite.

Discharges resulting from a general rise of the water table typically emerge at a break of slope and have been observed to expand across the toe flat and also upslope. Examples are widespread along fault scarps and similar ridges in the bedrock areas, such as the Whitelaw Fault scarp in the Axe Creek catchment, and at the junction of bedrock and alluvium at Kamarooka.

Water table seeps also emerge at the toes of terrace scarps and fans, as well as in depressions between dunes, in abandoned channels and in former lakes or swamps. A similar effect is often produced by farm dams and water supply channels, the water mound beneath them effectively raising the water table locally from which seepage water percolates downslope, or the general rise in water levels causes the seep to migrate both up and downslope. Seeps associated with farm dams are particularly common in the rolling uplands of north-central Victoria, while channel seeps are very frequent in the Mallee of north-western Victoria (Rowan, 1971), but are by no means confined to this area and often occur beside channels on subdued bedrock rises further to the east.

Regional groundwater discharge areas, the saline flats and the plains effluents, are of particular interest as they represent a continuation or reactivation of events which occurred in the geological past, and will have an important bearing on future salinisation trends. For example, the large areas in north-western Victoria which have gypsum at shallow depth are liable to resalinisation. Furthermore, the areas with saline subsoils, and thus subject to scalding, almost certainly owe their genesis to high saline water tables in the past and are possible sites for renewal of extensive salt seepage.

Concentration of salts at and near the surface is usually regarded as being a direct result of evaporation, the precise form this takes depending on whether or not the water table is at or below natural surface. In areas without surface salinisation but with the water table at shallow depth, near Kamarooka for example, the upper part of the profile indicates leaching, but concentration occurs between a level at or just above the water table, to a metre or more above. It is expected that this zone would be continuously saturated, either by free groundwater or capillary water and is characterised by the formation of clear crystalline gypsum. Where the capillary fringe reaches the surface in the alluvial valleys of the uplands, concentration occurs at two distinct levels: at the water table and at the surface, a high Na^+ and Cl^- content at each level being characteristic. These results are consistent with the observations of Lawrence (1966) in north-western Victoria.

The possible effect of plants on salt concentration should also be considered. It has been shown that forests, particularly in the maximum tree root zone, appear to concentrate soluble ions. A similar machanism may also apply with regard to shallow-rooting grasses in the cleared country topsoil. It has also been noticed that vegetation surrounding bare patches on Queensland seep areas appears to concentrate salt at its edge - the "fairy ring" effect (Hughes, personal communication, 1980). A similar phenomenon has been seen at Kamarooka, bare patches expanding and clumps of sea barley grass contracting, each with its trailing peripheral rim of precipitated salt.

Concentration of salts is more efficient in the warmer, drier areas of the north than in the more humid areas of the south. In general, salt seeps tend to contain larger areas of bare ground the further north their location. On the other hand some seeps, particularly in southern Victoria, have little if any bare ground and are recognised by a characteristic cover of salt tolerant species, particularly flat weeds and sea barley grass. Differences between seeps in different areas, and sometimes in the same area, also arise from the geomorphic situation of the seep, principally whether or not the accumulation area is a closed basin or is open to drainage, and whether the seep dries out periodically or is continually wet. For these reasons, groundwaters of similar salinity can produce different surface effects and those of different salinity, quite similar effects.

4 MANAGEMENT IMPLICATIONS

Although the quantities of soluble salts stored in the soil and substrate vary and affect salting phenomena in detail, the water recharge-discharge system is the controlling factor in salinisation. Any solution to the problem must therefore involve the control of water both at and below the surface. There are two aspects to this. Firstly, the additional recharge which follows clearing must be prevented, and secondly the pressures which have built up in confined aquifers, and water tables which have risen, must be reduced.

Owing to the varied and complicated conditions under which salting occurs, the details of reclamation schemes will necessarily vary from one place to another. The essential prerequisite is, therefore, a detailed knowledge of the conditions prevailing in each area in relation to the causitive factors involved. In some cases this would involve an integrated approach between widely separated areas, such as a recharge zone near the hills and discharge sites several hundreds of kilometres distant. At the other extreme, the local intake to a small perched seep can be cited.

4.1 Recharge Control

It has been shown that recharge may be uniform over a particular area. In some areas however, it can be more or less restricted to certain zones. The infiltration and salt profile studies near Bendigo suggest that recharge occurs over the whole area, although it is suspected that preferred paths of enhanced percolation probably exist. These are likely to be root holes left by grubbed or burnt out stumps, rocky outcrops with shallow stony soils, especially where these have been deep ripped, and areas of certain soils which are subject to tunnel and similar subsurface erosion processes, often augmented by rabbits.

Another possible preferred infiltration zone is found in the rubbly fans which form the lower slopes of rocky ridges in many parts of the Victorian uplands. These fans are often pitted by many small depressions which retard runoff and act as surface detention hollows. Recharge is undoubtedly enhanced by some agricultural procedures, such as long fallow, but the precise geohydrological effects of contour ploughing and banking as an erosion prevention measure, at least under Victorian conditions, are unknown (Jenkin, 1979).

Attempts are being made to design strategies for recharge control but are hampered by the lack of precise knowledge of the recharge paths and the quantities of water involved. Research in Western Australia (reviewed by Peck, 1977, 1978; Williamson and Bettenay, 1979, and several unpublished symposia), and in Victoria (Jenkin, 1979; Jenkin and Irwin, 1975, 1980; and others), is helping to overcome this deficiency. Equally important, but even more inadequately known, are the effects of various agricultural practices and the performances of particular plant species under different soil, climatic and geomorphic conditions. In Victoria, data have been collected in a few limited areas for some years, but the results have not yet been fully evaluated and are geared more to erosion control and water yield than to the recharge problem. Water relationships in several cropping situations, involving minimal tillage, various rotations and different fertiliser and gypsum applications are being studied in northern Victoria by the Soil Conservation Authority and the Department of Agriculture. In the Axe Creek basin near Bendigo monitoring programmes have recently been set up to determine the relative effects of unimproved and Phalaris pasture on soil moisture in duplex soils with medium to heavy clay B horizons, the typical profile for the district.

Although most of these investigations are at an early stage, it appears that recharge control could be achieved in some districts without fundamental changes in present farming practice. In the upland areas particularly, the establishment and maintenance of vigorous pasture may be the key. In addition, most areas suspected of particularly high water intake are in the least productive parts of the farm, for example on the rocky ridges and adjacent rubbly fans. Here, vigorous pasture would be difficult to establish, and planting or allowing the regeneration of deep-rooted and vigorously transpiring trees would be appropriate. It is emphasised that recharge control in the upland areas and its fringes does not necessarily involve the exclusion of large areas from agricultural production, enforcing non-viability on farms, but rather than production should increase.

On the plains the problems are more difficult in that the main recharge area may be many tens or even hundreds of kilometres distant. Superimposed upon this is the effect of local water intake, the local situation thus reinforcing the regional. This duel effect appears to be occurring in many parts of north-western Victoria, the problem being generated by artesian pressures built up in the main aquifers, recharged in the upland front zone, and compounded by local intake following removal of native vegetation (mainly Mallee), the imposition of long fallow periods, and the destruction of organic matter by stubble burning. In this case, it is necessary to control two entirely separate sources of water aquisition involving different farming routines and different farming traditions.

4.2 Discharge Control

Salinity control has been preoccupied with the terminal manifestation of salt concentration. This does not solve the problem but, in some cases, exaggerates it. Attempts to produce a vegetative cover on salt seeps results in a vulnerable pasture which only survives slight grazing pressure. Furthermore, research into salt-tolerant tree species is likely to end in frustration as planting trees in salted areas is likely to aggravate the situation. Firstly, although River Red Gum and Tamarix are relatively salt-tolerant, their long-term viability is doubted. Secondly, the planting of salt-tolerant trees on salt seeps to lower the water table and encourage leaching of salts, has several difficulties. Thus the trees, in transpiring large quantities of water, will concentrate salts in the root zone and probably contribute to their own eventual demise. Furthermore, many salt seep soils disperse readily when fresh water is applied, effectively reducing infiltration to zero. Finally, even if infiltration could be achieved through cultivation and the application of ameliorants, it is likely that the amounts of water required to produce leaching would keep water tables high, as quite small quantities are required to do this.

5 CONCLUSIONS

It has been shown that salinisation occurs in all but the highest and wettest areas in Victoria. Salt storage, except at its lowest limit, does not determine whether or not a salinisation problem will appear, but rather the upper levels of salinity which can be expected in ground and surface waters in a particular area. In turn, this is related to the time required to reach a new equilibrium when conditions change. Salt storage has two aspects, the level of salinity and the total volume. It is therefore legitimate to expect that storage related to perched waters of limited dimensions will have an equally limited life, that is, it will equilibrate in a finite, calculable number of years.

Such situations however, are rare in Victoria and it is suggested that the soluble ions stored in the weathered zone and in the groundwater provide, in practical terms, an inexhaustible salt supply. If the means of moving and concentrating these salts, particularly the former, are not removed the joint problems of soil and water salinisation will continue to grow.

The environmental salinity balance is extremely sensitive, as shown by fluctuations during the recent geological past, and the profound hydrological change imposed by the removal of forests is bound to have far-reaching effects, the magnitude of which has been demonstrated by comparative soil water and salinity studies. Not only are there gross changes such as these, but each form of land use superimposes its own particular effect and may partly compensate for, but more often aggravate, the initial environmental impact.

It is essential to be aware of the preferred paths of water movement in each terrain type, and to know in detail the effects of particular agricultural procedures under those conditions, if effective remedial strategies are to be developed.

The treatment of both soil and water salinisation problems can, but should not, be mutually exclusive. The long-term solution of soil salting, which must be based on control of the hydrological cycle, inevitably involves control of surface and subsurface waters. The corollary of a permanent reduction in soil salting therefore is a corresponding improvement in water supply quality.

6 REFERENCES

Blackburn, G., 1962. Stranded coastal dunes in north-western Victoria. Aust. J. Sci., 24: 388-389.

Bowler, J.M. and Magee, J.W., 1978. Geomorphology of the Mallee Region in semi arid northern Victoria and western New South Wales. Proc. Roy. Soc. Vic., 90: 5-25.

Bowler, J.M. and Harford, L.B., 1966. Quaternary tectonics and the evolution of the Riverine plain near Echuca, Victoria. J. Geol. Soc. Aust., 13: 339-354.

Cane, R.F., 1962. The salt lakes of Linga, Victoria. Proc. Roy. Soc. Vic. 75: 75-88.

Cope, F., 1958. Catchment salting in Victoria. Soil Cons. Auth. Vic., Tech. Comm. 1.

Currey, D.T., 1964. The former extent of Lake Corangamite. Proc. Roy. Soc. Vic., 77: 370-386.

Gibbons, F.R. and Downes, R.G., 1964. A study of the land in southwestern Victoria. Soil Cons. Auth. Vic., Tech. Comm. 3.

Hills, E.S., 1939. The physiography of north-western Victoria. Proc. Roy. Soc. Vic., 51: 297-323.

Hutton, J.T. and Leslie, T.I., 1958. Accession of non-nitrogenous ions dissolved in rainwater to soils in Victoria. Aust. J. Agric. Res., 9: 492-507.

Jenkin, J.J., 1976. Geomorphology. In: Geology of Victoria. Spec. Publ. Geol. Soc. Aust., 5: 329-348.

Jenkin, J.J., 1979. Dryland salting in Victoria. Water Res. Foundation and Soil Cons. Auth. Vic.

Jenkin, J.J. and Irwin, R.W., 1975. The Northern Slopes land deterioration project - some preliminary conclusions. Soil Cons. Auth. Vic.

Jenkin, J.J. and Irwin, R.W., 1980. Land deterioration on the Northern Slopes of Victoria. Soil Cons. Auth. Vic.

Lawrence, C.R., 1966. Cainozoic stratigraphy and structure of the Mallee region, Victoria, Proc. Roy. Soc. Vic. 79: 517-553.

Lawrence, C.R., 1975. Interrelationship of geology, hydrodynamics, and hydrochemistry of the southern Murray Basin. Mem. Geol. Surv. Vic., 30.

Lawrence, C.R., 1976. Groundwater. In: Geology of Victoria, Spec. Publ. Geol. Soc. Aust., 5: 411-417.

Legrand, H., 1979. Evaluation techniques of fractured-rock hydrology. In: Black and Stephenson (Eds.), Contemporary Hydrogeology. J. Hydrol. 43: 333-346.

Macumber, P.G., 1969. Interrelationship between physiography, hydrology, sedimentation and salinisation on the Loddon River Plains, Australia. J. Hydrol., 7: 39-57.

Macumber, P.G., 1978. Hydrologic change in the Loddon Basin: the influence of groundwater dynamics on surface processes. Proc. Roy. Soc. Vic., 90: 125-138.

Peck, A.J., 1977. Development and reclamation of secondary salinity. In: Russell, J. and Greacen E., (Eds.) Soil factors in crop production in a semi-arid environment. Univ. of Qld. Press: 201-319.

Peck, A.J., 1978. Salinisation of non-irrigated soils and associated streams: a review. Aust. J. Soil Res., 16: 157-68.

Rowan, J.N., 1971. Salting on dryland farms in north-western Victoria. Soil Cons. Auth. Vic., Tech. Comm. 7.

Sibley, G.T., 1967. A study of the land in the Grampians area. Soil Cons. Auth. Vic., Tech. Comm. 4.

Williamson, D.R. and Bettenay, E., 1979. Agricultural land use and its effect on catchment output of salt and water - evidence from southern Australia. In: The Agricultural Industry and its Effects on Water Quality. Int. Conf., Hamilton, New Zealand, May 1979.

GROUNDWATER SYSTEMS ASSOCIATED WITH SECONDARY SALINITY IN WESTERN AUSTRALIA

R.A. NULSEN and C.J. HENSCHKE
Division of Resource Management,
Department of Agriculture, South Perth, Western Australia, 6151.

ABSTRACT

Nulsen, R.A. and Henschke, C.J., 1981. Groundwater systems associated with secondary salinity in Western Australia. Agric. Water Manage., 1981.

Increased surface and stream salinity in Western Australia occurs after the native vegetation is replaced by agricultural species and is almost invariably associated with rises in water table levels and increases in the potentiometric head of the deep, semi-confined aquifers. The recharge zone of the deep aquifers can sometimes be identified from the geomorphology of the catchment and from infiltration rate measurements in this zone, which may be up to 200 times the infiltration rates of the lower slopes. Water transmission through the aquifers is governed by low hydraulic gradients and low hydraulic conductivities. However, there is evidence that a considerable quantity of the water is transported at much faster rates through discrete channels. These channels have been observed in discharge zones where they contribute to the rate of rise of saline water towards the soil surface.

1 INTRODUCTION

Writing in 1829 of the country east of the Darling Range, Nathaniel Ogle noted "The alternations of salt pools and fresh streams, even in the same channel, have puzzled exploring parties, but will hereafter be easily accounted for, and probably the cause made useful, like all the bountiful and various provisions of Providence, to mankind." Ogle's confidence in our ability to easily understand the causes of land and stream salinity in Western Australia has not yet been fully realised. Present understanding of the mechanisms of salinisation is largely qualitative and few quantitative data exist.

The association between removal of native vegetation and subsequent increases in salinity of surface soils and streams was first reported by Wood (1924). The association has subsequently been confirmed by a number of workers (Teakle and Burvill, 1945; Smith, 1962; Bettenay et al., 1964; Mulcahy, 1978). Similar responses to changes in land use have been reported in the eastern States of Australia (Abbott, 1880; Cope, 1958; van Dÿk, 1969; Rowan, 1971; Pauli, 1972) and in North America (Greenlee et al., 1968; Halvorson and Black, 1974; Doering and Sandoval, 1976).

While the mechanics of salinisation are not unique to Western Australia, the quantities of salt and water involved, the pathways they follow and the rates of movement are a unique function of the Western Australian landscape: a landscape characterised by a peneplain developed by subaerial weathering of Archean rocks. Except on the margins of the peneplain, surface drainage is limited to a series of interconnected salinas and subsurface drainage is restricted by a relatively shallow, impermeable bedrock and an absence of sedimentary aquifers.

2 THE HYDROLOGIC MODEL

Hydrologic models of catchments vary in complexity. The simplest model is that of a groundwater flow system consisting of an intake area, a flow path and a discharge area. More complex models detail each of these three major components and the most comprehensive models attempt to integrate all three into the one continuous system. It is not the purpose of this paper to present a comprehensive model of groundwater systems, but rather to examine what is currently known about each of the three major components of the basic hydrologic model.

Teakle and Burvill (1945), Smith (1962) and Bettenay et al. (1964) all proposed a similar simple hydrologic model for various locations in Western Australia. In all cases the emphasis was on a defined intake area, upon which a change in land use caused a hydrologic perturbation resulting eventually in increased soil salinity in the discharge zone.

2.1 The Recharge Area

Skeletal soils often are present immediately below rock outcrops and Bettenay et al. (1964) proposed that these areas were recharge zones for the confined aquifer in the Belka Valley (Fig. 1). Bligh (1980) measured infiltration rate of the soils of the Belka Valley and his data are summarised in Table 1.

TABLE 1

Relative infiltration rates of soil surfaces in the Belka Valley.

Infiltration rate of the alluvial flood plain (9.37×10^{-9} ms^{-1}) is taken as unity. (After Bligh, 1980).

Soil Surface	Relative Infiltration Rate
Upper pediment	181
Pediment slopes	0.6
Upper valleys	1.6
Lower valleys	188
Alluvial flood plain	1

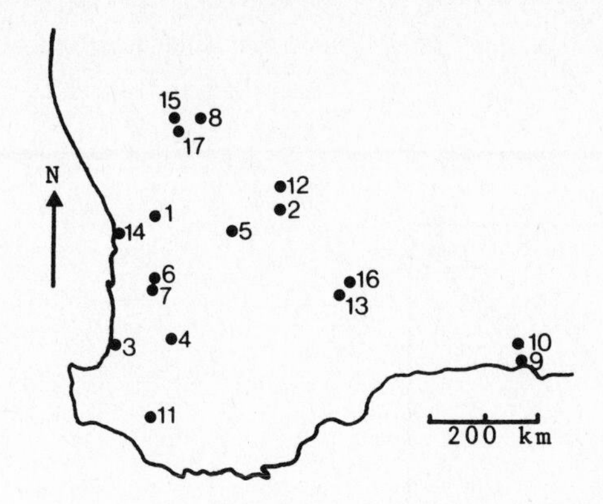

Fig. 1. Location map. 1, Bakers Hill; 2, Belka Valley; 3, Bunbury; 4, Collie; 5, Dangin; 6, Del Park; 7, Dwellingup; 8, East Wongan; 9, Esperance; 10, Esperance Downs Research Station; 11, Manjimup; 12, Merredin; 13, Newdegate; 14, Perth; 15, West Kondut; 16, Wheelock's; 17, Wongan Hills.

The upper pediments described by Bligh (1980) correspond to the upper parts of the Booraan association of Bettenay et al. (1964). The relatively high infiltration rates measured indicate that these areas may act as preferential intake areas for the seepages that occur downslope. High infiltration rates on the lower valley surfaces may be effective in directing rainfall and runoff from the upper valleys and pediment slopes to the groundwater system.

Potential intake zones below rock outcrops have been observed in the East Wongan catchment, a cross section of which is shown in Fig. 2. Here again, the soils immediately below the outcrop are sandy and do not contribute runoff. Booth (personal communication, 1979), has observed intense summer storm runoff from the northern outcrop infiltrating directly at the rock base.

On Wheelock's catchment, north-east of Newdegate, there is no outcropping rock but coarse textured surface and subsurface soils are associated with heath vegetation on the upper slopes. The saturated conductivity (K_s), measured by the single auger hole method, of the coarse soils was 20 times the K_s of the soils of the lower slopes. Infiltration rates of the coarse soils, measured with ring infiltrometers, were 150 to 200 times faster than rates for the soils of the lower slopes. The ratio of infiltration rates is similar to that measured by Bligh (1980) in the Belka Valley and hence differences in infiltration rate may be useful in defining recharge zones. However, it is imperative that measurement of soil physical parameters that will be used to help delineate recharge areas be done on soil surfaces that have received similar treatment. There are large differences in the hydraulic conductivity of, for instance, virgin soils and those that have been cleared and developed for agriculture. Bligh (1979, 1980) measured K_s of the same soil type in both the virgin and developed states. Sampling was done within 10 m, the results are summarised in Table 2.

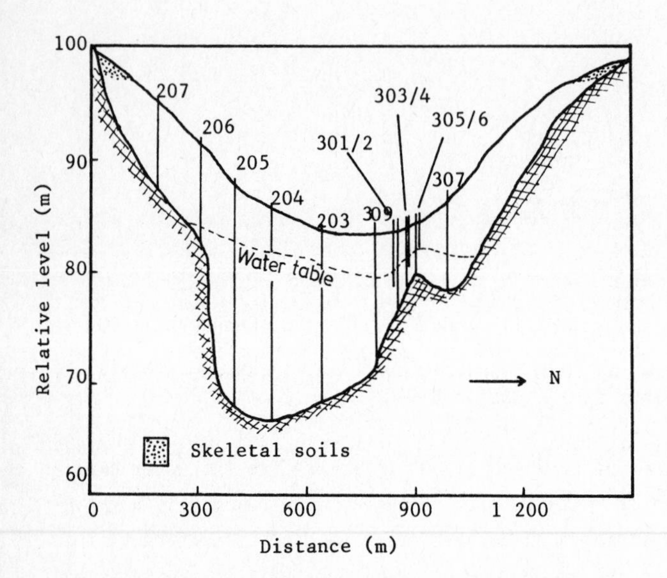

Fig. 2. Transverse cross section of East Wongan catchment.

TABLE 2

Hydraulic conductivity of virgin soils and adjacent soils developed for agriculture. (After Bligh, 1980).

Soil	Hydraulic Conductivity (ms^{-1}) Virgin	Developed
Clay loam	62.7×10^{-6}	9.71×10^{-6}
Sandy clay	46.8×10^{-9}	9.37×10^{-9}
Clay	671×10^{-9} - 532×10^{-9}	2.66×10^{-9} - 486×10^{-9}

Stoneman (1965) also observed differences in the infiltration rate of virgin and developed soils. At one site he found, after six hours infiltration, that the virgin soil had infiltrated three times the depth of water of the developed site.

Rapid water entry into the soil can occur down macropores in the recharge zone. Peck et al. (1980b) indicate that root channels and veins of quartz grains within the kaolinitic pallid zone contribute to groundwater recharge and lateral water movement. Observations of water flow in discrete channels, or preferred pathways, were first recorded by Lawes et al. (1882) who observed an almost immediate increase in the flow of the Rothamsted drain after rain. They proposed that this rapid response was due to flow through open channels and cracks in the soil profile. Thomas and Phillips (1979) in a brief review presented evidence showing the importance of macropores in soil and groundwater recharge and in salt transport through the soil.

2.2 The Flow Path

The flow path in most groundwater systems in Western Australia consists of a shallow unconfined aquifer and a deeper semi-confined aquifer. Conacher and Murray (1973) and Conacher (1974, 1975) consider the shallow, unconfined aquifer to be the most important contributor to salinity. Smith (1962, 1966); Bettenay et al. (1964); Williamson and Bettenay (1980) and Anon. (1979) have concluded that the deeper aquifer contributes the majority of the salts causing the salinity problem. Salt transport through the landscape is discussed in detail by Peck et al. (1981).

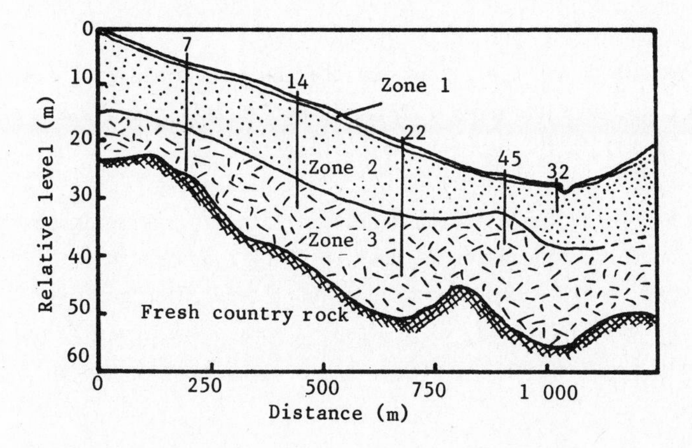

Fig. 3. Transverse cross section of Dangin catchment showing three hydrologic zones.

Fig. 3 is a cross section of the Dangin catchment. Based on morphology and texture three hydrologically different zones are discernible. Zone 1 consists of sands and gravels to a depth of generally less than 1 m. There is a distinct textural change to the sandy clays of zone 2. The clays are predominantly kaolinitic and the degree of weathering decreases with depth. The reduced weathering of zone 3 is characterised by an increase in the size and quantity of quartz and feldspar granules. Note that there is not a distinct textural change from zone 2 to zone 3. The layer above the fresh parent material is thought to contain numerous fractures and joints. Evidence for this is sparse but Furness (1978a) observed joints in hard rock cores and Peck et al. (1980a) cite other reports of the same.

2.2.1 Aquifer thickness

The thickness of aquifers varies with the surface topography and contours of the bedrock. Peck et al. (1980a) reported thicknesses of 10 to 30 m for the deep aquifers in the Darling Range catchments. Furness (1978a) measured a saturated thickness of 24 to 34 m for the Sutton catchments near Manjimup. Thicknesses of this order also apply to catchments further east, and Fig. 3 shows that the aquifer at Dangin is 20 to 30 m thick, while the maximum thickness in the East Wongan cross section (Fig. 2) is 16 m. The West Kondut catchment has an extremely undulating bedrock and the maximum aquifer depth measured was 32 m. George (1978) reported that the thickness of the aquifer at Esperance Downs averaged 8 to 10 m, which is significantly less than thicknesses reported elsewhere in the south-west of the State. George (1978) considered that the shallow bedrock at Esperance Downs was a major factor influencing saline encroachment on the catchment.

Little information is available on the depth of the shallow unconfined aquifers. Conacher (1974, 1975) and Conacher and Murray (1973) did not install piezometers at depths greater than 1.2 m and it is therefore impossible to define these aquifer limits. Shallow observation wells installed at Dangin have shown that there is an ephemeral perched water table in the pisolitic laterite. Williamson (1973) found permanent unconfined groundwater in deep spillway sands north of Meckering. The spillway sands could be as deep as 6 m and the perching bed was either fresh country rock or the kaolin clay of the pallid zone. There was evidence that the relatively fresh waters from these shallow aquifers leaked to the saline deeper aquifer.

2.2.2 Length of flow path and hydraulic gradients.

In most groundwater systems in Western Australia the length of the flow path is much greater than the thickness and thus it is reasonable to neglect vertical hydraulic gradients and assume horizontal flow between the intake area and discharge zone. Flow path length ranges from a few hundred metres for catchments in the dissected landscape of the Darling Range to several kilometres for catchments of the eastern wheatbelt.

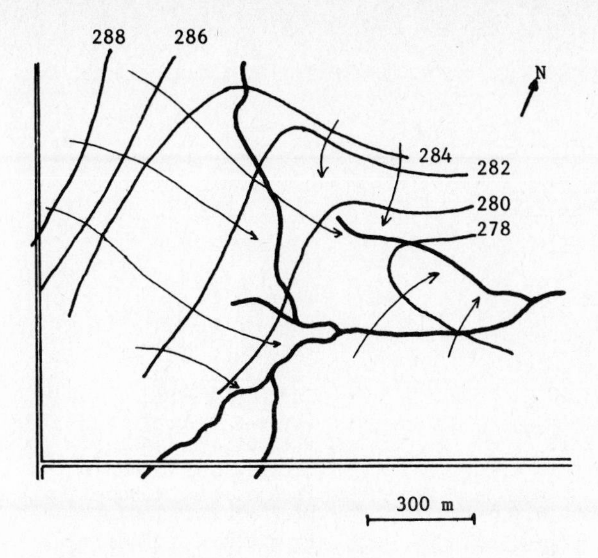

Fig. 4. Potentiometric surface (m) on September 5, 1979 for the semi-confined aquifer at Dangin.

Fig. 4 is a schematic of the flow for the semi-confined aquifer at Dangin and it illustrates that the underground flowpaths do not necessarily terminate at the first obvious surface watercourse. The west to east flowpaths bypass the first watercourse and converge on the second, less well defined watercourse. It is suspected that the second watercourse was the more active before clearing for agriculture changed the surface hydrology of the area.

The hydraulic gradients in the area range from 23×10^{-3} upslope to 7.1×10^{-3} near the watercourse. Larger gradients within the range 10×10^{-3} to 29×10^{-3} were measured by Furness (1978a, 1978b) for catchments near Manjimup.

In contrast to the decrease in hydraulic gradient near the watercourse at Dangin, George (1978) measured an increase in the gradient near the watercourse at Esperance Downs. The increased gradient occurred after the transition from agricultural land use to native vegetation and George considered that the higher annual transpiration of the native vegetation was responsible for maintenance of the higher gradient.

2.2.3 Aquifer transmission characteristics

Few measurements have been made of the transmission characteristics of aquifers associated with land and stream salinity in Western Australia. Table 3 summarises the available published data on hydraulic conductivities.

TABLE 3

Hydraulic conductivities of deep aquifers associated with land and stream salinity in Western Australia.

Location	Hydraulic Conductivity (ms^{-1})	Source
Del Park	409×10^{-9}	Bestow (1976)
Manjimup	75.2×10^{-9}	Furness (1977)
Manjimup	162×10^{-9}	Furness (1977)
Manjimup	255×10^{-9}	Furness (1978a)
Manjimup	463×10^{-9}	Furness (1978b)
Manjimup	359×10^{-9}	Furness (1978b)
Dwellingup	370×10^{-9}	Peck et al. (1980b)*
Bakers Hill	94.9×10^{-9}	Peck et al. (1980b)*
Collie West	13.9×10^{-3}	Peck et al. (1980b)*
Collie East	62.5×10^{-9}	Peck et al. (1980b)*
Upper Helena	1.39×10^{-6}	Peck et al. (1980b)*

* Values are the arithmetic means. See further discussion in the text.

Bestow (1976) estimated K_s from a groundwater study, Furness (1977, 1978 a,b) computed K_s from pump test data using the method of Kruseman and de Ridder (1970) and Peck et al. (1980b) calculated K_s from slug test data (Bouwer and Rice, 1976). Despite the different methods used there is a reasonable agreement of the order of magnitude of K_s. The Dwellingup data of Peck et al. (1980b) and the Del Park data of Bestow (1976) are from nearby areas and the values agree well. Peck et al. (1980b) argue that the arithmetic mean of a log-normal frequency distribution of spatially random values of K_s is the upper limit of K_s. The bulk value of K_s is believed to lie between the geometric and arithmetic means.

The slug test was used to determine K_s in 24 wells at West Kondut and in 22 at East Wongan. There was no difference in K_s between the two catchments. Arithmetic means were 871×10^{-9} and 840×10^{-9} ms^{-1} for West Kondut and East Wongan respectively. Corresponding geometric means were 326×10^{-9} and 182×10^{-9} ms^{-1}.

Hydraulic conductivities of aquifers in both the Darling Range and the Wongan Hills area are slow by Klute's (1965) criteria. Furness (1977) observed that the values of K_s he obtained for predominantly clay material were much greater than other published values for clays. He ascribed the difference to the effect of macropores. Fissures and joints in freshly weathered bedrock cores have shown iron staining which was taken as evidence of water flow through them (Furness, 1978a). Root channels with water flowing through them were observed at 6 m depth while excavating an earth tank in country that was originally vegetated with *Eucalyptus salubris* (Frith, J.L., personal communication, 1979).

These preferred pathways can transport a major proportion of the water moving through the soil. Nulsen (1980b) found that macropores were responsible for more than 99% of the water flow through soils from Kellerberrin and Newdegate in Western Australia.

2.2.4 Rates of water movement

Most computations of the rate of water movement use Darcy's Law. At Del Park Bestow (1976) calculated a velocity of 81.0 x 10^{-9} ms^{-1} normal to the equipotentials and a vertical velocity component of 225 x 10^{-9} ms^{-1}. The resultant vector velocity is 267 x 10^{-9} ms^{-1} in a direction 22° from vertical. He notes that deep in the aquifer the velocity would tend to 81.0 x 10^{-9} ms^{-1} in a direction normal to the equipotentials. Using the data of Furness (1978a,b) flow velocities for the catchments near Manjimup range from 2.5 x 10^{-9} to 13 x 10^{-9} ms^{-1} which are of the same order as the velocities derived by Bestow (1976).

The applicability of Darcy's Law to systems containing discrete channels is questionable. Atkinson (1978) suggested that for catchments, such as the one he studied in Wales, containing large preferred pathways for water flow variations on the Darcy-Weisbach equation would be appropriate for computation of fluxes. However, the size, extent and continuity of preferred pathways in aquifers in Western Australia are largely unknown and it is therefore not possible to identify which equation is appropriate for calculating flow velocities and dependent fluxes.

2.3 The Discharge Area

Discharge of saline water from the groundwater system can be either by flow into a stream system or by capillary rise and subsequent evaporation from the soil surface.

The net upward flux of water from the deep aquifer system to the discharge area is illustrated in Fig. 5 for the Dangin catchment. The hydrographs for the piezometer nest at site 32 are shown in Fig. 6 and indicate that throughout the period of observation there has been a potentiometric gradient forcing water towards the surface. For long periods the potentiometric head of well 32C was above the ground surface and the mean vertical potentiometric gradient from the deeper aquifer to the shallow aquifer was 0.132. This is greater than the hydraulic gradient between sites 45 and 32 of 7 x 10^{-3} and hence the potential vertical flow can cope with the horizontal flux.

Potentiometric heads above ground level are not uncommon in saline seepage areas and have been measured at West Kondut, East Wongan, Bakers Hill, Dangin and in the Collie catchments and observed during investigatory drilling in saline areas throughout most of the south-west of the State.

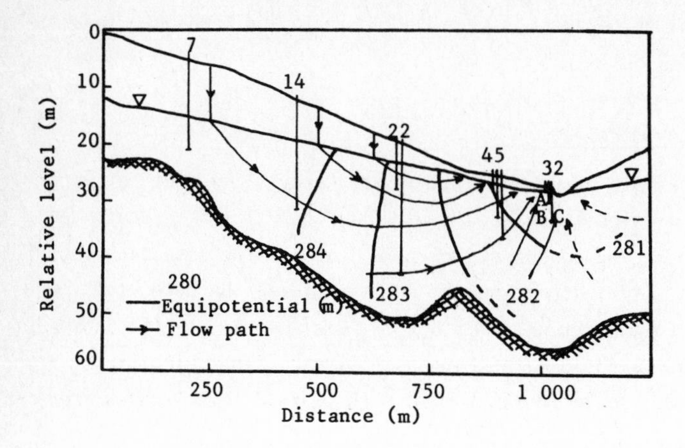

Fig. 5. Schematic of the groundwater flow in a transverse cross section of the Dangin catchment.

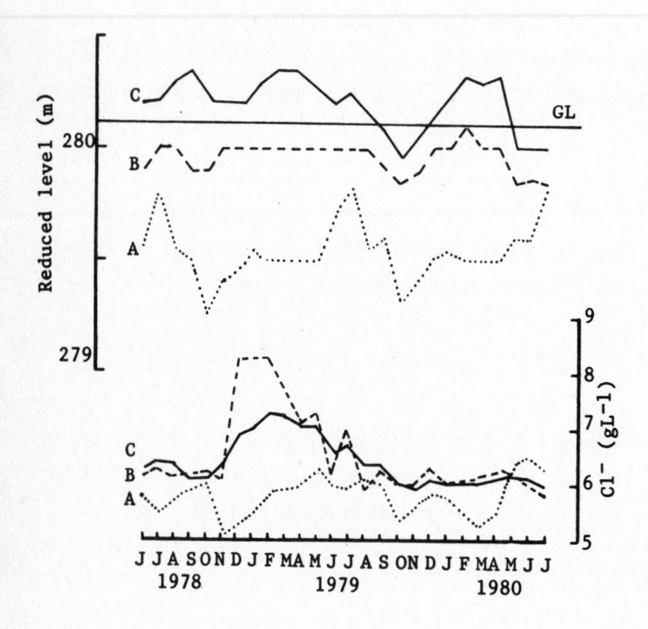

Fig. 6. Hydrographs and associated water Cl^- contents for the three piezometers at site 32 of the Dangin catchment. GL = ground surface.

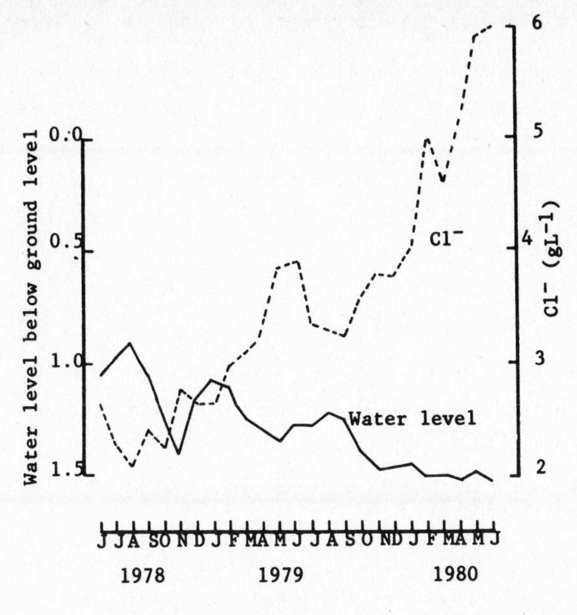

Fig. 7. Water levels and associated Cl^- concentrations for piezometer 304 of the East Wongan catchment.

Subsurface water flow in discrete channels in the discharge area is a common observation. Nulsen (1978) observed water flow through old root channels and interpedal voids of up to 2×10^{-3} m diameter and he calculated that such a channel could transport as much saline water as 37 m^2 of the clay matrix. The preferred pathways serve to increase the transport of saline groundwater towards the soil surface.

When the groundwater in the discharge zone comes to within 1.5 m of the surface, near surface soils become too saline to support most of the cereals and pastures grown in Western Australia (Nulsen, 1980a). The rate of water table rise can be rapid. Williamson and Bettenay (1980) measured a rise of 1.5 m yr^{-1} after clearing a catchment near Bakers Hill. However there is not always a rise in the water table associated with increased surface salinity. On the East Wongan catchment (Fig. 2) there has been a rapid expansion of surface expression of soil salinity over the last three years with a concomitant fall in the water table (Fig. 7). There has, however, been a dramatic rise in the water salinity. Fig. 7 shows the changes in water levels and water salinity in test well 304 which was 3.4 m deep and slotted over the bottom 2 m. A similar response occurred in piezometer 306 but little change in salinity was measured in 301 which was 10 m downslope of 304. Piezometer 301 is also downslope of the bedrock deformation which peaks under 305 and 306 (Fig. 2). The rapid rise in salinity may be due to concentration of the groundwater after three years of below average rainfall, or there may be a slug of more saline water moving through the aquifer.

The long term trends in water level and salinity (Fig. 7) at a rapidly expanding scald contrast with the relative constancy of these parameters at an established scald site as exemplified by Fig. 6 for the floor of the Dangin catchment. Although fluctuations are obvious in Fig. 6 there is no discernible trend in either the water levels or the water salinities. Only if there is some trend would the severity of the scald be expected to change.

3 CONCLUSIONS

Groundwaters associated with secondary salinity in Western Australia occur mainly in semi-confined aquifers which are characterised by low hydraulic conductivities and low hydraulic gradients. The implication is therefore, that the velocity of water movement through the aquifers is low, but this might not be so if the flow occurs in large, discrete channels.

The occurrence of secondary soil salinity is usually associated with a rise in the level of the water table and only by reversing the trend and significantly reducing the level of the water table can any amelioration of the surface salinity be anticipated.

4 REFERENCES

Abbott, W.E., 1980. Ringbarking and its effects. J. Roy. Soc. N.S.W., 14: 97-102.

Anon., 1979. Investigations of the Whittington interceptor system of salinity control. Water Resour. Inform. Note. Public Works Dept. West. Aust. 25 pp.

Atkinson, T.C., 1978. Techniques for measuring subsurface flow on hillslopes. In: Hillslope Hydrology. Kirby, M.J. (Ed.). John Wiley & Son, Chichester. 73-120.

Bestow, T.T., 1976. The effects of bauxite mining at Del Park on groundwater hydrology. 1975 Ann. Rep. Geol. Survey West. Aust., 13-31.

Bettenay, E., Blackmore, A.V. and Hingston, F.J., 1964. Aspects of the hydrologic cycle and related salinity in the Belka Valley, Western Australia. Aust. J. Soil Res., 2: 187-210.

Bligh, K.J., 1979. Runoff generation processes in a seasonally arid farmland catchment, a preliminary assessment. Hydrology Symposium, Perth. Inst. Engineers Aust., Canberra. 28-30.

Bligh, K.J., 1980. Estimating winter runoff from farmland. Agric. Engng. Conf., Geelong, Inst. Engineers Aust., Canberra. 161-165.

Bouwer, H. and Rice, R.C., 1976. A slug test for determining hydraulic conductivity of unconfined aquifers with completely and partially penetrating wells. Water Resour. Res., 12: 423-428.

Conacher, A.J., 1974. Salt scald: a W.A. case study in rehabilitation. Sci. Technol., Surrey Hills, Vic., 11: 14-16.

Conacher, A.J., 1975. Throughflow as a mechanism responsible for excessive soil salinisation in non-irrigated, previously arable lands in the Western Australian wheatbelt: a field study. Catena, 2: 31-67.

Conacher, A.J. and Murray, I.D., 1973. Implications and causes of salinity problems in the Western Australian wheatbelt: the York-Mawson area. Aust. Geograph. Stud., 11: 40-61.

Cope, F., 1958. Catchment salting in Victoria. Soil Conserv. Auth. of Victoria, Melbourne.
Doering, E.J. and Sandoval, F.M., 1976. Hydrology of saline seeps in the Northern Great Plains. Trans. Amer. Soc. Agric. Engnrs., 19: 856-865.
Furness, L.J., 1977. Aquifer tests on bores in the Yerraminnup Creek catchments. Proj. 2, Manjimup Woodchip Ind. Res., Mines Dept. West. Aust. Hydrology Rep. No. 1498: 10 pp.
Furness, L.J., 1978a. Groundwater and salt flows from the Sutton catchments. Proj. 2, Manjimup Woodchip Ind. Res., Mines Dept. West. Aust. Hydrology Rep. No. 1904: 16 pp.
Furness, L.J., 1978b. Groundwater and salt flows from the Lewin catchments. Proj. 2, Manjimup Woodchip Ind. Res., Mines Dept. West. Aust. Hydrology Rep. No. 1905: 13 pp.
George, P.R., 1978. The hydrogeology of a dryland salt seepage area in south-western Australia. Proc. Subcomm. Salt Affected Soils. 11th Int. Soil Sci. Congr., Edmonton. 3.1-3.13.
Greenlee, G.M., Pawluk, S. and Bowser, W.E., 1968. Occurrence of soil salinity in the drylands of south-western Alberta. Can. J. Soil Sci., 48: 65-75.
Halvorson, A.D. and Black, A.L., 1974. Saline seep development in dryland soils of north-eastern Montana. J. Soil Water Cons. 29: 77-81.
Klute, A., 1965. Laboratory measurement of hydraulic conductivity of saturated soil. In: Methods of Soil Analysis. Agron. 9: 210-221. Amer. Soc. Agron., Madison.
Kruseman, G.P. and de Ridder, N.A., 1970. Analysis and evaluation of pumping test data. Int. Inst. Land Reclamation and Improvement, Wageningen. Bull. No. 11.
Lawes, J.B., Gilbert, J.H. and Warington, R., 1882. On the amount and composition of the rain and drainage waters collected at Rothamsted. William Clowes and Sons, London.
Mulcahy, M.J., 1978. Salinisation in the south-west of Western Australia. Search, 9: 269-272.
Nulsen, R.A., 1978. Water movement through soil. J. Agric. West. Aust. 19: 106-107.
Nulsen, R.A., 1980a. Critical depth to saline groundwater in non-irrigated situations. Aust. J. Soil Res. (In press).
Nulsen, R.A., 1980b. Preferred pathway leakage of saline water from semi-confined aquifers. Int. Symp., Salt Affected Soils, Karnal, India, 1980. 227-232.
Ogle, N., 1839. The Colony of Western Australia. James Fraser, London. Reprinted in 1977 by John Ferguson, St. Ives, New South Wales. 298 pp.
Pauli, H.W., 1972. Relationships between land use and water pollution. In: Water Pollution Rep. No. 38: 5.1-5.13. Water Res. Found. Australia.
Peck, A.J., Hurle, D.H., Height, M.I. and Yendle, P.A., 1980a. Groundwater systems in deeply weathered materials in the Darling Range, Western Australia. CSIRO Div. Land Resour. Management Tech. Paper No. (In press).
Peck, A.J., Yendle, P.A. and Batini, F.E., 1980b. Hydraulic conductivity of deeply weathered materials in the Darling Range, Western Australia. Aust. J. Soil Res. 18: 129-138.
Peck, A.J., Johnston, C.D. and Williamson, D.R., 1981. Soil solute distribution and transport in relation to dryland salinity in Western Australia. Agric. Water Manage.4: 83-102
Rowan, J.N., 1971. Salting on dryland farms in north-western Victoria. Soil Conserv. Auth. of Victoria, Melbourne.
Smith, S.T., 1962. Some aspects of soil salinity in Western Australia. M.Sc. (Agric.) thesis. Univ. West Aust.
Smith, S.T., 1966. The relationship of flooding and saline water tables. J. Agric. West. Aust., 7: 334-340.
Stoneman, T.C., 1965. Soil structure studies in the Western Australian wheatbelt. M.Sc. (Agric.) thesis, Univ. of West. Aust.
Teakle, L.J.H. and Burvill, G.H., 1945. The management of salt lands in W.A. J. Agric. West Aust., 22: 87-93.

Thomas, G.W. and Phillips, R.E., 1979. Consequences of water movement in macropores. J. Environ. Qual., 8: 149-152.
Van Dÿk, D.C., 1969. Relic salt, a major cause of recent land damage in the Yass Valley, Southern Tablelands, N.S.W. Aust. Geographer, 11: 13-21.
Williamson, D.R., 1973. Shallow groundwater resources of some sands in south-western Australia. Hydrol. Symp. Nat. Conf. Publ. No. 73/3, 85-90.
Williamson, D.R. and Bettenay, E., 1980. Agricultural land use and its effect on catchment output of salt and water - evidence from southern Australia. In: 'Progress in Water Technology'. Vol. II, No. 6, Pergamon: London.
Wood, W.A., 1924. Increases of salt in soil and streams following the destruction of native vegetation. J. Roy. Soc. West Aust., 10: 35-47.

ROLE OF SOLUTE-TRANSPORT MODELS IN THE ANALYSIS OF GROUNDWATER SALINITY PROBLEMS IN AGRICULTURAL AREAS

L.F. KONIKOW
U.S. Geological Survey, Virginia, U.S.A.

ABSTRACT

Konikow, L.F., 1981. Role of solute-transport models in the analysis of groundwater salinity problems in agricultural areas. Agric. Water Manage., 1981.

Undesirable salinity increases occur in both groundwater and surface water and are commonly related to agricultural practices. Groundwater recharge from precipitation or irrigation will transport and disperse residual salts concentrated by evapotranspiration, salts leached from soil and aquifer materials, as well as some dissolved fertilizers and pesticides. Where stream salinity is affected by agricultural practices, the increases in salt load usually are attributable mostly to a groundwater component of flow. Thus, efforts to predict, manage, or control stream salinity increases should consider the role of groundwater in salt transport. Two examples of groundwater salinity problems in Colorado, U.S.A., illustrate that a model which simulates accurately the transport and dispersion of solutes in flowing groundwater can be (1) a valuable investigative tool to help understand the processes and parameters controlling the movement and fate of the salt, and (2) a valuable management tool for predicting responses and optimizing the development and use of the total water resource.

1. INTRODUCTION

Undesirable increases in salinity in soil or water is a problem that has confronted civilization for centuries. But it is not merely a problem of the past. Salinity problems are continuing to have a significant impact on society, primarily because of consequent damages to sources of water supply and to agricultural productivity.

Just as Man's disturbance of the hydrological cycle will affect the hydrologic budget of an area, so will a stress on the salt cycle (the full pattern of chemical precipitation, dissolution, and movement of salt) change the salt budget of an area (for example, see Holmes, 1971). Because the principal means of salt transport is as dissolved constituents in flowing water, hydrologic stresses can, in turn, affect salinity.

Undesirable salinity increases may occur in either groundwater or surface water. With few exceptions though, salinity problems in streams can be attributed mostly to a groundwater component of flow. Thus, most efforts to predict, manage, or control salinity increases have, by necessity, had to consider the role of salt transport in groundwater.

Research on the processes and parameters controlling the transport and dispersion of dissolved chemicals in flowing groundwater has been spurred by a variety of serious water-quality problems. These include agricultural problems, such as diffuse salinity increases caused by irrigation return flows, or the development of local salt seeps in dryland farming areas, as well as other types of solute-transport problems, including sea water encroachment in coastal aquifers, natural brine discharges from saline aquifers, disposal of saline wastewater, upconing from underlying saline aquifers because of overpumping, and migration of leachates from solid-waste disposal sites. Much relevant research has also been accomplished within the petroleum industry, where, for example, efficient secondary recovery efforts may require the capability to analyze and predict miscible displacement processes occurring in a petroleum reservoir.

Practical necessities and recently increased environmental awareness have resulted in a number of governmental programs to control or limit salinity increases in streams in the United States. Salinity control projects presently in various stages of planning by the U.S. Army Corps of Engineers and the Water and Power Resources Service (formerly the Bureau of Reclamation) may cost many hundreds of millions of dollars. These and other similar recent changes in legal and management aspects of water resources have increased the need for accurate technical evaluations of the sources of salinity and of salt transport in a variety of hydrogeologic settings. Such evaluations warrant a multi-disciplinary approach, utilizing expertise in hydraulics, hydrogeology, and geochemistry, as well as other disciplines. Although each specific problem is unique in terms of the origin of the salts, hydrogeologic framework, and impact on the environment, each problem can be analyzed with a common perspective of the basic physical, chemical, and hydraulic principles that govern salt transport in water.

2 SOLUTE-TRANSPORT MODELS

2.1 Governing Equations

Reliable quantitative predictions of salt transport in groundwater can only be made if we understand the processes and parameters controlling the convective transport, hydrodynamic dispersion and chemical reactions affecting the concentration of dissolved chemicals in groundwater and if we can quantitatively represent them in a systematic analytical model. A usable model for a variety of hydrogeologic situations must be accurate, functional and transferable. Aquifers generally have heterogeneous properties and complex boundary conditions; therefore, it is necessary to use a deterministic, distributed parameter, digital simulation model to solve the mathematical equations that describe the transport processes.

The purpose of a model that simulates solute transport in groundwater is to compute the concentration of a dissolved chemical species in an aquifer at any specified place and time. Because convective transport and hydrodynamic dispersion depend on the velocity of groundwater flow, the mathematical simulation model must solve at least two simultaneous partial differential equations. One is the equation of flow, from which groundwater velocities are obtained, and the second is the solute-transport equation, describing the chemical concentration in the groundwater.

The theory and development of the equations describing groundwater flow and solute transport have been well documented in the literature (for example, see Bredehoeft and Pinder, 1973; Freeze and Cherry, 1979; and Bear, 1979). By following the developments of Cooper (1966) and of Pinder and Bredehoeft (1968), the equation describing the transient flow of a homogeneous slightly compressible fluid through a nonhomogeneous anisotropic aquifer may be written in cartesian tensor notation as:

$$\frac{\partial}{\partial x_i}\left(K_{ij}\frac{\partial h}{\partial x_j}\right) = S_s\frac{\partial h}{\partial t} + W^* \tag{1}$$

where K_{ij} = the hydraulic conductivity tensor, LT^{-1}; h = the hydraulic head, L; S_s = the specific storage, L^{-1}; $W^* = W^*(x,y,z,t)$ = the volume flux per unit volume (positive sign for outflow and negative for inflow), T^{-1}; x_i, x_j = the cartesian coordinates, L; and t = time, T.

An expression for the average seepage velocity of groundwater can be derived from Darcy's law and can be written in cartesian tensor notation as:

$$V_i = \frac{-K_{ij}}{\varepsilon}\frac{\partial h}{\partial x_j}, \tag{2}$$

in which ε is the effective porosity of the aquifer.

Where salinity (and hence, fluid density) varies significantly, the relations among water levels, heads, pressures and fluid velocities are not as straightforward as in a system containing a homogeneous fluid. Calculations of flow rates and directions then require pressure, density, and elevation data, instead of just head measurements.

Grove (1976) presents a generalized form of the solute-transport equation in which he incorporates terms to represent chemical reactions and solute concentrations both in the pore fluid and on the solid surface, as follows:

$$\varepsilon\frac{\partial c}{\partial t} + \frac{\partial}{\partial x_i}(\varepsilon c V_i) - \frac{\partial}{\partial x_i}(\varepsilon D_{ij}\frac{\partial c}{\partial x_j}) + C'W^* = \text{CHEM} \tag{3}$$

where CHEM equals:

$-\rho_b \frac{\partial \bar{c}}{\partial t}$ for linear equilibrium controlled ion-exchange reactions,

$\sum_{k=1}^{s} R_k$ for s chemical rate controlled reactions, and

$-\lambda(\varepsilon c+\rho_b\bar{c})$ for decay,

and where:

c = the concentration of the solute (mass of solute/volume of liquid), ML^{-3};

D_{ij} = the coefficient of hydrodynamic dispersion, L^2T^{-1};

C' = the concentration of the solute in the source or sink fluid, ML^{-3};

ρ_b = the bulk density of the solid (mass of solid/volume of sediment), ML^{-3};

$\bar{c}$ = the concentration of the species adsorbed on the solid (mass of solute/mass of sediment);

R_k = the rate of production of the solute in reaction k of s different reactions, $ML^{-3}T^{-1}$; and

λ = the decay constant (equal to ln 2/half life), T^{-1}.

2.2 Numerical Methods

Perhaps the most important technical advance in the analysis of groundwater salinity problems during the past 10 years has been the development of numerical simulation models that efficiently solve the governing flow and transport equations for the properties and boundaries of a specific situation. These numerical simulation models offer a tool to evaluate conceptual models and to test hypotheses concerning complex field problems. The models do not replace field data, but they can guide toward developing a better understanding of a problem and designing a more efficient data collection network. The development of numerical models applicable to groundwater salinity problems is still in a state of flux; no single model is available yet that is equally suitable for the entire spectrum of possible problems. Particularly difficult numerical problems arise if the chemical reaction terms are highly nonlinear or if the concentration of the solute of interest is strongly dependent on the concentration of numerous other chemical constituents.

Three general classes of numerical methods have been used to solve the solute-transport equation: finite difference methods, finite element methods and the method of characteristics. Each method has some advantages, disadvantages, and special limitations for applications to field problems. Each method also requires that the area of interest be subdivided by a grid into a number of smaller subareas.

The method of characteristics was originally developed to solve hyperbolic equations. If solute transport is dominated by convective transport, as is common in many field problems, then equation (3) may closely approximate a hyperbolic equation and be highly compatible with the method of characteristics. Although it is difficult to present a rigorous mathematical proof for this numerical scheme, it has been successfully applied to a variety of field problems. The method solves a system of ordinary differential equations that is equivalent to equation (3). The numerical solution is achieved by introducing a set of moving points that are traced with reference to the stationary coordinates of a finite difference grid. Each point has a concentration associated with it and is moved through the flow field in proportion to the flow velocity at its location. The change in concentration due to dispersion, fluid sources, and chemical reactions can be computed with an explicit finite difference equation. The development and application of this technique to problems of flow through porous media have been presented by Garder et al. (1964), Pinder and Cooper (1970), Reddell and Sunada (1970), and Bredehoeft and Pinder (1973), and a general computer program is documented by Konikow and Bredehoeft (1978).

Finite difference methods solve an equation that approximates the partial differential equation. Although problems of numerical dispersion, overshoot, and undershoot may induce significant errors for some problems, these methods can efficiently provide accurate answers, particularly when dispersive transport is large compared to convective transport. In the latter case, the computation time required may be 2 to 5 times less than for the method of characteristics. In general, the finite difference methods are the simplest mathematically and the easiest to program for a digital computer. A three-dimensional, transient, finite difference model that simultaneously solves the pressure, energy-transport, and mass-transport equations for variable density fluids is described by INTERA (1979).

Finite element methods use assumed functions of the dependent variables and parameters to evaluate equivalent integral formulations of the partial differential equations. Pinder (1973), Segol and Pinder (1976), and Gupta et al. (1975) have indicated that Galerkin's procedure is well suited to solve solute-transport problems. Pinder and Gray (1977) present a comprehensive analysis and review of the application of finite element methods to groundwater flow and transport problems. These methods generally require the use of more sophisticated mathematics than the previous two methods, but for many problems may be more accurate numerically and more efficient computationally than the other two methods. A major advantage of the finite element methods is the flexibility of the finite element grid, which allows a close spatial approximation of irregular boundaries of parameter zones. However, Gupta et al. (1975, p. 69) report that in problems dominated by convection, the finite element methods may also have difficulties.

The selection of a numerical method for a particular problem depends on several factors, such as accuracy, efficiency/cost, and usability. The first two factors are related primarily to the nature of the field problem, availability of data, and scope or intensity of the investigation. A trade-off between accuracy and cost is frequently required. The usability of a method may depend more on the availability of a documented program and on the mathematical background of the modeler. Greater efficiency is usually attainable if the modeler can modify and adapt a program to the specific field problem of interest.

3. MODEL APPLICATION TO IRRIGATED ALLUVIAL VALLEY

Groundwater and surface water are interrelated in stream-aquifer systems in which groundwater in the flood-plain alluvium is in hydraulic connection with the stream. In the arid to semiarid regions of the western United States, the fertile flood-plain soils are commonly irrigated with both diverted surface water and pumped shallow groundwater. Much of the applied irrigation water is lost by evapotranspiration, but some applied water recharges the alluvial aquifer and provides return flow to the stream. Dissolved solids become concentrated in the recharged water because evapotranspiration consumes some of the applied water but has little effect on the total mass of chemical constituents dissolved in the water. The down-valley reuse of water causes a buildup of salts to levels that may approach concentrations intolerable to many crops.

The increase in salinity of groundwater and surface water in the Arkansas River Valley of south-eastern Colorado, U.S.A., is primarily related to irrigation practices. To demonstrate the feasibility of modeling groundwater flow and salinity changes that occur in such an irrigated river valley, an 18 km reach of the valley was selected for a detailed investigation. The hydrogeologic framework and water use patterns within the study reach are representative of most of the Arkansas River valley in south-eastern Colorado and, in fact, are typical of many other irrigated river valleys in semiarid to arid areas.

The study was designed to simulate a 1-year period of record that includes one complete irrigation season. During this time, a network of 4 surface water stations and 63 observation wells was maintained to determine all inflows, outflows, and changes in aquifer storage of both water and dissolved solids within the study area. These observed data were used as a basis for calibrating the simulation model. A more complete description of the study is presented by Konikow and Bredehoeft (1974a).

3.1 Description of Study Area

The alluvial valley is about 2.4 km wide and occupies an area of about 44 km^2. The climate is semiarid, having an average annual precipitation of approximately 330 mm. The average annual temperature is 10^oC, and the mean monthly temperature ranges from -1.1^oC in January to 26^oC in July. The average discharge of the Arkansas River at La Junta is about 7.4 m^3s^{-1}. Highest flows normally occur between May and September; during 1960 to 1970, daily mean discharges have ranged from less than 0.03 to about 570 m^3s^{-1}. A major diversion from the river to the Fort Lyon Canal is made near the upstream end of the study area.

The valley is cut into relatively impermeable bedrock consisting of shale and limestone of Cretaceous age. The alluvium consists of inhomogeneous deposits of gravel, sand, silt and clay. Within the study reach the Arkansas River has a sandy channel and is in good hydraulic connection with the aquifer.

3.2 Model Calibration

Many input data are required for the model; the accuracy of these data affect the reliability of the computed results. The data requirements are summarized in Table 1 and are discussed in more detail by Konikow and Bredehoeft (1974a). In this study area it proved reasonable to assume that the chemical parameter of interest (dissolved solids concentration) is not affected by chemical reactions. In cases where reactions must be considered, additional parameters must obviously be defined to describe the nature and rates of governing reactions; this may not be a trivial task.

Variations of hydrologic and water-quality parameters with time were simulated on a monthly basis. The model was calibrated by adjusting most system characteristics within a narrow range of values until a best fit was obtained between the

TABLE 1

Summary of data requirements

System characteristics	Stresses
Transmissivity	Surface water inflow (quantity & quality)
Specific yield	Groundwater withdrawals
Saturated thickness	Precipitation
Boundaries	Surface water applications
Effective porosity	Groundwater applications
Dispersivity	Consumptive use
Initial water table elevations	Recharge (quantity and quality)
Initial dissolved-solids concentrations	Stream stage
Streambed leakance	Tributary inflow

observed data and the simulation results. In general it is more difficult to calibrate a solute-transport model of an aquifer than it is to calibrate a groundwater flow model. Fewer parameters need to be defined to compute the head distribution with a flow model than are required to compute concentration changes with a solute-transport model.

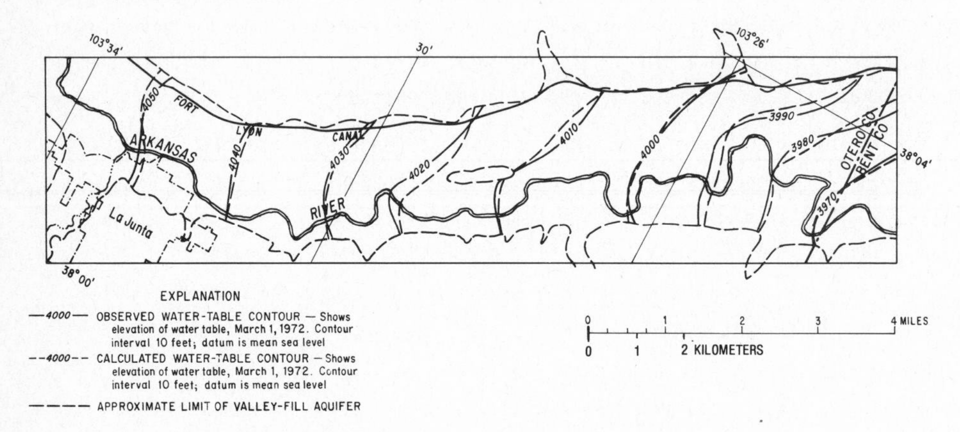

Fig. 1. Observed and calculated water table configurations at the end of the study period, March 1, 1972.

A satisfactory working model of the flow of water in the system is essentially a prerequisite for a successful water-quality model. The hydrologic model of this study area was calibrated in part by comparing the observed and calculated water table configurations after 1 year (Fig. 1), and in part on the basis of observed and calculated water level fluctuations over time (Fig. 2) (Konikow and Bredehoeft, 1974a). During the study period, the measured water levels in 23 observation wells varied by an average of about 1 m. The model reproduced the water table elevation within 0.3 m of the observed value more than 90% of the time. The two largest sources of error were probably the monthly pumpage data and the estimated rates of recharge.

The dissolved-solids concentration observed in the aquifer during the 1-year study period ranged from 800 mg L^{-1} to over 3 800 mg L^{-1}. Comparisons were made between the observed and calculated spatial water-quality patterns, areas of change, and variations of dissolved-solids concentration with time. (See Figs. 3,4 and 5). These data indicate that the model can successfully reproduce observed changes in water quality in the aquifer. The dissolved-solids concentration was computed within 10% of the observed value approximately 80% of the time. The differences are related to both data deficiencies and modeling errors.

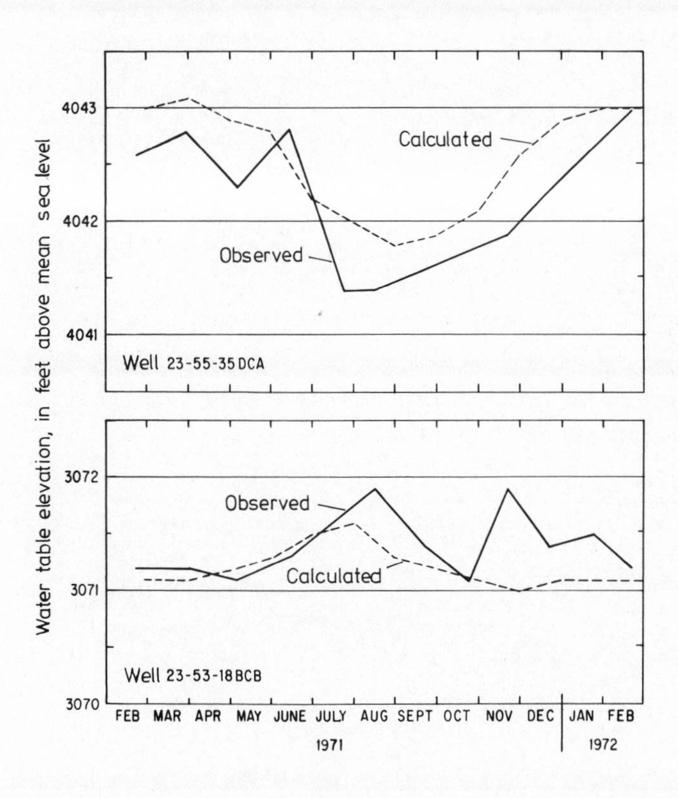

Fig. 2. Observed and calculated water levels in two wells during study period.

Changes in surface water quality that occur within the study reach were also analyzed (Fig. 6). Changes in flow and salinity of the river are determined for each time step by simple downstream routing in which the total inflow to each successive stream cell is adjusted for the flux at that node computed by the model. The flux between the river and the aquifer at each stream node is represented as a head-dependent leakage function. The observed change in the dissolved-solids concentration in the river was an increase that averaged 475 mg L^{-1}. This represents an increase of about 40% over the 1 180 mg L^{-1} average value observed in the river entering the study reach at La Junta. The calculated increases are in close agreement with the observations. Approximately one half of the total increase is attributable to the dissolved-solids load contributed by tributary inflow, and the other half is attributable to the discharge to the river of high-salinity groundwater.

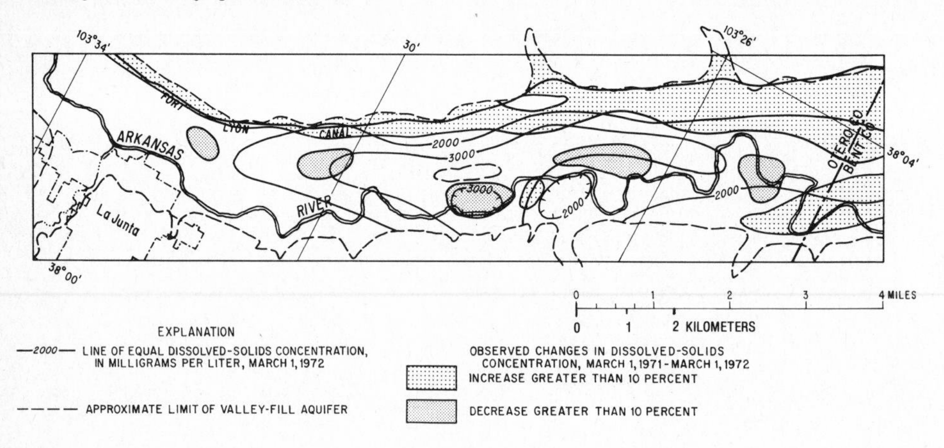

Fig. 3. Observed dissolved-solids concentration in aquifer on March 1, 1972, and areas of observed changes since March 1, 1971.

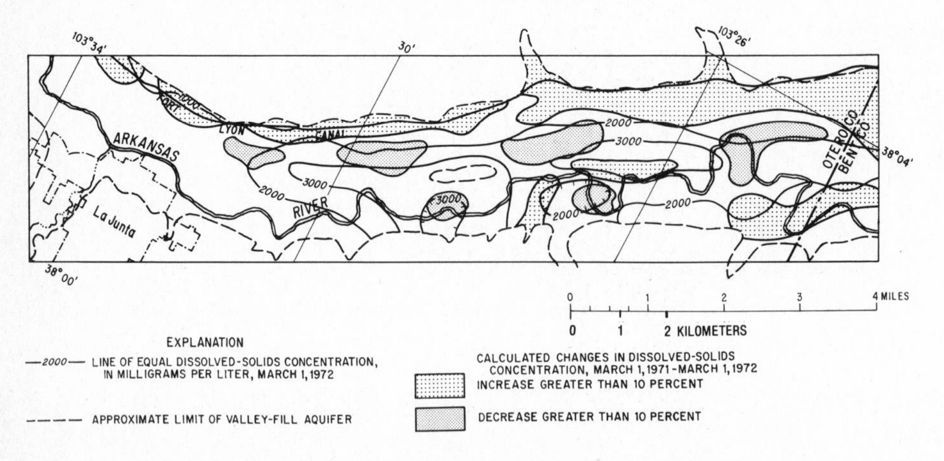

Fig. 4. Calculated dissolved-solids concentration in aquifer on March 1, 1972, and areas of calculated changes since March 1, 1971.

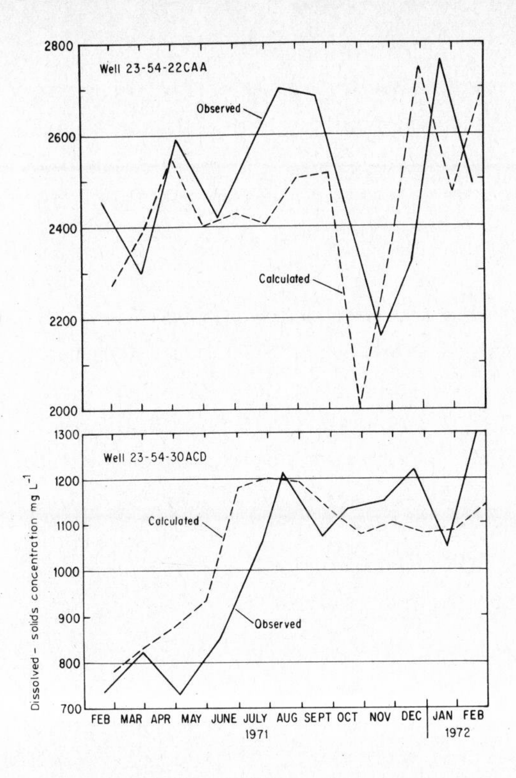

Fig. 5. Observed and calculated dissolved-solids concentrations of groundwater at two wells during the study period.

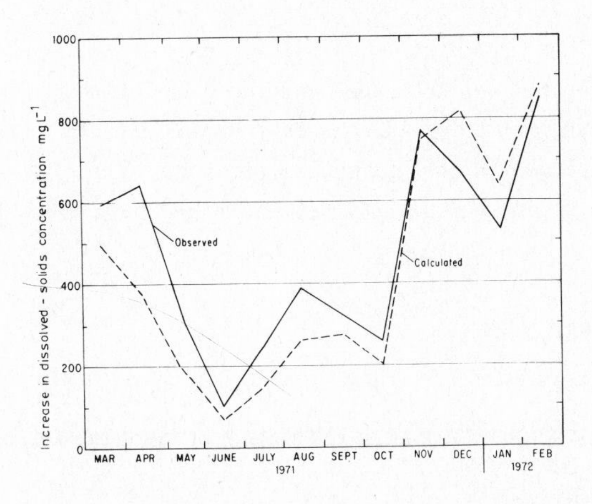

Fig. 6. Observed and calculated monthly average changes in dissolved-solids concentration in the Arkansas River in the study reach.

3.3 Predictive Capability of Model

After the simulation model is calibrated, it can be used to predict future responses if future stresses can be estimated. The model's predictions can be utilized to help evaluate the impact of alternative decisions or policies regarding problems of water planning, water management, and water-quality control, or to study the effects of an occurrence of an extreme hydrologic event such as a flood or a drought.

The application of the model to management problems in the Arkansas River valley was demonstrated by evaluating the impact of several possible changes (Konikow and Bredehoeft, 1974b). The calibrated model was used to predict long-term (5-year) changes that might occur within the stream-aquifer system, both with and without any changes in irrigation practices. Predicting long-term trends without any changes was based on the assumption that all stresses which occurred during the first year would recur in each succeeding future year. Extending the simulation period on the basis of the stresses and irrigation practices observed during the 1-year study period provides a basis for comparing the results of changing management decisions.

These evaluations of possible water-management changes resulted in predictions that increasing the proportion of groundwater use in the study area would cause significant salinity increases in the aquifer, water table declines and streamflow losses. However, increasing the proportion of surface water use would generally produce the opposite effects. Improving irrigation efficiency reduced the overall requirement for irrigation water but induced a long-term buildup of salts in the aquifer. The model also indicated that if the Fort Lyon Canal could be lined to prevent seepage losses, then significant groundwater salinity increases and water table declines would occur in the area between the canal and the river. In general, the analyses of the predictive tests for this study area indicate that initial or short-term responses to changes in irrigation practices are strongly related to antecedent conditions and do not necessarily reflect actual long-term responses to these changes.

4. MODEL APPLICATION TO NONIRRIGATED AREA

Significant damage to parts of a wheat field on a dryland farm in the South Platte River valley, near Denver, Colorado, occurred in the early 1970's. Fig. 7 presents a progressive series of closeup photographs that illustrate the nature of the damage. Clearly the crop damage is related to the development of a salt crust at the soil surface in parts of the field. The observed phenomenon and its impacts are similar to those at other salt seep problem areas, but in this case, the occurrence of the salt seep problem is related largely to industrial activities at the adjacent (and upgradient) Rocky Mountain Arsenal.

Fig. 7. Progressive series of closeups showing salt crust development in South Platte River valley, Colorado.

The Rocky Mountain Arsenal has been operating since 1942; its operations have produced liquid wastes that contain complex organic and inorganic chemicals, including a characteristically high chloride concentration that apparently ranged up to about 5 000 mg L^{-1}. From 1943 to 1956 the high-chloride wastes were discharged to unlined disposal ponds A, B, C, D and E, shown in Fig. 8. Fig. 8 also shows that after about 14 years the definable high-chloride plume had extended to a length of nearly 10 km. Interpretation of the hydraulic, chemical and geologic data indicates that liquid wastes seeped out of the unlined disposal ponds, infiltrated into the underlying alluvial aquifer and migrated downgradient toward the South Platte River (Konikow, 1977).

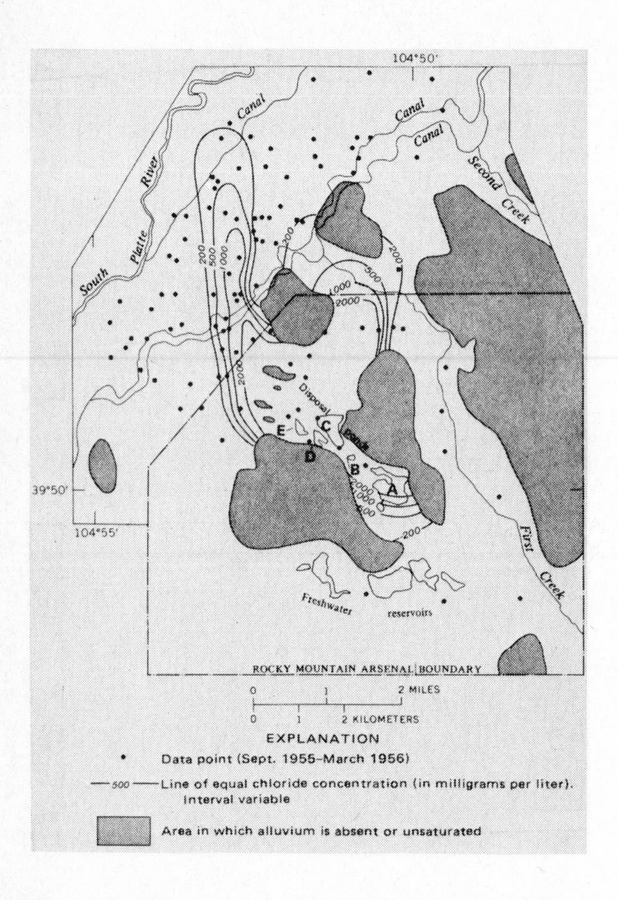

Fig. 8. Observed chloride concentration, 1956.

The rate and direction of movement of the dissolved salts through the aquifer is closely related to the groundwater flow pattern, as reflected by the water table configuration shown in Fig. 9. The hydrogeologic characteristics of the alluvium in this area indicate that the aquifer is nonuniform in thickness, sloping, discontinuous and heterogeneous (Konikow, 1975). The areas in which the alluvium is either absent or unsaturated most of the time form internal barriers that significantly affect the rates and directions of groundwater flow within the alluvial aquifer, and hence affect solute transport.

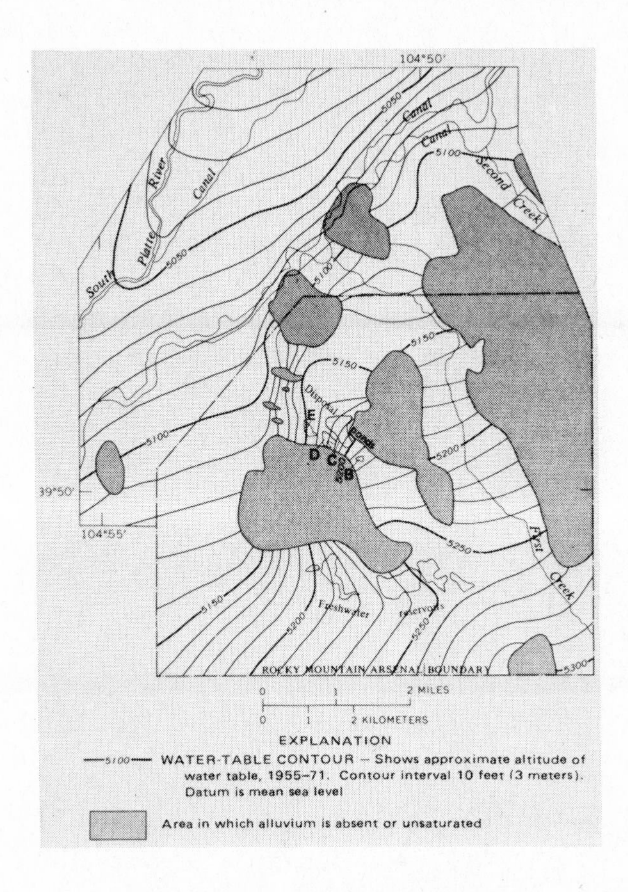

Fig. 9. General water table configuration in the alluvial aquifer in and adjacent to the Rocky Mountain Arsenal, 1955 to 1971.

Since 1956, disposal has been into an asphalt-lined reservoir, which thereby contributed to a subsequent decrease in the extent and magnitude of groundwater contamination. Fig. 10 illustrates that by 1972 chloride concentrations greater than 1 000 mg L^{-1} had become limited to just two small parts of the main area of contamination. Both are areas of relatively low hydraulic conductivity. The area of salt crust development shown in Fig. 7 occurs within the area of the northernmost of these two remaining high-chloride zones. In 1968 or 1969 an apparent effort to help flush the contaminants from the aquifer system was begun.

The apparent policy was to keep the largest of the unlined reservoirs (pond C) filled with relatively good quality water diverted from nearby freshwater reservoirs. Subsequent infiltration from 1968 to 1972 was estimated to average 0.03 m^3s^{-1}. This artificial recharge caused increases in water table elevations, hydraulic gradients and flow velocities in adjacent downgradient areas.

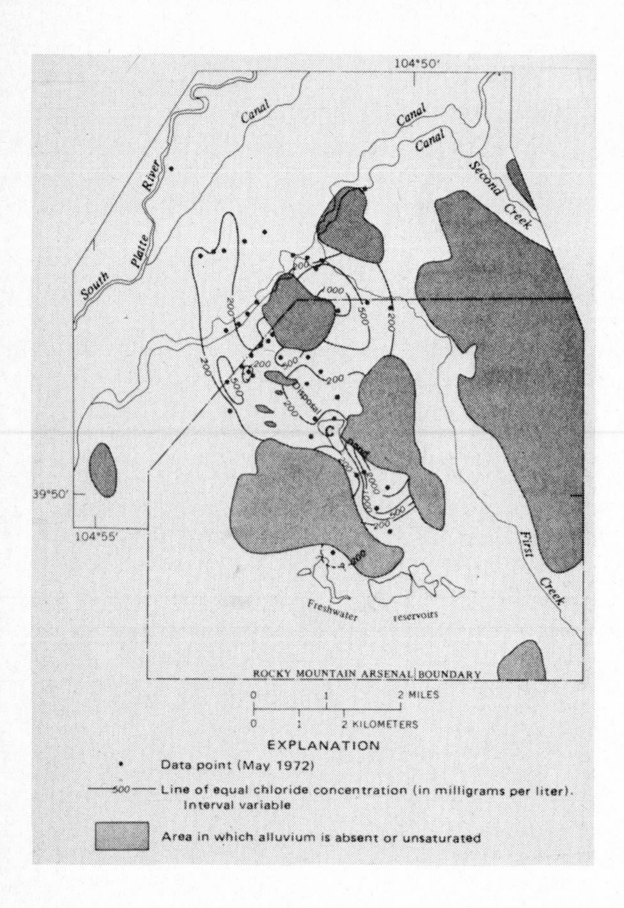

Fig. 10. Observed chloride concentration, May 1972.

The solute-transport model described by Konikow and Bredehoeft (1978) was used to simulate the movement of dissolved salts in the alluvial aquifer. The results indicated that the simulation model quantitatively integrated the effects of the major factors that controlled changes in chloride concentration and accurately reproduced the 30-year history of chloride contamination. The pattern of contamination computed for 1972 is shown in Fig. 11. The computed pattern agrees fairly well with the observed pattern, although the former shows somewhat longer plumes. After the 30-year simulation period, the model has identified (1) the two areas where high chloride concentrations were still present, (2) the reduction in the size and strength of the plume since 1956, and (3) changes in water table elevations in response to changes in water input to the unlined ponds.

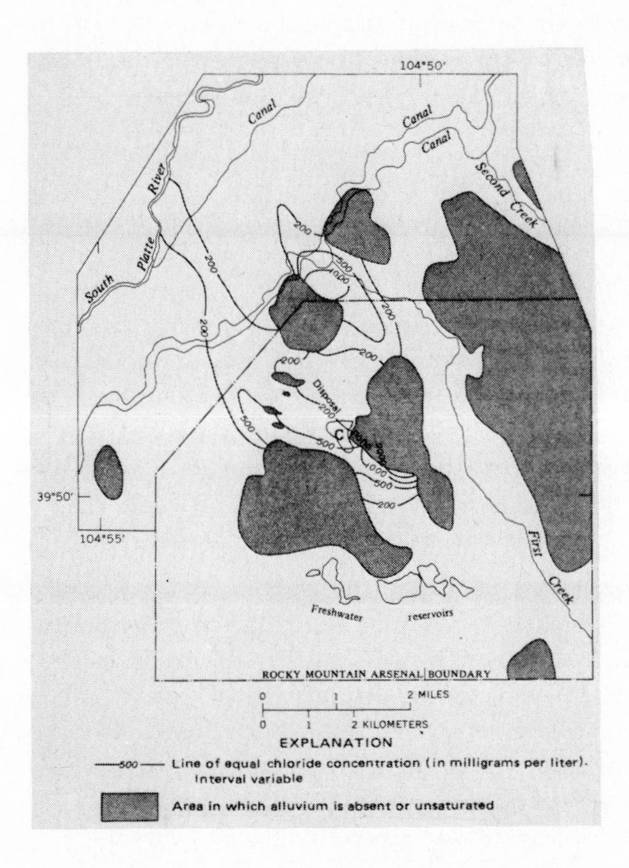

Fig. 11. Computed chloride concentration at the start of 1972.

Field examinations indicated that the surficial salt problems were restricted to local topographic lows, where the water table was relatively close to the land surface (less than 1 m). The soil surface was very moist within the damaged areas, even after several weeks of dry weather. Analyses by X-ray diffraction of a soil sample containing the white salt crust indicated that it contained 38% water soluble salts. After dehydration, these salts consisted of nearly equal amounts of halite and bassanite ($CaSO_4 \cdot \frac{1}{2}H_2O$) (H.A. Tourtelot, written communication, 1974). The coincidence of the salt crust with the area of high chloride content in groundwater and with areas of shallow water table indicates that the salts have accumulated at the soil surface because of evaporation of saline groundwater transmitted to the land surface by capillary flow.

The calibrated model was also used to test some possible water management alternatives, including the coupling of artificial recharge at pond C with construction and operation of hydraulic sinks along the northern boundary of the arsenal. The results indicated that drainage at the postulated hydraulic sinks would produce water table declines in the adjacent salt seep area of sufficient magnitude (1 to 2 m) to prevent the soil salinity problem from recurring.

5. CONCLUSIONS

Although the origins and causes of the salinity problems in the two examples discussed in this report are assuredly different from those that are most common in south-western Australia, the solute-transport principles and investigative approaches are general and transferable. The common thread that links the analysis of all of them is the universal nature of the physical and chemical laws governing fluid flow, transport processes and chemical solubility. The two examples illustrate (1) the crucial role of groundwater in transporting and dispersing salts through a local or regional aquifer to the land surface or into a surface water body, and (2) the value of a model as an investigative tool to help understand the processes and parameters controlling the movement and fate of the salt.

A complex engineering solution, such as proposed for the Rocky Mountain Arsenal, might be feasible when the salt problem is predominantly related to an isolated and specific industrial activity, but such types of control measures would likely be impractical where salt seeps become pervasive within a region because of agricultural activities. Nevertheless, any comprehensive investigation of the latter type of problem could be greatly aided by the application of solute-transport models, because the model analysis provides a disciplined format to assess the consistency within and between (1) concepts of the governing processes and (2) data describing the relevant coefficients. Feedback from preliminary models should help the investigator to set improved priorities for the collection of additional data, as well as to help test hypotheses concerning governing processes and hence develop an improved conceptual model of the salt seep problem. More advanced and calibrated models should then evolve from the improved field data and refined conceptual models. These then would enable a prediction and evaluation to be made of the range of effects on the quality and quantity of ground and surface water that might ensue from implementing a new land or water management procedure that would be economically and physically feasible in the local environment.

6 REFERENCES

Bear, J., 1979. Hydraulics of Groundwater. McGraw-Hill, New York, 567 pp.
Bredehoeft, J.D. and Pinder, G.F., 1973. Mass transport in flowing groundwater. Water Resour. Res. 9: 194-210.
Cooper, H.H., Jr., 1966. The equation of groundwater flow in fixed and deforming coordinates. J. Geophys. Research. 71: 4785-4790.
Freeze, R.A. and Cherry, J.A., 1979. Groundwater. Prentice-Hall, Inc., Englewood Cliffs, 604 pp.
Garder, A.O., Peaceman, D.W. and Pozzi, A.L., Jr., 1964. Numerical calculation of multidimensional miscible displacement by the method of characteristics. Soc. Petroleum Engineers J. 4: 26-36.
Grove, D.B., 1976. Ion exchange reactions important in groundwater quality models. In: Advances in Groundwater Hydrology, American Water Resour. Assoc., pp. 144-152.
Gupta, S.K., Tanji, K.K. and Luthin, J.N., 1975. A three-dimensional finite element groundwater model. California Water Resour. Center, Contribution 152, 119 pp.
Holmes, J.W., 1971. Salinity and the hydrologic cycle. In: Talsma, T. and Philip, J.R. (Eds.) Salinity and Water Use. MacMillan, London, pp. 25-40.
INTERA Environmental Consultants, Inc., 1979. Revision of the documentation for a model for calculating effects of liquid waste disposal in deep saline aquifers. U.S. Geol. Survey Water-Resour. Inv. 79-96, 73 pp.
Konikow, L.F., 1975. Hydrogeologic maps of the alluvial aquifer in and adjacent to the Rocky Mountain Arsenal, Colorado. U.S. Geol. Survey Open-File Rept. 74-342.
Konikow, L.F., 1977. Modeling chloride movement in the alluvial aquifer at the Rocky Mountain Arsenal, Colorado. U.S. Geol. Survey Water-Supply Pap. 2044, 43 pp.
Konikow, L.F. and Bredehoeft, J.D., 1974a. Modeling flow and chemical quality changes in an irrigated stream-aquifer system. Water Resour. Res., 10: 546-562,
Konikow, L.F. and Bredehoeft, J.D., 1974b. A water quality model to evaluate water management practices in an irrigated stream-aquifer system. In: Flack, J.E. and Howe, C.W. (Eds.), Salinity in Water Resources. Merriman, Boulder, pp. 36-59.
Konikow, L.F. and Bredehoeft, J.D., 1978. Computer model of two-dimensional solute transport and dispersion in groundwater. U.S. Geol. Survey Techniques of Water-Resources Inv., Book 7, Chap. C2, 90 pp.
Pinder, G.F., 1973. A Galerkin-finite element simulation of groundwater contamination on Long Island, New York. Water Resour. Res., 9: 1657-1669.
Pinder, G.F. and Bredehoeft, J.D., 1968. Application of the digital computer for aquifer evaluation. Water Resour. Res., 4: 1069-1093.
Pinder, G.F. and Cooper, H.H., Jr., 1970. A numerical technique for calculating the transient position of the saltwater front. Water Resour. Res., 6: 875-882.
Pinder, G.F. and Gray, W.G., 1977. Finite element simulation in surface and subsurface hydrology. Academic Press, New York, 295 pp.
Reddell, D.L. and Sunada, D.K., 1970. Numerical simulation of dispersion in groundwater aquifers. Colorado State Univ. Hydrology Pap. 41, 79 pp.
Segol, G. and Pinder, G.F., 1976. Transient simulation of saltwater intrusion in south-eastern Florida. Water Resour. Res., 12: 65-70.

RIVER BASIN HYDROSALINITY MODELING

K.K. TANJI
Department of Land, Air and Water Resources,
University of California, U.S.A.

ABSTRACT

Tanji, K.K., (1981). River basin hydrosalinity modeling. Agric. Water Manage., 1981.

Hydrologic-salinity modeling at the river sub-basin and basin spatial scales in the Western U.S.A. are evaluated. In the past two decades, significant advances have been made in basin hydrosalinity modeling. These models are providing increased understanding of the complexity of water and salt flows in large-scale systems. Most models, however, have been developed for specific problems and do not lend themselves for application to a wide variety of salinity problems. Much more work is needed in the areas of field verification of models and increasing their utility for management and policy-making decisions.

1. INTRODUCTION

One approach to integrating and synthesizing a developing body of knowledge on a problem, as in the case of stream-salinity problems in Western Australia, is mathematical and computer simulation modeling. Such modeling efforts may have diverse objectives; obtaining greater conceptual understanding of a complex problem, identifying information and data collection needs, obtaining more quantitative evaluation of observed phenomena and measured data, and predicting probable future trends for alternative management options.

This paper reports on representative hydrosalinity modeling efforts for the past two decades in the Western U.S.A. and evaluates their potential application to the land-stream salinity problem in Western Australia.

2. HYDROSALINITY MODELS

2.1 Representative Models from Western U.S.A.

Table 1 presents a partial listing of hydrosalinity models for large-scale systems that represent a variety of problem situations, modeling approaches, theory and assumptions, and applications. Not included are the finely-tuned modeling efforts for small-scale subsystems (e.g., field plots) as well as those dealing exclusively with reservoir or groundwater subsystems since they are addressed elsewhere in this Workshop.

TABLE 1

Representative large-scale hydrosalinity models for Western U.S.A.

Model	Temporal and Spatial Units	Salinity Parameter and Considerations	Applications
Woods and Orlob (1963)	Monthly for one year, 13 sub-basins and 5 reservoirs; 809 km^2	Hypothetical pollutant, e.g., pesticides TDS, etc; assumed pollutant is conservative and affected by use — factor of water	Water quality investigation in irrigated lands in the Klamath - Tule Lake area in Oregon and California
Woods (1967)	Monthly for 1 year, 21 irrigation units, reservoir and 5 streams, $8.09 \times 10^3 km^2$	EC and other quality parameters treated as conservative but subject to consumptive use and reuse concentration changes	Water quality control and management investigation in irrigated lands of the Sacramento River Basin, California
California Department of Water Resources (1969)	Monthly for 25 years; 9 study areas and streams; $9.71 \times 10^3 km^2$	EC treated as a conservative parameter with change in concentration related to a combination of linear and multiple regression analyses in historical data	Model based on 1951-65 period and projections made from 1970 to 2020 for agricultural return flows in the Lower San Joaquin River Basin
Water Resources Engineers (1969)	Yearly for 55 years, 247 polygonal nodes, $1.44 \times 10^3 km^2$	TDS treated as a conservative parameter; TDS sources include agricultural, industrial, and municipal waste loadings	Salt and water balance study focused on groundwater salinity projections in the Upper Santa Ana River Basin, California from the 1965 to 2015 period
Hyatt et al. (1970)	Monthly for 2 years; 40 sub-basins and major rivers and tributaries; $285 \times 10^3 km^2$	TDS treated as a conservative parameter with considerations for salt pickup from natural geologic sources	Simulated water and salt flows for the 1964-65 period in the Upper Colorado River Basin in Colorado, Utah, Wyoming, and New Mexico
Thomas et al. (1971)	Monthly for 2 years; 5 major soil types and up to 19 soil layers each; irrigated valley; 13.2 km^2	Salinity treated as a nonconservative parameter, six major solute species considered are involved in cation exchange, solubility of gypsum and lime	Predicted water and TDS (Ca, Mg, Na, HCO_3, Cl, and SO_4) flows in the Little Bear River Basin in Utah for the 1967-68 period
Konikow and Bredehoeft (1973)	Monthly for 1 year; 44 km^2 river valley subdivided into large number of cells	TDS treated as a conservative parameter and subject to concentrating effects of evapotranspiration.	Simulated surface and groundwater salinities in the irrigated Arkansas River Valley in Colorado
Bay-Valley Consultants (1974)	Monthly for 1 year; $112 \times 10^3 km^2$ area subdivided into 56 sub-areas	TDS computed by irrigated agriculture model that considered chemical weathering and residual fertilizers as additional sources of salts	Calculated water and salt flows based on projected 1970, 1980, 1990, and 2000 levels of development in the Sacramento River Basin, San Joaquin River Basin and the Delta in California
Tanji (1977)	Irrigation season or annually; irrigation projects of 1977 and 663 km^2 areas	TDS considered salt pickup — salt deposition of irrigation water salts as a function of water quality and leaching fraction, and salt pickup from native soil gypsum	Model applied to the Glenn-Colusa Irrigation District in the Sacramento River Basin and the Panoche Drainage District in the San Joaquin River Basin, California
U.S. Bureau of Reclamation (1977)	Monthly for variable periods; up to 20 river basin nodes, each node with up to 10 operational sequences	TDS treated as a nonconservative parameter and simulated by chemistry model that involves cation exchange and gypsum and lime solubility	Model has been verified with small-scale systems and used to project water quality inputs in large-scale planned and existing irrigation projects

Model 1, the Woods and Orlob (1963) Lost River System Model from the University of California at Berkeley is one of the first water-quality, computer-based modeling efforts. Model 1 was originally developed to appraise pesticide residues that appeared to be causing substantial kills of water fowl but could be applied to other water pollutants. The strength of their model was in dynamic simulation of the complex hydrologic flows in the U.S. Bureau of Reclamation Klamath Project, including recycling of irrigation return flows which contributes towards buildup of pollutants. The weakness of this model was in assuming each irrigation use or reuse increased the pollutant concentration by one unit (use-factor).

The Lost River System Model was followed by Model 2 (Woods, 1967) to demonstrate the wider applicability of the former model to low flow conditions in the Sacramento River system. This second effort extensively evaluated model sensitivity to changes in some of the more important physical parameters such as groundwater storage and elevations as well as simulation of various water quality parameters. Instead of use-factor, concentrating effects on pollutants were predicted by a consumptive use model. The simulations indicate comparatively good fit with observed data for water flow, electrical conductivity (EC), total hardness, bicarbonate-alkalinity and sodium but not for nitrogen, phosphorus and boron.

Model 3 (California Department of Water Resources, 1969) was used to evaluate conditions of deteriorating water quality in the lower reaches of the San Joaquin River and their potential salinity impacts on irrigated agriculture. Hydrologic and salinity responses were projected for present and future possible operational schemes, including the construction of the Peripheral Canal and the San Joaquin Drainage Facility. The projections on flow and EC were based on linear regression and/or combination of linear and multiple regressions obtained from historical data. The EC's were also used to correlate with sodium, total hardness, chloride, bicarbonate, and boron. Good to fair agreement was obtained for most of the river reaches, except the model overestimated historic flows and underestimated historic EC's during the months of June and November as well as extremely dry years.

Model 4 (Water Resources Engineers, Inc., 1969) was focused mainly on salt balance in the groundwaters of the Upper Santa Ana River Basin in Southern California with some considerations on several surface water subsystems. The floor of this basin was subdivided into polygonal modes representing distinct hydrologic entities, consisting of an unsaturated soil and a saturated zone from which pumping takes place as well as a deeper zone unaffected by pumping. Besides identifying and estimating annual TDS waste loadings from numerous agricultural and nonagricultural activities, this model emphasized the importance of travel time of salts through the unsaturated zone. The model was verified for groundwater elevation with a mean standard deviation of about 2 m as well as predicting the trends in total dissolved solids (TDS) for the 1951 through 1960 period. Sensitivity analyses were also carried out, for example, on initial concentration of TDS in the unsaturated zone.

Model 5 (Hyatt et al.,1970) from Utah State University was used to estimate the hydrology and TDS in the Upper Colorado River Basin. Unlike the others listed herein, the mathematical model was executed on an analog computer. This macroscopic model is a modified version of earlier hydrologic models (e.g., Riley and Chadwick, 1967) developed at Utah that included snowmelt and unmeasured surface-water inflows. In addition to considering salt flows in the irrigated valley floors, it was necessary to estimate the natural within-the-basin salt contributions. The latter includes saline mineral springs as well as chemical weathering of salt-affected shales formed in a marine environment. Detailed verification of the model was made with the White River Sub-basin, one of 40 sub-basins. Reasonably good agreement of monthly water and salt out-flows were reported for most of the sub-basins.

Model 6 (Thomas et al.,1971), also from Utah State University, is a hybrid computer model i.e., hydrology is simulated with an analog computer while salinity is simulated with a digital computer. This model is one of the first large-scale models that simulated TDS as a nonconservative parameter by explicitly considering the major soil chemical reactions. TDS is considered to be the sum of six major solute species (Ca^{2+}, Mg^{2+}, Na^{+}, HCO_3^-, Cl^-, SO_4^{2-}). Model 6 successfully simulated measured outflows of water and solute species, except for Na^+, from a portion of the Little River Basin in Utah.

Model 7 (Konikow and Bredehoeft, 1973) simulated the flow and salinity in a stream-aquifer system along a 17.7 km reach of the Arkansas River system in Colorado. This U.S. Geological Survey model used extensive hydrogeologic and groundwater pumping records. Calculated water table elevations in the aquifer were within 0.3 m of the observed values about 90% of the time and calculated dissolved solids (same as TDS) were within 10% of the observed values about 80% of the time in both the aquifer and stream. This model appeared to be sensitive to potential evapotranspiration and canal leakage and relatively insensitive to variations in precipitation, pumpage, and underflow through the alluvium.

Model 8 (Bay-Valley Consultants, 1974) was developed for the California State Water Resources Control Board to estimate TDS and N waste loadings from irrigated agriculture over a large area. This mass-balance model focuses on the crop root zone portion of irrigated lands. The mass emission of TDS considered salts contributed by soil chemical weathering and residual fertilizers as well as uptake of salts by crop plants, in addition to salts present in the applied irrigation water. The utility of this simplified macroscopic model is its applicability to areas with limited field data.

A similar version of the Bay-Valley Consultants' model was formulated by Tanji (1977) and subjected to model calibration and prediction in two irrigation districts in California. Model 9, as in the former model, focuses on the crop root zone. Besides the concentrating effects of the soil solution from evapotranspiration losses of water, salt pickup-salt deposition of the applied water was considered on the basis of detailed water chemistry and leaching fraction. Moreover, salt pickup from native soil gypsum was also considered. This conceptual hydrosalinity model was verified with data from the Glenn-Colusa Irrigation District (GCID), which contains little or no salt-affected soils, in the Sacramento River Basin and then applied to the Panoche Drainage District (PDD), which contain salt-affected soils, in the San Joaquin River Basin. The calculated seasonal water flow and concentration and mass of TDS in the collected surface return flows were, respectively, -33%, +23%, and -18% of measured values in GCID and, respectively, +17%, -7%, and +9% in PDD for typical irrigation seasons.

Model 10 (U.S. Bureau of Reclamation, 1977) is a comprehensive macroscale model widely used by this agency to evaluate environmental impacts of current and planned irrigation projects. Salinity is treated in a manner similar to that of the Thomas et al. (1971) model and hydrology, similar to that of the Hyatt et. al. (1970) and other models. The U.S. Bureau of Reclamation (now known as the U.S. Water and Power Resources Services) has, however, extended past hydrosalinity modeling efforts. Among others, Model 10 simulates the land-stream-reservoir system, is capable of handling tile-drainage outflows, and has several bypass options built into the model depending upon availability of information and data or for site-specific boundary and initial conditions. The Bureau of Reclamation model has been verified in part with observed data from smaller-scaled studies with good results, e.g. Dutt et al. (1972) as well as application by an independent party (Gelhar and McLin, 1979).

The hydrosalinity models developed in the Western U.S.A. all involve irrigated agriculture since it is the largest consumptive user of water. In nearly all modeling efforts, the primary motivation was to project the quantity and quality of irrigation return flows. These models generally compute at time scales ranging from monthly to annually but may involve smaller time intervals for simulation of particular processes. These models consider comparatively large spatial units and stream nodes in contrast to the more refined research-oriented models for smaller-scaled problems.

The term hydrosalinity model explicitly implies that a hydrologic and a salinity submodel are coupled. Because the flow of salts is directly dependent upon the flow of water, salinity simulation requires outputs from the former. The modeling approaches taken for simulation of water and salt flows, however, may be quite diverse.

2.2 Hydrologic Submodel

Fig. 1 from Woods (1967) is an example of an idealized hydrologic unit consisting of three basic elements or subsystems; surface water (stream), soil water (land), and groundwater (subsurface). Each hydrologic unit in a river basin contains two or three of these subsystems between which water (as well as salts) are transferred according to physical relationships or to some operational plan. Each of these subsystems may be further subdivided into pools and fluxes. The pools (storage compartments) contain the state variable water (and salts) while the fluxes describe the pathways and rates of transfer from one pool to another. For each hydrologic element, Woods and Orlob (1963) and Woods (1967) invoked the continuity equation such that:

$$\frac{dS}{dt} = \Sigma Q \tag{1}$$

where S is the total volume of stored water; Q is the rate of flow into or out of the element; and t is time.

The hydrologic unit is viewed by Model 2 to have subsystem inputs of precipitation (P, P') and land surface (Q_{si}), stream (Q_{ci}) and groundwater (Q_{gi}) rim inflows with subsystem outputs of evapotranspiration (U, E), exported water (Q_p), and stream outflows (Q_{co}). To make this scheme more complete one can add land surface (Q_{so}) and groundwater rim outflows (Q_{go}), imported water (Q_i), and deep percolation beyond the reach of groundwater pumpage (Q_{dp}).

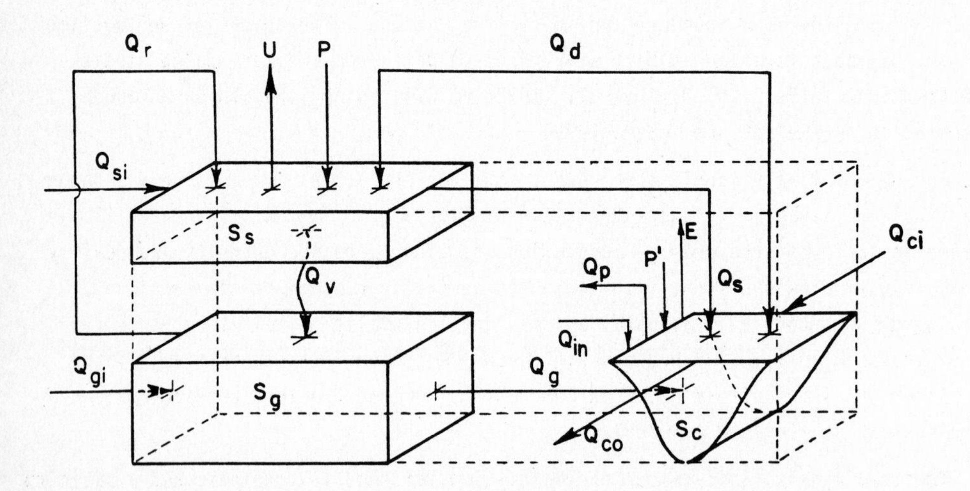

Fig. 1. A proposed hydrologic model consisting of soil water, groundwater, and stream subsystem (Woods, 1967)

By including the above additions to the Woods (1967) idealized hydrologic unit, the changes in the internal water balance in the stream, soil water, and groundwater elements may be respectively given by:

$$\frac{dS_c}{dt} = Q_{ci} - Q_{co} + Q_s + Q_g - Q_d + Q_{in} - Q_p + P' - E \qquad (2)$$

$$\frac{dS_s}{dt} = Q_{si} - Q_s + Q_d + Q_r - U + P - Q_v - Q_{so} + Q_i \qquad (3)$$

$$\frac{dS_g}{dt} = Q_{gi} - Q_g - Q_v - Q_r - Q_{go} - Q_{dp} \qquad (4)$$

where the subscripts c, s, and g refer respectively to the stream, soil water and groundwater storage elements; i and o respectively denote inflow and outflow from the element; d refers to diversion for irrigation use; p and i refer respectively to export from stream and import to land surface; v and dp denote respectively vertical seepage from soil to groundwater and deep percolation; r denotes recirculation or pumpage and may include rising groundwaters to the land surface; P and P' denote respectively precipitation on land and free water surfaces; and U and E refer respectively to evapotranspiration from land surface and evaporation from free water surface (Fig. 1). Other schematic versions of the hydrologic unit or river subsystem are found in Utah State University Foundation (1969), Water Resources Engineers, Inc. (1969), More (1969) for the Stanford Watershed model (Crawford and Linsley, 1966), and Hyatt et al. (1971).

Fig. 2 illustrates the usual block building involved in the development of systems level modeling. Model 10 (U.S. Bureau of Reclamation, 1977) is composed of four system elements each of which are further defined, as in the case of soil hydrology system, into subsystem and further into interactive processes.

In the state-of-the-art dynamic simulation models, the underlying theoretical basis for the mathematical models are the principles of continuity of mass, momentum, and energy. Continuity of momentum is important if the model is to describe fast-flowing waters in channels while continuity of energy is important if the model is to describe radiant energy transfers, e.g., snowmelt. Continuity of mass is essential for all models.

The state variable of interest in the hydrologic submodel is water in the liquid, gaseous and solid forms but only the first-named phase is usually simulated and the other two are treated as either output or input. Occurrence of precipitation, rain and snow, is a stochastic phenomenon but is treated deterministically in these models by reading in temporal and spatial values, i.e, they are not simulated. Applied irrigation water is also treated as input and may be also variable in both time and space. The usual processes simulated are

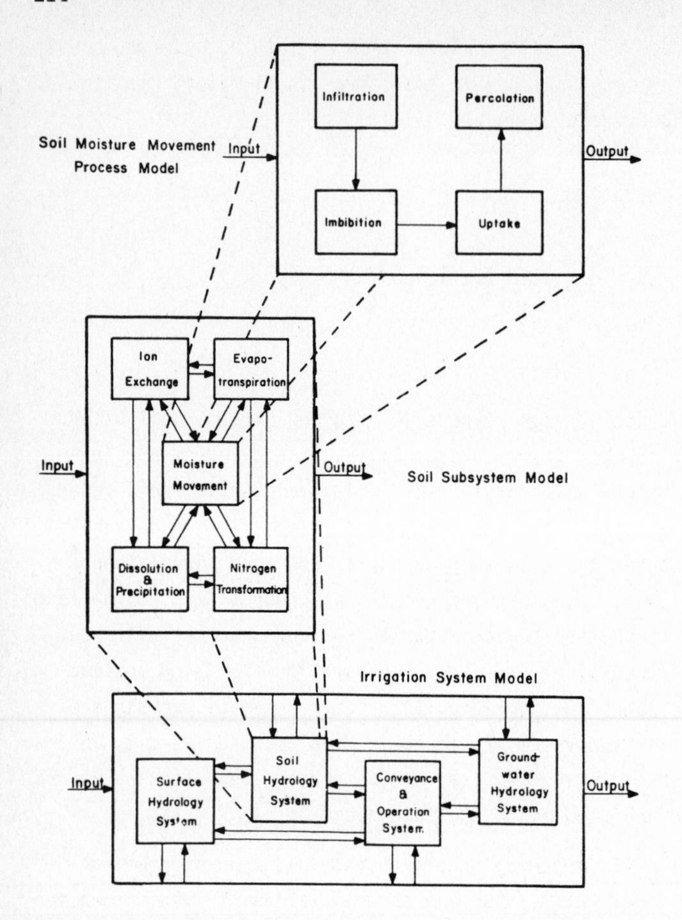

Fig. 2. A conceptual diagram of block building in irrigation return flow modeling (Walker, 1978).

infiltration, water movement in soil and other porous geologic materials, runoff, transpiration, and evaporation; snowmelt and channel flow are considered in some models. Most of the mathematical models for water flow in hydrologic simulation may be reduced to:

$$q = -\frac{1}{R} \cdot \frac{d\psi}{dz} = -K\frac{d\psi}{dz} \tag{5}$$

where q is flux of water; R is resistivity; K is conductivity; ψ is potential; z is distance; and $d\psi/dz$ is the potential gradient, the driving force. In addition to the driving force and transfer coefficient, the state variable in the system is constrained by capacity, i.e., volumetric water storage in a given pool. Transient soil water flow, a major process, was described by Nimah and Hanks (1973) as:

$$\frac{\partial \theta}{\partial t} = \frac{\partial}{\partial z} [K(\theta) \frac{\partial H}{\partial z}] + A(z,t) \quad (6)$$

where θ is the volumetric soil water content; $K(\theta)$ is hydraulic conductivity; H is the hydraulic head; and A is the sink term for root water extraction.

Hornsby (1973) reviewed state-of-the-art of modeling irrigation return flows with particular reference to salinity control. He concluded that simulation modeling will play an increasingly important role in water resources management, especially for examining alternative approaches to irrigation management practices and to improve control of salinity. Walker (1978) has made a detailed appraisal on system and subsystem as well as process models for irrigation return flows. His evaluations include scope of the model, input requirements, spatial and time scales, structure of the code, basic mathematical or analytical approach and general comments for about 40 models. In addition, Vansteenkiste (1975, 1976) has edited proceedings of computer simulation of water resources systems which contain many reports of interest that are beyond the scope of this paper.

2.3 Salinity Submodel

The sources and sinks of salinity in waters are manifold. All models consider the dissolved mineral salts in applied waters and few in rainwater. As indicated in Table 1, some models consider salinity to be conservative while others simulate it as a non-conservative parameter. The lumped salinity parameter chosen is either TDS or EC. The former is an extensive gravimetrically-measured parameter while the latter is an intensive electrically-measured parameter. The former is used to calculate mass of salts from the product of salt concentration and water volume. The latter gives a better estimate of the impact of salinity to biological organisms, including plants, because it is related to osmotic or solute potential. Many convert from one to the other using empirical conversion factors.

Other sources of salinity may include dry fallout of airbourne salts, which is not usually considered, chemical weathering (dissolution) of minerals from soils, rocks and other geologic materials, decomposition of organic compounds and wastes such as animal manures, fertilizer residues, soil amendments like gypsum, saline seeps, etc. Nonagricultural sources may include rising connate waters and mineral springs, municipal and industrial discharges, and mining, gas and geothermal production operations.

The sinks for dissolved mineral salts include mineral precipitation, sorption of certain solute species by soils and other earth materials including phosphorus and ammonium clay fixation, ion exchange and adsorption, uptake by plants, and wind erosion. Of these sink processes, mineral precipitation and plant uptake appear to be most frequently modeled while cation exchange may be considered. This exchange process affects only the relative proportions of cations and not salt concentration per se but such changes may have direct influence on other chemical processes leading to salt concentration changes, e.g., mineral solubility.

In addition to the solid-liquid phase interaction for salts, the dissolved solute species, both mineral and organic in nature, may interact and associate with each other forming ion pairs and complexes so that they no longer act as free ions. Such ion association affects mineral solubility (e.g., Tanji, 1969a) and EC (Tanji, 1969b).

In considering salinity as a reactive parameter, the solute species usually considered are Na^+, Ca^{2+}, Mg^{2+}, Cl^-, HCO_3^-, CO_3^{2-}, and SO_4^{2-}, and for the solid phases the minerals gypsum ($CaSO_4 \cdot 2H_2O$) and lime ($CaCO_3$). Not considered are other minerals that may be significant contributors such as those present in salt crusts or efflorescence (halite [NaCl], mirabilite [$Na_2SO_4 \cdot 10H_2O$], blödite [$Na_2Mg(SO_4)_2 \cdot 4H_2O$], hexahydrate [$MgSO_4 \cdot 6H_2O$], etc.) and oceanic or marine aerosols.

The modeling approach taken for treating salinity as a reactive parameter is chemical equilibrium (thermodynamic). This theory only considers the difference between initial and final states and does not consider time as a variable. Most of the chemical reactions occur at rates rapid enough so that invoking chemical equilibrium does not appear to be a serious problem, especially in systems in which water flow rates are comparatively small.

The mathematical model for mineral solubility and precipitation is:

$$K_{sp} = a_C a_A = m_C r_C m_A r_A \tag{7}$$

where K_{sp} is the solubility product constant; a_C and a_A are respectively the ionic activities of the dissociated cationic and anionic species C and A; and m and r are the respective analytical concentrations and ionic activity coefficients of the subscripted ion species.

The stoichiometric (total) solubility of the mineral, however, is described not only by the activity concentration of the dissolution products but also by any possible ion association between the mineral dissolution products as well as with other solute species that may be present.

The formation of ion pairs and complexes is described by:

$$K_d = \frac{a_C a_A}{a_{(CA)^{V_+ + V_-}}} \tag{8}$$

where K_d is the dissociation constant and the denominator contains the activity of the ion pair CA having a net charge of $V_+ + V_-$.

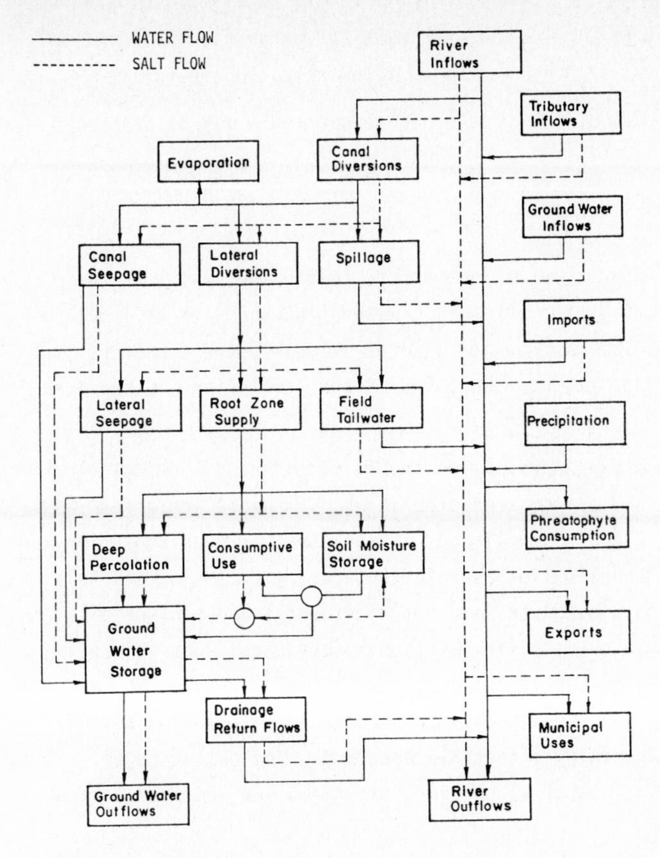

Fig. 3. A conceptual diagram of a generalized hydro-salinity model. (Hornsby, 1972).

Cation exchange involves the adsorption and desorption of positively charged ions to the negatively charged sites of the soil exchange complex. For instance, the exchange between adsorbed Ca^{2+} and solution Na^+ may be described by:

$$K_{Ca^{2+}\,Na^{+}} = \frac{(E_{Na})^2\ (Ca^{2+})}{(E_{Ca})\ (Na^+)^2} \tag{9}$$

where $K_{Ca^{2+}\,Na^{+}}$ is the equilibrium exchange constant and E denotes the exchangeable cationic species.

Soil water plays a major role in salinity considerations since it affects both salt concentration as well as salt movement. In the former, salt concentration does not usually change in direct proportion to change in soil water contents because chemical processes such as mineral solubility, cation exchange, and ion association are affected. This phenomenon has been quantitatively handled by Tanji et al. (1967). In the latter, the principal media by which soluble salts are transported is by liquid water. Associated with the water flow pathways identified in Fig. 1 are the salt flows. Fig. 3 illustrates this point.

The transport of an adsorbing solute species in porous media may be described (Bresler, 1973) by:

$$\frac{\partial(\theta C)}{\partial t} = \frac{\partial}{\partial z}\left[D(v,\theta)\frac{\partial C}{\partial z}\right] - \frac{\partial(qC)}{\partial z} - \frac{\partial(\rho z)}{\partial t} \qquad (10)$$

where D is the apparent diffusion coefficient of the solute species; v is the average pore water velocity; q is the volumetric water flux;and ρ is soil bulk density. The first term on the RHS of equation (10) is for diffusive transport, the second is for convective transport, and the third is for adsorption. Other sink or source terms may be added to equation (10).

Biggar and Tanji (1977) made an appraisal on the difficulties of predicting and evaluating the discharge of salts from irrigated lands. For most irrigated lands, salinity control cannot be adequately analyzed without considering dissolution and precipitation of earth minerals, ion association, cation exchange, and salt (and water) transport in both unsaturated and saturated soils and substrata. These processes also should be assessed in relation to the usual spatial variability encountered in the field. It is not entirely clear what processes, and under what conditions can be ignored or how approximate the description of the processes can be without sacrificing acceptable accuracy. The dominance of NaCl type of salts in Western Australia, however, may allow for some simplification in soil chemistry.

3 COMPARATIVE SALT PROBLEM SCENARIOS

There is a need first to consider the similarities and contrasting features of the salt problem between Western U.S.A. and the south-west of Western Australia before attempting to evaluate the utility of hydrosalinity models developed in the former.

There appears to be a consensus opinion (Public Works Department, 1979) that the salt problem in Western Australia is mainly caused by clearing of forested land wherein a change occurs in both water and salt balance. The major source of salts is said to be oceanic salts present in rainwater. Under natural conditions of deep-rooted trees and other plants, most of the precipitation is lost back to the atmosphere via evapotranspiration (ET) and salts tend to accumulate in the soil. But upon conversion to shallower-rooted vegetation such as pastures and crops, ET is comparatively smaller and there is a tendency for salts which have accumulated for centuries to be leached out of the soil. In other words, more salts and more water are drained off, and the net effect of clearing is a rise in stream salinity. The subsurface pathways by which water and salts enter the stream system may be quite complex and highly dependent upon local hydrogeologic conditions. The salt problem is manifested in diverse ways including saline seepage and perched water tables, water salinity in confined as well as unconfined aquifers, and dryland salt (Malcolm and Stoneman, 1976). Saline seepage has been reviewed in depth by Peck (1978) while other manifestations of the salt problem have been discussed by Mulcahy (1978) and Conacher (1979).

In contrast to the above scenario, salt problems in Western U.S.A. are somewhat different, in that the primary source of dissolved salts is chemical weathering of minerals in soils, sediments, and rocks (Richards, 1954). The major source of salinity in waters appears to be soils derived from marine shales, sandstones, and conglomerates. In areas of lower precipitation, about 250 mm or less, there is usually an accumulation of soluble salts (similar to the composition of seawater) present in surface soils. In areas of higher precipitation or where irrigation is practiced, the soluble salts are usually leached out of the surface soil and accumulate in the subsoil or perched water table and eventually in deeper groundwaters and surface water bodies (Skogerboe and Law, 1971).

Of the salts present in salt-affected soils and other geologic materials, the unweathered solid phase minerals appear to be the major contributor of dissolved salts. Salt-affected soils contain an abundance of gypsum ($CaSO_4 \cdot H_2O$), calcium carbonate, mainly calcite ($CaCO_3$), and dolomite ($CaMg(CO_3)_2$). The products of chemical weathering and cation exchange are mainly SO_4^{2-}, Na^+, and Mg^{2+} in many areas, and SO_4^{2-}, Cl^-, Na^+, Mg^{2+} and Ca^{2+} in some localized areas. This array of solute species differs from those found in Western Australia, where Cl^- and Na^+ dominates. The Western U.S.A. streams as they become highly saline, especially in closed basins or streams traversing through arid lands, have chemical compositions similar to or approaching that of sea water, i.e. 31% Na^+ and 55% Cl^- (% of total mg L^{-1} of TDS or 77% Na^+ and 90% Cl^-, respectively, of sum of meq L^{-1} of cations and anions). A secondary source of salts is the dissolved salts present in the irrigation water diverted from streams and pumped from groundwater basins as well as water imported from outside the river basin.

Salts from salt-affected land are discharged into stream systems mainly through natural (rainfall and snowmelt) runoff and seepage and by irrigation return flows. Although the original source(s) of salts and the nature of the salts may differ between Western U.S.A. and the south-west of Western Australia, the manifestations of salt problems are similar, i.e., presence of saline seeps in localized areas, elevated salinity in the perched water table and groundwater, and rise in stream salinity.

4 APPLICABILITY OF MODELS

The question posed here is, are any of the models presented in Table 1, or others, applicable to the salinity problems in the south-west of Western Australia? The response could range from negative to positive depending upon a number of considerations:

a. How much information and data are available?

b. What are the objectives of modeling?

c. Will existing models simulate adequately water and salt flows?

d. What is the computer requirement for the model?

e. Has the model been sufficiently documented and tested?

One of the first considerations that need to be addressed is the availability of information and data base, i.e., for hydrology, geology, soils, land use, water distribution and use, vegetation and cropping patterns, irrigation and drainage, and sources and discharges of salinity. These will be addressed by others in this conference. The extent of such information and data needs are in part dictated by the next question.

What are the objectives of hydrosalinity modeling? Four illustrative modeling objectives were previously noted in Section 1 of this paper. A fifth objective is, modeling fosters co-ordination and inter-disciplinary efforts. If the modeling objective involves scientific questions and refined simulations, the requirements of data base are very demanding. If the modeling objective, however, concerns obtaining a first-cut approximation and simulation results for planning purposes, then the information and data requirements may not be as great.

It has been the author's experience that a firmer approximation could be obtained going from adequate problem characterization to simplified modeling (e.g., Model 9) and not vice versa. What to simplify cannot be previously ascertained until sufficient knowledge is available. There is some tendency to scale-up finely-tuned models developed from small experiments to large-scaled systems (e.g., Model 10). This approach causes some difficulties since the cost of obtaining input data for initial conditions and model parameters and coefficients are exorbitant. Some aggregation and simplification is usually desirable in tackling regional problems but in so doing, there may be a potential danger of masking over important processes and events in a particular problem situation.

Hydrologic data requirement for Models 3, 5, 8 and 9 are judged to be less demanding than the others. On the other hand, salinity data requirements in Models 6 and 10 are very extensive and in Models 8 and 9, moderate. Model 10, however, has an option to bypass details of soil chemistry. The models presented in Table 1 offer a number of combinations of data requirements for hydrology and salinity.

The third question is, will the model adequately simulate water and salt flows? If the chemistry of salinity is complex, a simplified chemistry submodel coupled to a detailed hydrology submodel probably would not provide satisfactory simulations. In contrast, if the chemistry dominates over physical processes, a detailed chemistry submodel coupled to a simplified hydrologic submodel may yield satisfactory simulations.

In most instances, and in contrast to subsurface flows, adequate data are available or reasonable estimates can be made of water and salt flows on the land surface. One of the major and most frequent unknowns is the splitting of deep percolation vertically as well as laterally. Most models are of the one-dimensional type which may not be entirely applicable when two-dimensional assessments are required. Models 4 and 7 appear to have strengths for subsurface water flows while Models 3 and 5 are weak in this regard.

Another problem frequently encountered is rising groundwaters, the simulation of which usually requires considerable hydrogeologic data. A third problem is the nature of the salt mixing phenomenon at the interface between the percolating water and resident water in the saturated zone.

Since the salt problem in the south-west of Australia is mainly caused by a change in water balance, triggered by a shift from forested to cleared land, simulating evapotranspiration losses is of prime importance. Although all models consider this sink for water, the modeling approaches taken vary widely from assumed time-invariant rates in most models to those attempts to simulate consumptive water use based on climatic, plant, and soil factors over small time intervals (e.g., Model 10).

For those conditions in which earth minerals play a major role in salinity, it is generally necessary to model mineral solubility and other associated chemical reactions. These processes may proceed not only in the soil zone but also in the underlying sub-stratum as well as in surface water bodies.

Since the carbonates (dissolved CO_2, HCO_3^-, and CO_3^{2-}) are indigeneous to all natural waters, even of the NaCl type waters and represents a sparingly soluble mineral, it is desirable to consider this portion of the chemistry. In the absence of detailed soil chemistry, a simplified treatment of the role of carbonates (and sulfates) in soil salinity is found in Model 9 which lumps together net salt pickup (dissolution) or salt deposition (precipitation) of lime and gypsum.

Many models are formulated and applied to site-specific problems, and are not amendable to more generalized problem situations. If a model is to be applied from one problem situation to another, the mathematical and/or computer model generally needs modification. Moreover, most hydrosalinity models do not lend themselves to providing guidelines for decision-making policies and management because optimization is not explicitly considered (Horner and Dudek, 1980).

A fourth question is, what is the computer system requirement? All of the models given in Table 1 can be run on medium-sized computers, except Model 10 which requires the capability of a CDC 7400 system. Some models, like Model 9, can be run on a programmable desk calculator having 8 K bytes of memory.

A fifth question is, has the model been sufficiently documented and tested? It is highly desirable to have a user's manual, including documentation. The documentation package may consist of source listing, flow chart(s), list of instructions needed to operate the program, sample input and output, and list of error conditions. Without such information, the user frequently encounters difficulties deciphering the model Models 2, 4, 5, 6, and 10 appear to be adequately documented.

Since a program written and debugged on one computer usually is not operational when fed to a second computer, documentation and adequate comments in the program are highly desirable. This problem of getting a model operational occurs even if the computers are identical systems because of local modifications at the respective computer centres.

The testing of a computer model may comprise of mathematical validation of the numeric scheme, sensitivity analyses by varying one model parameter while keeping all others constant, calibration of model parameters and coefficients for fitting simulations closer to field data, and verification of the model using a set of data different from that used for calibration. Most of the models in Table 1 have been calibrated and for some, sensitivity analyses have been conducted (e.g., Models 2, 4, and 10). To the best of this author's knowledge, none of the models cited in Table 1 have been rigorously tested against all of the aforementioned tests. Rigorous verifications are constrained by the high cost of aquiring field data.

Even though hydrosalinity modeling was initiated about two decades ago, most models are seldom widely utilized by other modelers. Although the mathematical modeling approach and some computer subroutines are similar, there seems to be a tendency to formulate yet another model each time a salinity problem is encountered. This is mainly due to existing models being too site-specific and not having sufficient generality. Other reasons may be lack of documentation or inadequate verification.

5 CONCLUSIONS

Over the past two decades, significant advances have been made in various aspects of the basin hydrosalinity model. The complexity of these models ranges from simple mass balance to state-of-the-art computer simulation models. These models are providing increased insight and understanding of the complexity of water and salt flows in large-scale systems. However, it is clear that much more work is needed in the areas of field validation of models and contributions of guidelines to planning and management.

A review of ten representative hydrosalinity models developed in the Western U.S.A. indicates there does not exist a single model that has strength in both water and salts for all three subsystems (soil, groundwater, stream). Most models have been developed for site-specific problem situations and for a variety of modeling objectives, and do not lend themselves to universal application or to other site-specific problems. A more serious deficiency is the lack of adequate field data to verify existing models and the cost of acquiring such information and data. This is particularly true for the finely-tuned models simulating over small time and space intervals. On the other hand, there is some question about the applicability of overly simplified models to complex salt problems.

Recognizing these deficiencies and limitations, it is suggested that a first-cut approximation be initially made with more simple models to obtain a fix on probable flow pathways and salt loadings. This objective appears to have been achieved in the land and stream salt problem in Western Australia. In some cases, no further modeling effort is required to evaluate and forecast potential effects of management alternatives. In other cases, more complex and refined modeling may be in order. Continued modeling efforts are recommended for a variety of reasons. Modeling fosters co-ordination and co-operation among disciplinary groups as well as agencies. Modeling may indicate what additional information and data are needed for more quantitative appraisals. Finally, modeling gives increased insight and better conceptual understanding because it forces one to examine the relative importance of different interacting physical, biological, and chemical processes.

6 REFERENCES

Bay-Valley Consultants, 1974. Recommended Water Quality Management Plan. Sacramento River Basin, San Joaquin River Basin, Sacramento-San Joaquin Delta. Report to the California State Water Resources Control Board, pp. 1-1 through 7-11.

Biggar, J.W. and Tanji, K.K., 1977. Soil-salt interactions in relation to salt control. Trans. Am. Soc. Agric. Engnrs. 20: 68-75.

Bresler, E., 1973. Simultaneous transport of solute and water under transient, unsaturated flow conditions. Water Resour. Res. 9: 975-986.

California Department of Water Resources, 1969. Lower San Joaquin River Water Quality Investigation, Bull. No. 143-5. Department of Water Resources, Sacramento, 207 pp.

Conacher, A., 1979. Comment on salinity problems in Southwestern Australia. Search 10: 162-164.

Crawford, N.H. and Linsley, R.K., 1966. Digital simulation in hydrology: Stanford Watershed Model IV, Tech. Rep. No. 39,, Department of Civil Engineering, Stanford University, Palo Alto, California, 210 pp.
Dutt, G.R., Shaffer, M.S. and Moore, W.S., 1972. Computer simulation model of dynamic bio-physico-chemical processes in soils, Agricultural Experiment Station, Tech. Bull. 196, University of Arizona, Tucson, 101 pp.
Gelhar, L.W. and McLin, S.G., 1979. Evaluation of a hydrosalinity model of irrigation return flow quality in the Mesilla Valley, New Mexico, EPA-600/2-79-173, U.S. Environmental Protection Agency, 192 pp.
Horner, G.L. and Dudek, D.J., 1980. An analytical system for evaluation of land use and water quality policy impacts upon irrigated agriculture. In: Yaron D. and Tapiero, C.S. (Eds), Proceedings, ORAGWA International Conference, Jerusalem, Israel, North-Holland Publishing Company (in press).
Hornsby, A.G., 1973. Prediction modeling for salinity control in irrigation return flows, EPA-R2-73-168, U.S. Environmental Protection Agency, 55 pp.
Hyatt, M.L., Riley, J.P., McKee, M.L. and Israelson, E.K., 1970. Computer simulation of the hydrologic-salinity flow system within the Upper Colorado River Basin, PRWG 54-1, Utah Water Research Laboratory, Utah State University, Logan, 124 pp.
Konikow, L.F. and Bredehoeft, J.D., 1973. Simulation of hydrologic and chemical-quality variations in an irrigated stream-aquifer system. Colorado Water Resources Circular 17, Colorado Water Conservation Board, Denver, 43 pp.
Malcolm, C.V. and Stoneman, T.C., 1976. Salt encroachment - the 1974 Saltland survey. J. Dep. Agric., West Australia 17: 42-49.
More, R.J., 1969. The basin hydrological cycle, In: Chorley, R.J. (Ed), Water, Earth and Man, Methuen, London, pp. 67-76.
Mulcahy, M.J., 1978. Salinisation in the south west of Western Australia, Search 9: 269-272.
Nimah, M.N. and Hanks, R.J., 1973. Model for estimating soil water, plant, and atmospheric inter-relations: I. Desorption and sensitivity. Soil Sci. Soc. Am. Proc. 37: 522-527.
Peck, A.J., 1978. Salinisation of non-irrigated soils and associated streams: A review. Aust. J. Soil Res. 16: 157-168.
Public Works Department, Western Australia, 1979. Clearing and stream salinity in the South West of Western Australia. Document No. MDS 1/79.
Richards, L.A. (Ed.), 1954. Diagnosis and improvement of saline and alkali soils. U.S. Dept. Agric. Handbook 60, U.S. Government Printing Off., Washington, DC, 160p
Riley, I.P. and Chadwick, D.G., 1967. Application of an electric analog computer to the problems of river basin hydrology. Utah Water Research Laboratory, Utah State University, Logan, 199 pp.
Skogerboe, G.V. and Law, J.P., Jr., 1971. Research needs for irrigation return flow quality control. EPA 13030-11/71, U.S. Environmental Protection Agency, 98 pp.
Tanji, K.K., 1969a. Solubility of gypsum in aqueous electrolytes as affected by ion association and ionic strengths up to 0.15 M and 25°C. Env. Sci. and Technology 3: 656-661.
Tanji, K.K., 1969b. Predicting specific conductance from electrolyte properties and ion association in some aqueous solutions. Soil Sci. Soc. Am. Proc. 33: 887-890.
Tanji, K.K., 1977. A conceptual hydrosalinity model for predicting salt load in irrigation return flows. In: Dregne, H.E. (Ed.), Managing saline water for irrigation, Texas Tech. University. Lubbock, pp. 49-65.
Tanji, K.K., Dutt, G.R., Paul, J.L. and Doneen, L.D., 1967. Quality of percolating waters II. A computer method for predicting salt concentrations in soils at variable moisture content. University of California, Hilgardia 38: 307-318.
Thomas, J.L., Riley, J.P. and Israelson, E.K., 1971. A computer model of the quantity and chemical quality of return flow. PRWG 77-1, Utah Water Research Laboratory, Utah State University, Logan, 94 pp.
U.S. Bureau of Reclamation, 1977. Prediction of mineral quality of irrigation Return Flow, Vo. 111, simulation model of conjunctive use and water quality

for a river system or basin. User's Manual, EPA-600/2-77-179c. U.S. Environmental Protection Agency, 285 pp.
Utah State University Foundation, 1969. Characteristics and pollution problems of irrigation return flow. Report to Federal Water Pollution Control Administration, Ada, Oklahoma, 237 pp.
Vansteenkiste, G.C. (Ed.), 1975. Computer simulation of water resources systems. North-Holland Publishing Company, Amsterdam, 686 pp.
Vansteenkiste, G.C. (Ed.), 1976. System simulation in water resources. North-Holland Publishing Company, Amsterdam, 417 pp.
Walker, W.R., 1978. Identification and initial evaluation of irrigation return flow models. EPA-6001 2-78-144, U.S. Environmental Protection Agency, 124 pp.
Water Resources Engineers, Inc., 1969. An invesigation of salt balance in the Upper Santa Ana River Basin. Final Report to the California Department of Water Resources, 198 pp.
Woods, P.C., 1967. Management of hydrologic systems for water quality control. Water Resources Center Contribution No. 121, University of California, 121 pp.
Woods, P.C. and Orlob, G.T., 1963. The Lost River System. A water quality management investigation. Water Resources Center Contribution No. 68, University of California, 54 pp.

PREDICTING STREAM SALINITY CHANGES IN SOUTH-WESTERN AUSTRALIA

I.C. LOH and R.A. STOKES,
Public Works Department,
West Perth, Western Australia, 6005.

ABSTRACT

Loh, I.C. and Stokes, R.A., 1981. Predicting stream salinity changes in south-western Australia. Agric. Water Manage., 1981.

The hydrologic characteristics of the south-west of Western Australia are described with particular emphasis on the effect of permanent removal of forest vegetation on the variability of stream salinity. Stream salinity can change dramatically within hours during flood periods and very gradually over decades in response to land use change. A range of approaches to predicting these different scales of stream salinity variation are suggested. A regional prediction model is presented which simulates the annual flow and salinity of a large river system subject to a long clearing history. The model provides an estimate of the long-term effects of permanent clearing and can provide useful guidance to planners developing catchment management strategies. Preliminary results are presented but further calibration and development work is required.

1 INTRODUCTION

Increases in stream salinity and land salination following permanent clearing of deep-rooted native forest for agricultural development is now accepted as one of the major water supply and land management issues of the south-west of Western Australia. Similar problems following dryland agricultural development have occurred in other parts of southern Australia, the United States and in Canada (Peck, 1978) but nowhere else in the world has the deterioration of the available water resources been as great as in Western Australia.

Prior to European settlement virtually all the conventionally divertible surface water resources of the south-west (Fig. 1) were believed to be fresh (less than 500 mg L^{-1} Total Soluble Salts). Some 20% of these resources are now considered of marginal salinity (between 500 and 1000 mg L^{-1}) and a further 34% are considered brackish or saline (over 1000 mg L^{-1} TSS). The larger river systems draining the main agricultural areas of the State are particularly saline. The Swan-Avon system has a salinity of about 5000 mg L^{-1} TSS and the Murray and Frankland Rivers are in excess of 1900 mg L^{-1} TSS and still rising.

Recent legislative control over further clearing on 5 river systems has been enacted to minimise further deterioration of the most sensitive water resources (Public Works Department, W.A., 1979). In some cases minimisation of further deterioration is insufficient to protect the resource in the long term and remedial actions are required. A partial reforestation programme has already commenced on the catchment of Wellington reservoir (Fig. 1); a major source of irrigation and town water supply.

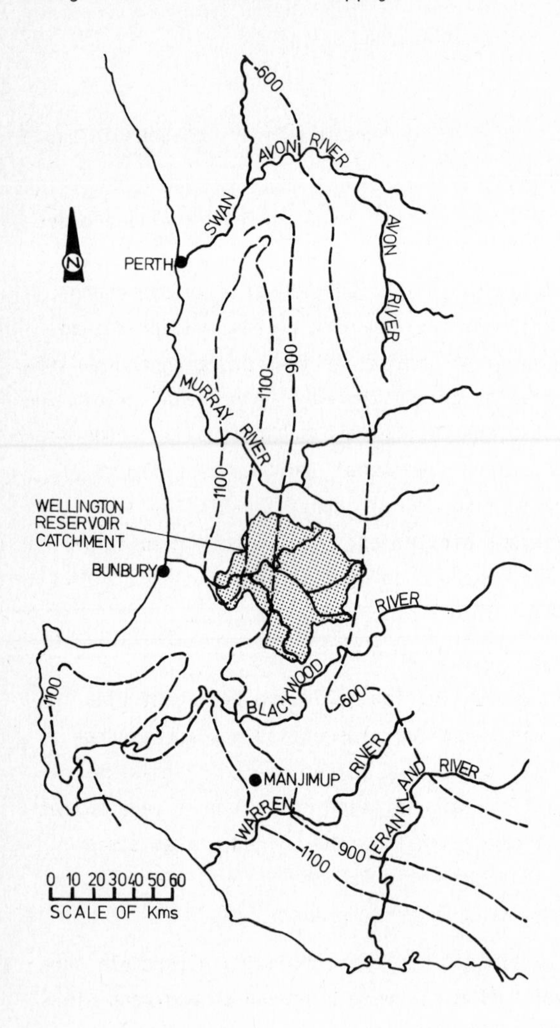

Fig. 1. Location map showing major rivers of the south-west of Western Australia.

A thorough understanding of the causes of and techniques for predicting changes in stream salinities is required to assist in developing cost effective remedial solutions.

Predictions of stream salinity variations are required:-

(i) to assist long-term catchment management and land use planning,

(ii) to assist in the design of engineering solutions such as improved reservoir operation or saline diversion schemes,

(iii) to assist in understanding the mechanisms of stream salinity increase following clearing in order to develop appropriate techniques for rehabilitation.

The approaches to prediction of stream salinity variations for these purposes varies greatly with scale.

(a) Decision makers require predictions of the salinity of major river systems over decades to assist them in formulating long term planning strategies.

(b) Engineers need to study variations in river flows and salinities over days, months and years to adequately simulate the effects of engineering schemes to improve the final salinity of water from complex supply systems.

(c) Scientists may need to study variations in stream quality and associated runoff generation mechanisms over time scales less than an hour on small research catchments if the redistribution of salts following land use change is to be understood and appropriate rehabilitation strategies developed.

This paper describes the range of hydrologic responses and salinity variations of rivers and streams which drain forested and agricultural land in the Darling Range of south-western Western Australia. The paper also discusses appropriate approaches to modelling stream salinity changes to solve different problems and describes a model to predict the annual salinity and flow of a large river system subject to a long history of permanent clearing.

2 HYDROLOGIC CHARACTERISTICS OF THE REGION

Over the past 10 years an extensive programme of research and investigation has improved our quantitative understanding of the hydrologic and related soil salinity characteristics of the region. The work has included studies relating land use and stream salinity (Collins, 1974; Williamson and Bettenay, 1979), catchment chloride balances (Peck and Hurle, 1972), rainwater chemistry (Hingston and Gailitis, 1977), soil solute storage (Johnston, 1980; Stokes et al., 1980), soil and landscape studies (Bettenay and Mulcahy, 1972) and groundwater characteristics (Peck et al., 1979), together with intensive hydrologic studies on small research catchments (Williamson and Johnston, 1979; Conservation and Environment, 1980). The following is a general summary of current knowledge.

Rainfall through the study region decreases with increasing distance from the coast. Rainfalls range from 1400 mm yr^{-1} in the west and south to less than

600 mm in the east and is winter dominant. Monthly rainfall only exceeds monthly evaporation in approximately 4 to 6 months each year. The major rivers rise on the relatively flat topography of the Darling Plateau and drain to the south and west through increasingly incised valleys to a narrow coastal plain and the Indian or Southern Oceans.

The geology of the Darling Range is dominated by the Archean Shield, a granitic basement which underlies the lateritic soil profiles of the region. Soil associations have generally formed in situ and are characterised by lateritic gravels and duricrust overlying pallid zones of sandy clays or clay and a zone of weathered bedrock. Soil profiles are generally deep, typically being 20 m, with the major portion being the pallid zone subsoils. Depth of weathering is variable but in the high rainfall regions (above 1100 mm yr^{-1}) greater dissection has taken place, slopes are steeper and bedrock frequently outcrops at the valley invert. The natural forest vegetation is characterised by a variety of Eucalyptus species with deep roots which extend well into the subsoils and weathering profile.

2.1 Small Forested Catchments (Generally 1st Order Streams Less Than 5 km^2)

Under natural forested conditions in the inland low rainfall region (less than 900 mm yr^{-1}) over 95% of streamflow occurs in the wet months of May to October and is generated from perched aquifers which develop seasonally in the shallow soils close to the streamline above the subsoil pallid zones. As evapotranspiration is high, virtually all water which infiltrates into the subsoils is returned to the atmosphere through transpiration by the deep-rooted native forest. Groundwaters, if present are localised and often occur more than 10 m below the valley inverts. Streamflow yields are small (less than 5% of rainfall) and may be zero in 30% of years in areas with less than 700 mm annual rainfall. As the streamflow is generated from perched aquifers which have no connection with the saline subsoils or groundwater, stream salinities are low (order 100 mg L^{-1} TSS).

Under natural forested conditions in the western and southern high rainfall areas (above 1100 mm yr^{-1}) there is sufficient infiltration to form extensive groundwater systems in the subsoils and weathering zone which discharge into the surface stream. Streamflow yields are much higher (15% of rainfall) although flows are still winter dominant and streams are rarely perennial. Much of the winter streamflow is generated by seepage from shallow soils and direct rainfall on saturated areas close to the streamline. Annual stream salinities are typically 150 to 250 mg L^{-1} TSS.

There is mounting evidence that catchments in the higher rainfall zones are discharging a greater quantity of salts in streamflow than they are receiving from the atmosphere while those in low rainfall areas are slowly accumulating salts in their landscape (Williamson and Johnston, 1979; Conservation and Environment, 1980). In low rainfall areas, salts which enter the subsoil clays are retained as there appears to be no groundwater outlet to the stream system and therefore under present climatic conditions salts must accumulate. Certainly, soil salt storages increase dramatically in rainfall areas below 900 mm yr^{-1} (Stokes, et al., 1980) where groundwaters rarely discharge to the surface stream.

The rainfall zone between the 1100 mm and 900 mm isohyet represents a transition between the high and low rainfall regions. Streams within this zone display intermediate but variable hydrologic characteristics which depend, to a large extent, on whether the subsoil groundwater system contributes to the surface hydrology (Conservation and Environment, 1980).

Streams in this rainfall zone tend to have the highest salinities of forested catchments in the region as some groundwater can contribute to streamflow unlike lower rainfall catchments, but have less rainfall and shallow subsurface flow to dilute this groundwater compared with higher rainfall catchments.

The above discussion has emphasised the variation in hydrologic characteristics of small forested catchments across the 100 km wide region. Large variations can also occur on very localised scale (order 100 m and less) and even at the soil structure scale, within these catchments.

Groundwater salinities have been observed to vary over an order of magnitude between bore holes 100 m apart (Batini et al., 1977). Localised seepages from areas less than 5% of a small forested research catchment (Salmon) have been observed to contribute over 50% of the total flow from the catchment. Preferred pathways for water movement through relatively impermeable soils have been identified on a scale of metres (Hurle and Johnston, 1979; Nulsen, 1978).

2.2 The Effects of Clearing on Small Catchments

Increases in stream salinity following clearing for agriculture have been observed since the turn of the century (Wood, 1924). Records of rising water tables have been recorded through the years (Teakle, 1938) but only recently have quantitative figures of increases in the water and salt discharge from a catchment been related directly to the intersection of saline subsoil groundwaters with the soil surface (Williamson and Bettenay, 1979).

The most detailed information on the effects of clearing on the hydrology of small catchments is available from three research catchments on Wellington Reservoir catchment (Wights, Lemon and Dons) as shown in Fig. 2. These have been calibrated with two control catchments and were recently partially cleared (1976/

77 summer). Results to date are therefore restricted to two to three years following clearing.

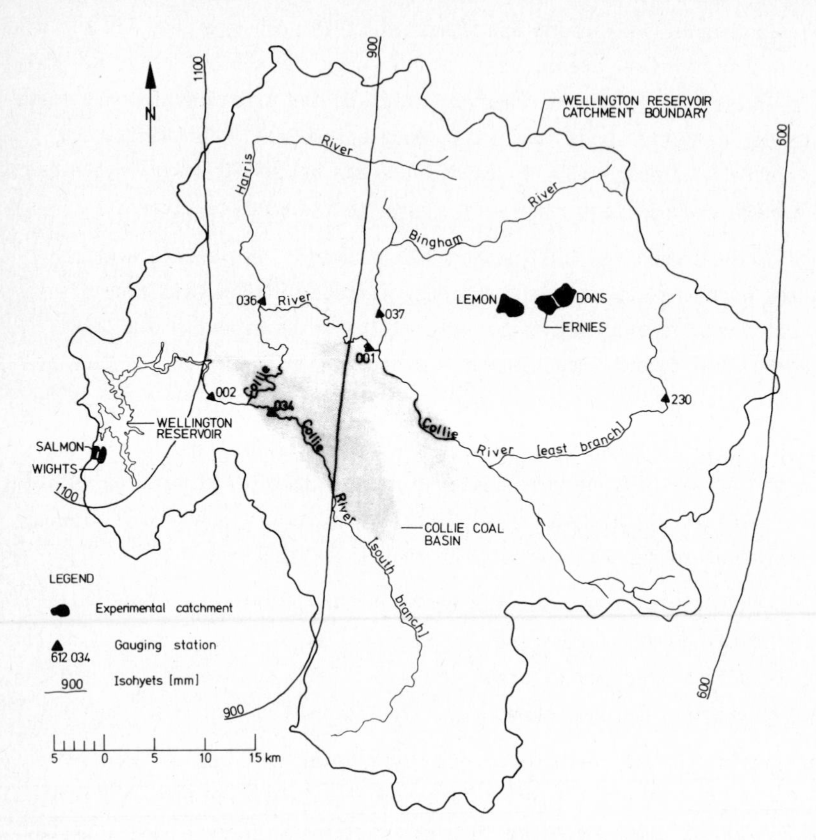

Fig. 2. Wellington Reservoir Catchment showing the location of experimental catchments and range of annual average rainfall across the catchment.

Increases in the subsoil groundwater have occurred in about 80% of bores which intersect permanent aquifers in the deeply weathered profiles (subsoil groundwaters) of the region (Peck et al., 1979). In the 750 mm rainfall area (Lemon catchment) rates of groundwater rise were about 0.8 m yr^{-1} although groundwaters still remain well below the stream bed (order 15 m) and do not contribute to the stream. No significant changes in stream salinity have occurred and salinities remain low (order 100 mg L^{-1} TSS).

In the 1150 mm rainfall zone (Wights catchment) rates of groundwater rise have been about 2 m yr^{-1} and seepage from the groundwater now continues throughout the year. Preliminary estimates of changes in the annual flow volume and chloride following clearing indicate increases by factors up to 3.1 and 4.9 respectively. While salinities have remained acceptable (order 300-350 mg L^{-1} TSS since clearing) the greater increase in chloride flow over streamflow suggests a consequent increase in salinity (Williamson and Johnston, 1979). The major increase in streamflow has occurred during the winter months and is related to greater moisture in the shallow soils and associated shallow aquifers. Changes have also occurred in soil hydraulic properties following clearing and regrowth control activities which have contributed to significant overland flow and much greater flood response.

3 STREAM SALINITY VARIATIONS OF LARGE CATCHMENTS (ORDER 100 km^2 AND GREATER)

The hydrologic response of large scale catchments is a function of the proportion of the catchment in different hydrologic regions (rainfall zones) and the complex distribution of land use both in time and space throughout the catchment.

As the hydrologic response across rainfall regions varies so dramatically, small areas of clearing in salt sensitive regions can have a large effect on stream salinity. This, together with the fact that natural hydrologic processes are so variable, has made it difficult to identify changes in flow and salinity from large catchments.

Long periods of good quality flow and salinity data are required to isolate long term salinity trends from highly variable short-term variations. An understanding of factors affecting short-term variability are important to enable long term salinity trends to be identified from the shortest possible length of record and to enable development of methods for predicting future salinities.

3.1 Changes in Salinity with Catchment Clearing Across the Region

Table 1 summarises annual salinity statistics of streamflow from 15 catchments which have more than 15 years of stream salinity data. The data are based on chloride ion concentrations collected at routine visits to gauging stations established under the States Water Resources Assessment Programme.

Catchments are grouped within three rainfall zones, namely greater than 1100 mm yr^{-1}, between 900 and 1100 mm yr^{-1} and less than 900 mm yr^{-1}. The effect of clearing on stream salinity varies considerably across these zones. Catchments in the group with average rainfalls in excess of 1100 mm yr^{-1} have relatively low salinities (below 260 mg L^{-1}) even though considerable clearing has taken place on three of the five catchments. Variations in the annual salinity are low relative to other catchments in the region and there is no significant difference between mean salinities prior to 1970 and during the 1970's.

Higher salinities occur in the forested catchments in the 900 to 1100 mm annual rainfall zone and again there is no evidence of higher salinities during the

1970's than prior to 1970. By contrast, the two catchments with clearing in excess of 30% (Wilgarup and Thompson) show either a high salinity (Wilgarup) or a significant increase between the pre and post 1970 mean salinity (Thompson).

The final group of catchments (with average catchment rainfalls below 900 mm yr^{-1}) all have significant clearing and have salinities in excess of 500 mg L^{-1} TSS prior to 1970, salinities in excess of 750 mg L^{-1} TSS during the 1970's and with significant differences in the mean in all cases.

TABLE 1

Annual flow weighted salinity statistics

Catchment					Annual Flow Weighted Salinity Statistics			
Gauging Station Name	Gauging Station Number	Area km^2	Average Rainfall mm	Permanent Clearing	Mean Prior to 1970 mgL^{-1} TSS	Mean Post 1970 mgL^{-1} TSS	Coeff. of Variation	Number of Years of Record
Barlee	608001	165	1170	0%	153	168 (NS)	19%	18
Shannon	606185	350	1190	3%	134	160 (NS)	17%	16
Clarke	613146	17.1	1150	15%	197	199 (NS)	16%	17
Dombakup	607155	115	1420	16%	199	225 (NS)	15%	19
Brunswick	612152	213	1240	25%	256	255 (NS)	25%	16
Yarragil	614044	72.5	1010	0%	478	380 (NS)	23%	26
Chalk	614123	104	970	0%	360	318 (NS)	17%	18
Harris	612036	383	955	8%	250	264 (NS)	25%	25
Wilgarup	607144	450	915	32%	810	977 (NS)	36%	19
Thompson	611111	102	925	40%	240	425 (1%)	39%	21
Collie South	612034	668	840	27%	592	891 (5%)	41%	24
Warren	607220	4035	865	32%	539	750 (5%)	25%	19
Kent	604053	1590	850	40%	571	1060 (1%)	49%	22
Murray	614006	6840	660	60%	1560	2220 (5%)	47%	37
Frankland	605012	5800	610	85%	1090	1990 (1%)	45%	38

Notes: (1) Annual salinities were calculated from the $\Sigma S_iF_i/\Sigma F_i$ where S_i is the chloride concentration of a grab sample taken at flow rate F_i. The salinities quoted have been derived from this annual series after conversion to Total Soluble Salts (TSS).

(2) The figures in brackets indicate the level of significance in differences in the mean salinity during the 1970's compared with the mean salinity prior to 1970.

(3) The Coefficient of variation quoted is based on the pooled estimate of variance of data pre and post 1970 divided by the mean for the complete record.

Although no long term data for fully forested catchments in this rainfall zone are listed in Table 1 the preclearing salinities of the catchments included are believed to have been between 100 and 350 mg L^{-1} TSS. This is based on experience from small research catchments (Section 2), broad scale sampling programmes throughout low rainfall forested areas (Batini and Selkirk, 1978) and salinity monitoring of less than 10 years at gauging stations in the region (Public Works Department, W.A., 1976).

3.2 Variation Between Years

Annual salinities are strongly influenced by annual streamflow. Different sequences of flood and drought years have caused major salinity fluctuations on partially cleared catchments which have far out-weighed the long-term increases caused by clearing. For example, increases in salinity of the Warren River (Gauging Station 607220) from 1957 to 1976 averaged 16 mg L^{-1} yr^{-1} while successive flow weighted annual salinities differed by as much as 300 mg L^{-1} on five occasions through the same period.

Table 1 shows that the variability of annual salinity increases with increasing catchment clearing, and increasing mean salinity.

A general equation of the form:

$$S = A\left[\frac{x}{x_m}\right]^n \qquad (1)$$

where S is the yearly salinity mg L^{-1}; x is the year streamflow in mm; x_m is the flow in a medium year in mm; A is the salinity in a median flow year and n is an exponent,

has been fitted to both forested and partially cleared catchments which do not display salinity time trends. Representative values for the two parameters of equation (1) and associated correlation coefficients are shown in Table 2. Between 50% and 90% of salinity variations can be explained by annual flow variations. Much higher exponents occur for the partially cleared catchment (average of -0.72) compared with forested catchments (average of -0.28). Variations by factors of 5 in the annual salinity in successive years for saline streams are quite possible.

TABLE 2

Parameters of equation relating stream salinity with annual flow.

Catchment			Median Flow	Permanent Clearing	Equation Parameters		r^2	Years of Record
Name and Gauging Station Number	Area km^2	Mean Rainfall mm	X_m mm	% of Catchment	A mgL^{-1}	n		
Barlee 608001	165	1170	204	0%	166	-.36	.70	16
Shannon 606185	350	1190	256	3%	170	-.25	.51	15
Harris 612136	383	955	112	8%	229	-.24	.62	25
Wilgarup 607144	450	915	73	32%	827	-.57	.82	19
Collie East 612230	169	650	33	61%	3290	-.86	.90	12
Weenup 609005	87	550	19	90%	4280	-.74	.91	5

Notes: (1) Parameters of Equation are defined by $S = A\frac{X}{X_m}^n$ where S is the annual salinity, X is the annual flow

(2) r^2 is coefficient of determination.

3.3 Variations Within Years

Fig. 3 shows a plot of salinity versus instantaneous flow rate at time of sampling for a forested and predominantly cleared catchment. While there is a general reduction in salinity with increasing flow rate, a wide range of salinities can occur for a given flow rate on the cleared catchment. Higher salinities occur during the early winter period when large quantities of salts are transported down the river system. These come from salts which have accumulated in valley depressions and along dry stream beds over the previous summer months. Salinities at highflows reduce in late winter and early spring after the surface accumulation of salts have been removed.

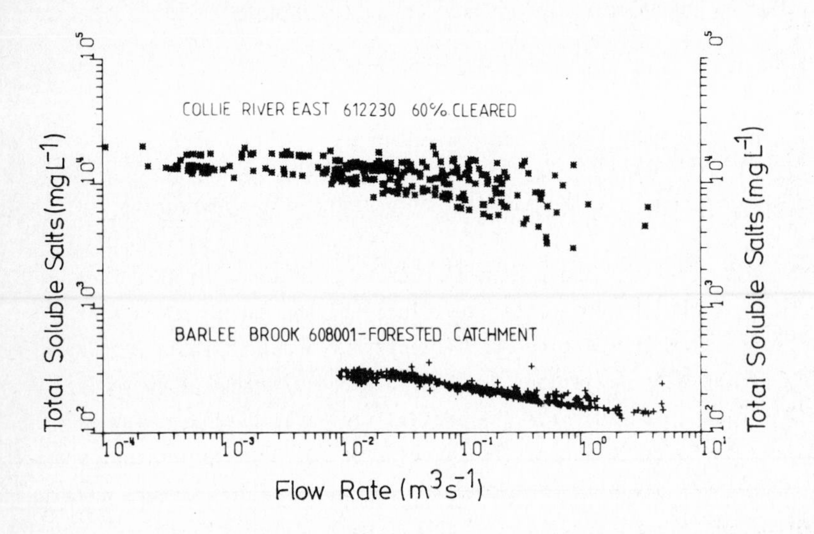

Fig. 3. Variation of salinity with flow rate for a forested and cleared catchment.

Fig. 4 shows the flow and salinity pattern during 1976 for a catchment which has its upper reaches (lower rainfall areas) predominantly cleared while its lower reaches remain forested. Early low flows from the higher rainfall forested areas are relatively fresh but become highly saline as the first major flows from the inland region contribute.

As streamflows decrease through spring and early summer saline groundwaters contribute a larger proportion of the flow and salinities rise. Salinities may reduce as saline seepage from the inland areas dry up and groundwater from forested areas become the primary contribution to streamflow. The complexity of the relationship between salinity, flow rate and time of year is highlighted in Fig. 5.

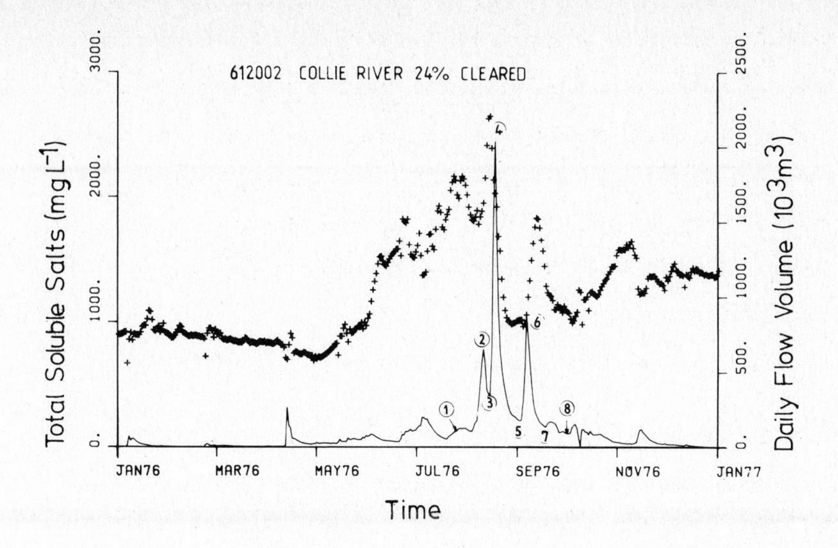

Fig. 4. Variations of flow and salinity with time for a partly cleared large catchment (area 2250 km^2).

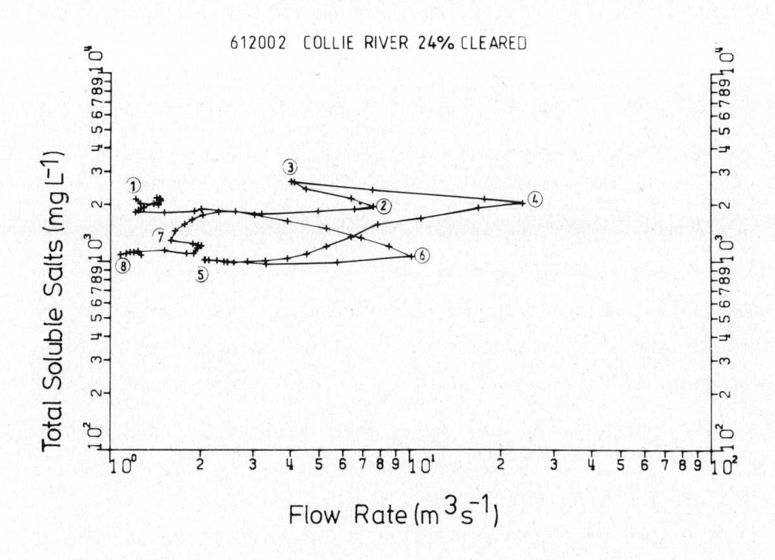

Fig. 5. Variation of salinity with flow between 24/7 and 30/8 from Fig. 4.

In the first major flow event salinities are higher on the rising side of the hydrograph while in the second, the high salinities occur well after the peak.

The actual pattern of flushing surface salts from catchments in any one year is an extremely complex function of:

(i) the extent of catchment clearing, its temporal and spacial distribution,

(ii) the level of groundwater tables over the previous summer,

(iii) the evaporative demand causing capillary rise of salts over the summer,

(iv) any summer or early autumn streamflow which redistributes the accumulated surface salts through the catchment,

(v) the sequence and magnitude of early winter flows,

(vi) and other unknown factors.

Modelling this behaviour on large complex catchments will be difficult.

4 STREAM SALINITY MODELLING

The discussion in the two previous sections has highlighted the many complex factors which affect stream salinity. All factors are not relevant to all problems. For example, the local scale variations in soil structure, groundwater salinity and flows are unimportant to predicting long-term salinity trends on major river systems but may be very relevant to developing cost-effective rehabilitation strategies on specific sites within these river basins. The following discussion highlights approaches to predicting stream salinities considered relevant to different aspects of the salinity problem in Western Australia.

4.1 Stream Salinity Prediction at the Small Research Catchment Scale

To adequately explain observed changes in the response of a small research catchment to land use change a knowledge of the processes of water and solute movement within the catchment is required. It is in this area where the science of soil water movement, plant water relations and evapotranspiration are most relevant and the need to take into account local variations in hydrologic characteristics are important.

Traditional engineering approaches of scaling catchment response by catchment area (lumped parameter modelling) may have proved satisfactory for predicting variations in flow but are unlikely to adequately account for variations in both salinity and flow without more realistic simulation of flow within the catchment. Models which couple sub-surface flow and channel flow and simulate the effect of saturated wetland variation within catchments are likely to be necessary for this purpose. However, the difficulty of adequately defining the surface topography, soil properties and groundwater characteristics of even a small catchment may mean that such models will be used with simplified catchment properties to produce conceptual changes in catchment response following land use change rather than being applied directly to a research catchment data set.

Observations of processes occurring within such catchments are important when formulating simple-parameterised models which account for distributed catchment response that can be readily applied to larger scale catchments.

4.2 Prediction of Daily Sequences of Flow and Salinity from Catchments Greater than 100 km^2

The complexity of daily distributions of flow and salinity have been highlighted in Section 3. Modelling of the distributed nature of catchment reponse is essential if the pattern of flushing surface salts from large catchments is to be adequately simulated. Parameterised mass balances of the surface salt accumulation and depletion, related to models of groundwater leakage is likely to be a critical component. The models developed for these purposes should build on results from the small research catchment studies and associated process modelling work but must be simpler and more appropriate to the data available from catchments of the order of 100's of square kilometres.

Until satisfactory modelling techniques are developed direct monitoring of flow and salinity variability of major streams will continue to provide a sound basis for evaluation of resource management strategies based on engineering solutions (Public Works Department, W.A., 1980).

4.3 Predicting Long Term Salinity Trends

Evaluation of the long-term effect of past clearing, further agricultural development and other land uses such as bauxite mining, forest clear felling and subsequent rehabilitation on stream salinity has been a major requirement of water resource planners in recent years.

It is in this area that most of the modelling effort has been directed in Western Australia. Peck (1975) developed the first techniques for estimating increased stream salinity from land use change. The model assumed an increased groundwater recharge following clearing, zero input-output lag and transmission losses through the groundwater system, a salinity of discharging groundwater based on soil solute concentrations and determined the salinity following land use change by adding the additional salt load and groundwater discharge to the average existing salt load and streamflow respectively. Improvements to this simple model have developed and include the allowance for a lag between the recharge and discharge of groundwater, and the variation of groundwater discharge volume and salinity across a catchment (Peck et al., 1977).

Limitations in the predictions based on these models include:

(i) neglect of both the increased surface runoff during winter and the portion of increased groundwater seepage which is lost to evaporation during the summer following land use change.

(ii) neglect of variations in flows and salinities from year to year,

(iii) the sensitivity of the predictions to the estimates of groundwater salinity following clearing and

(iv) the fact that no independent data set is available to test the predictions.

Attempts to overcome some of these limitations are discussed in the formulation of a model to predict the salinity and flow from the Wellington Reservoir Catchment.

5 PREDICTING ANNUAL SALINITY AND FLOWS FROM WELLINGTON RESERVOIR CATCHMENT

The Collie River rises in the dissected lateritic landscape of the Darling Range in areas with an annual rainfall of approximately 600 mm and drains westward through higher rainfall country reaching Wellington Reservoir where annual rainfall is approximately 1200 mm (Fig. 2).

Generally, the hydrologic characteristics, climate and soil salt storage across the catchment reflect the characteristics described in Sections 2 and 3. The exception is the Collie Coal Basin where a substantial groundwater resource of good quality exists in the sandy sediments of the basin.

5.1 Model Structure

To adequately model salinities and flows over such a variable catchment a complete clearing history, characteristics of the relationship between rainfall and streamflow, and a knowledge of the additional flow and salt load added to the river system following clearing across the catchment, are required.

The total catchment has been divided into 8 zones of differing rainfall and one which represents the Collie Coal Basin. For each zone an annual rainfall and area cleared has been determined from the year 1900. An annual model has been developed which estimates the flows and salt loads (flow times salinity expressed as kilograms of Total Soluble Salts) of the Collie River at Wellington Reservoir from these inputs of rainfall and historic clearing.

Each year a "forested" flow is generated for each zone from the annual rainfall of that zone, by a functional relationship relating millimetres of rainfall to millimetres of streamflow. Salinities for this flow are derived from equations of the form of equation (1). The catchment flow volume and salt load is simply calculated by multiplying the flow per unit area by the area of the zone and the flow volumes by its salinity, respectively and adding the zones.

The additional flow volume following clearing is another function of rainfall times the area cleared and is a combination of increased surface and shallow sub-surface flow and any increased flow from groundwater seepage.

The functional relationships between both rainfall and flow and rainfall and additional flow have been expressed in terms of the probability of non exceedance of annual rainfall to make comparisons across zones simpler. Annual rainfalls in the region are usually normally distributed (Loh and King, 1978).

Additional salt load comes from two sources. Firstly additional salt is transported by additional surface and shallow sub-surface flow, and secondly by increased groundwater discharge. Experience on Wights catchment indicates that by far the greatest increase in flow volume occurs from the shallow sub-surface flow and overland flow (see Section 2). In the lower rainfall zones addition of groundwater seepage is generally less than the evaporation rate. As a good approximation then, the addition of salts from the surface and sub-surface soils has been assumed a product of an average shallow groundwater salinity times the total increase in flow volume.

The additional groundwater salt load discharge is a product of the discharging groundwater volume prior to evaporation times the groundwater salinity. Groundwater discharge is related to the additional groundwater recharge by a time function which delays the discharge behind the recharge. This function is governed by the hydraulic and geometric properties of the groundwater aquifer. The model is constructed so that different zones have different time delays between recharge and discharge and different salinities of discharging groundwaters.

Effective salt discharge is also a function of the levels of groundwater. This in turn is related to the quantity of rainfall in both the preceding and current years. No quantification of variations of salt discharge from year to year at a particular saline seep have, as yet, been made. For simplicity discharge has been assumed a constant from year to year.

Final calculation of the total flow is given by:

$$F_t = \sum_{i=1}^{9} FF_i \times A_i + \sum_{i=1}^{9} AF_i \times AC_i \qquad (2)$$

where FF_i is forested flow in mm for zone i; A_i is Area of zone i in km^2; AF_i is additional flow in mm; AC_i is area cleared in km^2; and F_t is total flow in $10^6 m^3$.

The final salt load is given by:

$$SL_t = \sum_{i=1}^{9} FF_i \times A_i \times SF_i + \sum_{i=1}^{9} AF_i \times AC_i \times SS_t + \sum_{i=1}^{9} GR_i \times F_i \times AC_i \times GS_i \quad (3)$$

where SF_i is the forested salinity in kg m^{-3}; SS_i is the salinity of shallow subsurface groundwater in kg m^{-3}; GS_i is the salinity of discharging groundwater in kg m^{-3}; GR_i is the groundwater recharge rate in mm; F_i is a proportion of the groundwater recharge which is currently being discharged and in turn is a function of past clearing and the ground response; and SL_t is the Total Soluble Salt load in 10^6 kg.

The salinity of inflow is given by:

$$SAL_t = SL_t \times 1000/F_T \quad (4)$$

where SAL_t is in mg L^{-1} TSS.

For trial comparison with the historic salinity data in Wellington Reservoir the inflow salt load and flow were used as input to a simple two season reservoir model and a reservoir salinity at April each year calculated.

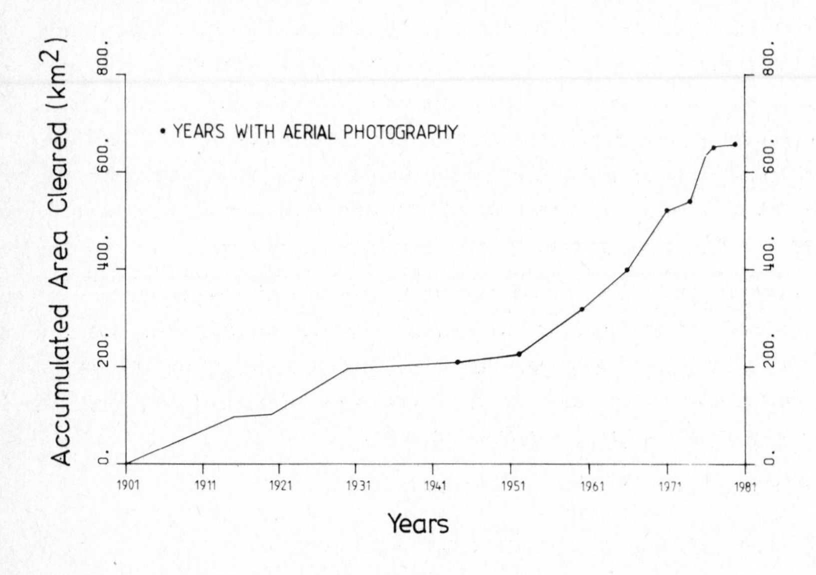

Fig. 6. Clearing history of Wellington Reservoir Catchment

5.2 Rainfall and Clearing History Since 1900

The historic rainfall data for the 9 different zones have been obtained from over 150 stations including both daily read and pluviometer data. Fig. 6 represents the pattern of total clearing on the catchment and is based on air photography of the catchment taken at approximately 6 to 8 year intervals from 1943/44. Prior to 1943 the pattern of development has been estimated from clearing statistics by Shires and records of early history in the district (Spencer, 1966).

Table 3 lists the mean and standard deviation of rainfall, total areas and area cleared as at March 1980 for each of the 9 zones of the catchment.

TABLE 3
Rainfall, clearing and forested salinity characteristics of zones within Wellington Reservoir Catchment.

Zone	Rainfall 1900 to 1979		Total Area km^2	Area Cleared 1980 km^2	Median Forested Salinity mgL^{-1}	Exponent in Equation (1) n
	Mean mm	S. Dev. mm				
1	1200	268	74	5.6	150	- .35
2	1100	242	141	20.2	175	- .35
3	1020	221	235	17.6	200	- .35
4	940	217	324	45.0	350	- .35
5*	883	199	277	57.0	200	- .25
6	835	207	447	118	250	- .60
7	741	181	754	172	100	- .14
8	672	163	389	109	100	- .14
9	615	141	189	114	100	- .14

*Zone 5 is the Coal Basin.

5.3 Flow and Salinity Characteristics when Fully Forested

Annual rainfall-streamflow relationships for the different zones were developed from a growing body of rainfall and streamflow data on gauged catchments throughout the Darling Range. Small forested catchments which had similar rainfall to particular zones of the Wellington catchment were selected to represent that zone and their annual streamflow (expressed in mm) plotted against the probability of non-exceedance of the annual catchment rainfall for the years of available record. This technique enabled a number of gauged catchments covering a particular zone to be included thereby extending the definition of the relationship over a greater range of years. Fig. 7 summarises the resulting annual rainfall-streamflow relationships of forested catchments across the region.

Annual streamflow is shown as a function of the probability of non-exceedance of rainfall for a given average rainfall. As indicated, forested catchment flows reduce dramatically with decreasing average rainfall across the region.

The salinity of streamflow in the forested state has been described in Section 2.1 and characterised in Tables 1 and 2.

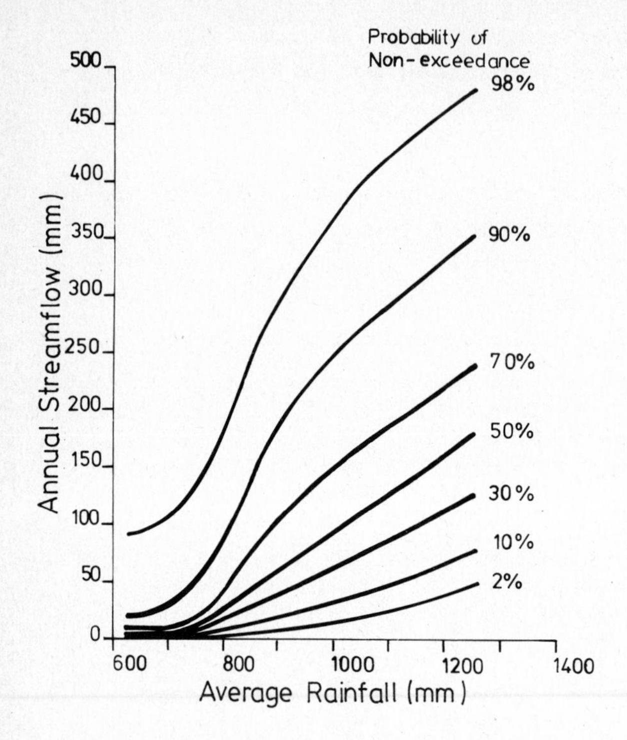

Fig. 7. Annual streamflow for forested catchments with different average catchment rainfalls.

Equation (1) was used to calculate the annual salinity from the annual flow for each zone. Table 3 includes the adopted parameters of the equation.

5.4 Increases in Streamflow Following Clearing

As noted in Section 2.2 research catchment data on the increases in streamflow following clearing are only available for the first 3 years after treatment. However, increases of up to 3 times the original forested flows in below average years have been noted on Wights Catchment. Data from the research catchments have been supplemented by longer records from the Collie East Branch (612230) (see Fig. 2). Increases were estimated by comparing observed streamflow from this 60% cleared catchment with the forested flow expected for zone 9. In this way, increases in streamflow were determined for the 1150 mm and 625 mm rainfall areas for a range of above and below average years. Increases in other rainfall areas have been determined by linear interpolation as shown in Fig. 8.

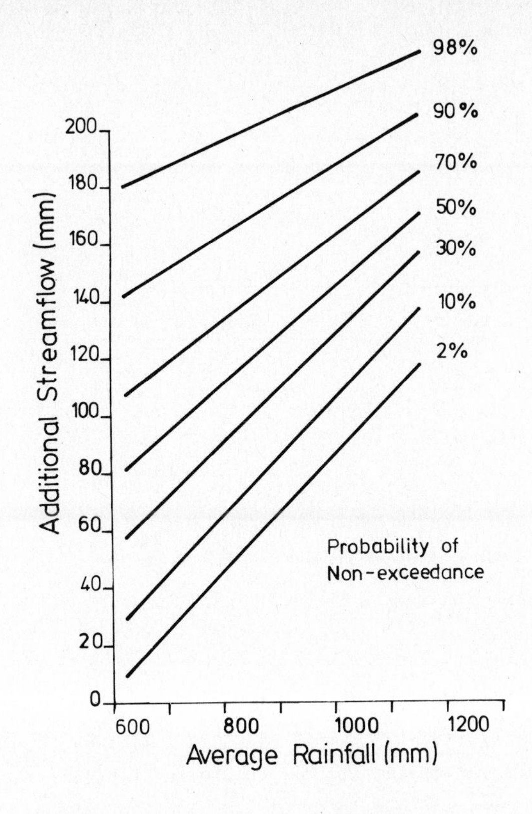

Fig. 8. Additional streamflow generated following clearing of catchments with different average annual rainfall.

5.5 Additional Solute Discharge by Groundwater Following Clearing

The additional salts discharged by groundwater, either directly or through capillary rise of salts from a saline water table is a function of the additional groundwater recharge, the relationship between groundwater recharge and discharge, and the salinity of the discharging groundwater.

5.5.1 Additional groundwater recharge

Increased groundwater recharge can be estimated from rates of groundwater rise and hydraulic properties of the groundwater aquifer prior to groundwaters intersecting the stream system. Table 4 summarises available data on groundwater level increases since clearing in the region. Data quality varies from detailed observations of groundwaters at research sites over relatively short periods (less than 5 years) to historic records of individual bores which indicate large increases over 20 or more years. Table 4 includes two sets of recharge rates based on the observed rates of groundwater rise for two assumed values of aquifer specific yield.

TABLE 4

Rates of groundwater rise following clearing.

Location	Average Annual Rainfall mm	Rate of Water Level Rise mm yr^{-1}	Recharge for Specific Yield of 0.05		0.04		Data Notes
			mm	% of Rainfall	mm	% of Rainfall	
Wights	1150	2000	100	8.7	80	7.0	1
Lemon	750	800	40	5.3	32	4.3	1
Bingham River	725	1200	60	8.3	48	6.6	2
Batalling	650	700 to 1500	35 to 75	5.4 to 10.0	28 to 60	4.3 to 9.2	2,3
Bakers Hill	590	600	30	5.1	24	4.1	1
Lake Toolabin	410	240 to 540	12 to 27	2.9 to 6.6	10 to 22	2.4 to 5.4	3,4
Salmon Gums	390	300	15	3.8	12	3.1	4

Notes: (1) Based on measurements of averaged groundwater rises from experimental sites. Referenced in Peck et al (1979) and Williamson and Betteney (1979)

(2) Elevation change estimated for position of water table relative to soil solute distribution

(3) Considerable uncertainty exists in the time taken for the groundwater rise. The known possible range is included.

(4) Based on individual bores (unpublished data from Department of Agriculture and Department of Mines)

5.5.2 The ratio of groundwater discharge to recharge

As additional recharge is added to the groundwater system water tables rise, and may take many years before they discharge to the stream. Only after further delay, when increased groundwater gradients have developed will a stable groundwater level be reached and the additional groundwater recharge be equalled by the groundwater discharge.

The boundary conditions assumed in the model for bauxite mining and rehabilitation (Peck et al., 1977) are not appropriate for the agricultural problem particularly where initial groundwater levels are well below the river bed.

Digital simulation of groundwater responses to recharge rates (shown in Table 4) are required for typical initial groundwater geometries characteristic of the different zones. Such groundwater modelling is planned for the near future. In the interim, only empirically derived relationships between the ratio of discharge to recharge, based on observed delays between clearing and the development of seeps have been adopted and are shown in Fig. 9.

5.5.3 The salinity of discharging groundwater

As the results from research catchments have not progressed sufficiently, other sources must be used to assess the degree of mobilisation of stored salts by increased groundwater levels following clearing. Stokes et al. (1980) have summarised the available data on soil solute and associated groundwater salinity data in the region. Drilling programmes in cleared areas have shown that the ratio of groundwater to soil solute concentration ranges between 0.5 and 1. This contrasts with forested sites in low rainfall areas where the ratio is often less than 0.25. By adopting a ratio of groundwater salinity to soil solute concentration for cleared sites and patterns of soil solute concentration across the region a first estimate of the discharging groundwater can be made whether an area has been cleared or not. The soil solute concentrations adopted for the 9 zones are based on data from Stokes et. al. (1980) and recent drilling activity in the Collie Coal basin and are shown in Table 5.

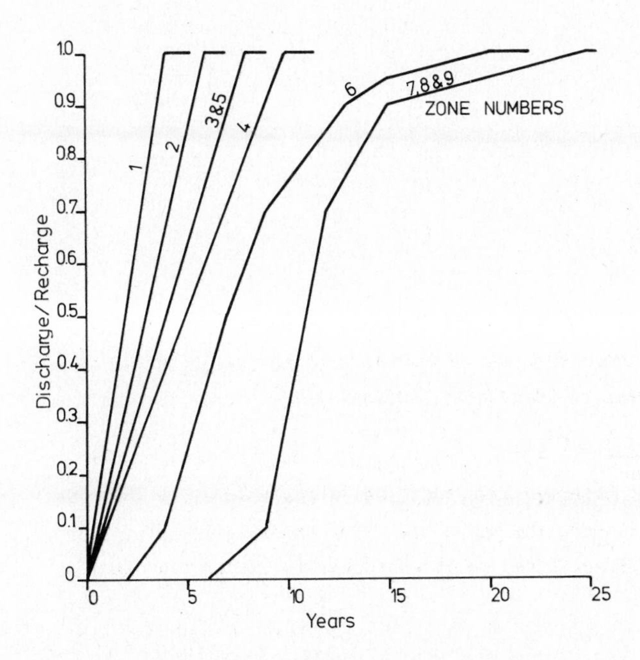

Fig. 9. Relationships used to define the delay between groundwater recharge and discharge for the zones defined in Table 3.

5.6 Model Calibration

Although estimates can be made of groundwater recharges, the relationship between recharge and discharge and the salinity of groundwater discharge following clearing as described above, it is in these variables where the greatest uncertainty in the input data exist. Modifications to these inputs are considered justifiable to improve the correlation between the observed and predicted reservoir salinities.

TABLE 5

Groundwater characteristics following clearing.

Zone	GR % of Rainfall	SS mgL^{-1}	Soil Solute Concentrations	Ratio of Groundwater to Soil Solute Concentrations
1	8.0	150	460	0.75
2	7.5	150	700	0.75
3	7.0	150	1110	0.75
4	6.5	150	1840	0.75
5	6.3	150	400	0.75
6	6.0	150	3220	0.75
7	5.5	150	6040	0.75
8	5.3	150	10270	0.75
9	5.0	150	12670	0.75

GR is groundwater recharge
SS is shallow subsoil groundwater salinity.

6 PRELIMINARY RESULTS

The model has so recently been developed that only limited checking and some initial simulation runs have been completed. The results presented should be considered preliminary only, as further work is planned.

6.1 Observed and Predicted Flows

Table 6 lists statistics of observed and predicted inflow volumes to the reservoir from 1945 to 1978. Although the means over the period agree well, the standard deviation of the observed sequence is considerably larger than the predicted sequence.

TABLE 6

Annual streamflow statistics 1945 to 1978

Statistic	Observed $10^6 m^3$	Predicted $10^6 m^3$
Mean	187	186
Std. Dev.	142	119
Max.	652	551
Min.	33	70

The inability to predict the extreme wet and dry years is primarily a consequence of using annual rainfall data to predict annual streamflows. The yearly distribution of rainfall can have a marked effect on streamflow yield and is particularly important in extreme years (either flood or drought).

6.2 Salinity Predictions

Initial predictions have been carried out using input groundwater data as summarised in Table 5. Fig. 10 shows the observed pattern of reservoir salinities compared with the predicted reservoir salinities from 1900. Also included in Fig. 10 are the estimated salinities if the catchment had not been cleared.

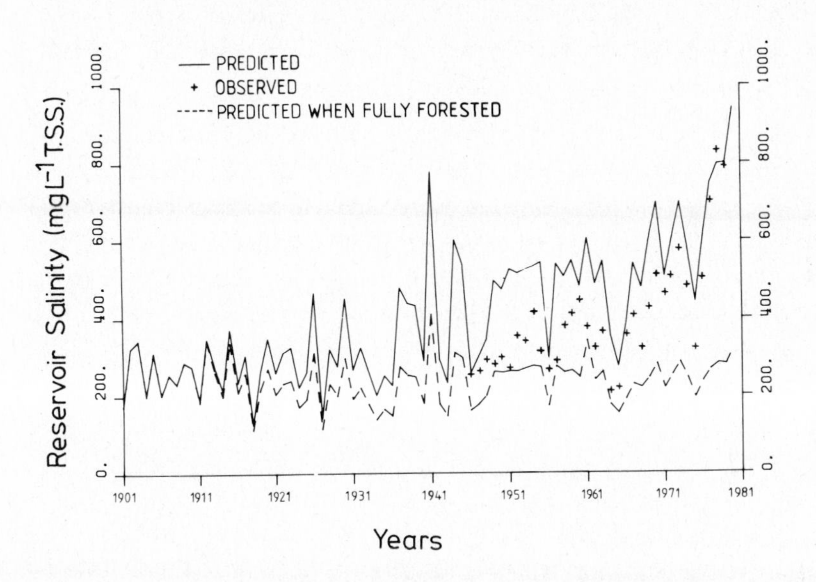

Fig. 10. Observed and predicted reservoir salinities.

Although reasonable predictions were obtained for the later 1970's significant overprediction occurred during the 1950's. A number of factors contribute to the discrepancy and include:

(i) the assumed clearing history prior to 1943/44

(ii) characteristic valley clearing on alluvial soils during the early phase of farm development which may have a different effect from more extensive clearing later.

(iii) overestimation of salinity in drought years by assuming that groundwater discharge is constant from year to year.

(iv) limitations in the flow generation technique from annual rainfall

(v) the assumed delay between groundwater recharge and discharge

(vi) and limitations in the early rainfall, streamflow and reservoir salinity data.

The extension of the lag between groundwater recharge and discharge to a maximum of 50 years reduced salinities in the 1950's but caused underestimates during the 1970's.

By modelling the additional groundwater discharge as a linear function of annual rainfall, increased salinities in flood years and reduced salinities in drought years were obtained, but resulted in only minor improvements in the predictions during the 1950's. The best time distribution of predicted salinities was obtained by assuming an exponential growth in clearing between 1900 and 1951 rather than the assumed pattern prior to 1943/44 (Fig. 6).

A general change in the predicted salinities can be obtained by changing either the groundwater recharge or the ratio between groundwater salinity and soil solute concentration. Direct comparison between the observed and predicted inflow salinity to Wellington Reservoir between 1974 and 1978 indicated that the best calibrations occur with groundwater recharge varying between 5% and 8% of rainfall and soil solute concentrations equalling groundwater salinities. These inputs have been used in the subsequent simulations discussed below.

Perhaps the most encouraging feature of the model is that it adequately predicts the distribution of salt loads across the catchment. Observed streamflow data between 1974 and 1978 show that some 68% of the total salt load comes from the area east of the 900 mm isohyet excluding the Coal Basin. Over the same period the model predicted that 72% of the salt load comes from this region. By contrast, when the catchment was fully forested only some 21% of the salt load came from this region.

The flow volumes from the eastern portion were slightly overestimated (40% compared with 34%) and this was most noticeable in dry years. Predictions of the fully forested state show that only 25% of the water originally came from the eastern portion of the catchment.

As discussed in Section 4 the primary aim of a model of this type is to provide general information to planners to assist with long term catchment management decisions. Examples of the type of possible predictions using the model are presented in Figs. 11 and 12.

Fig. 11 shows the pattern of inflow salinities for a wet, median and dry rainfall year from 1900 to 2010 assuming no further clearing takes place, and thereby provides an estimate of the ultimate inflow salinities without catchment management aimed at reducing inflow salinities. Fig. 12 is an estimate of the likely effect of clearing 90% of the area to the west of the 1000 mm isohyet to increase water yield in this low salt hazard zone. In this example the total inflow volume

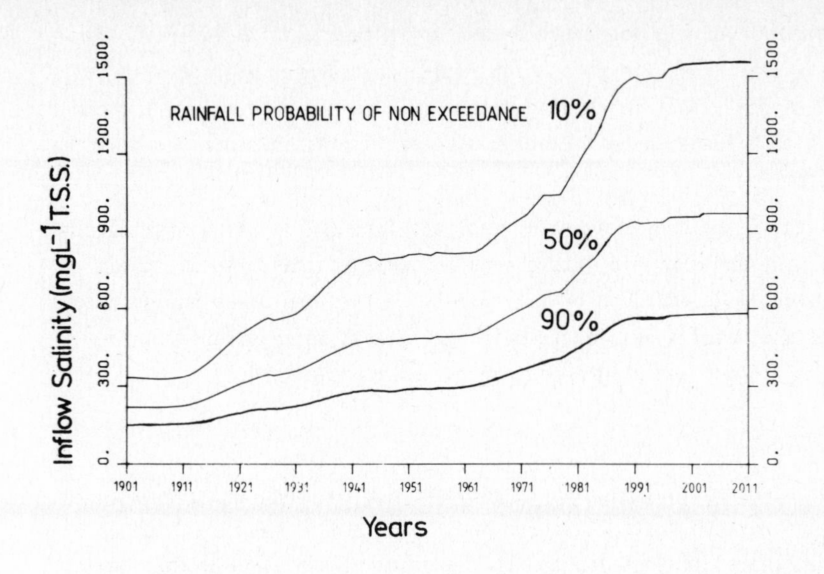

Fig. 11. Predictions of future inflow salinities assuming clearing remains static from 1980. Groundwater and soil solute concentrations are assumed equal.

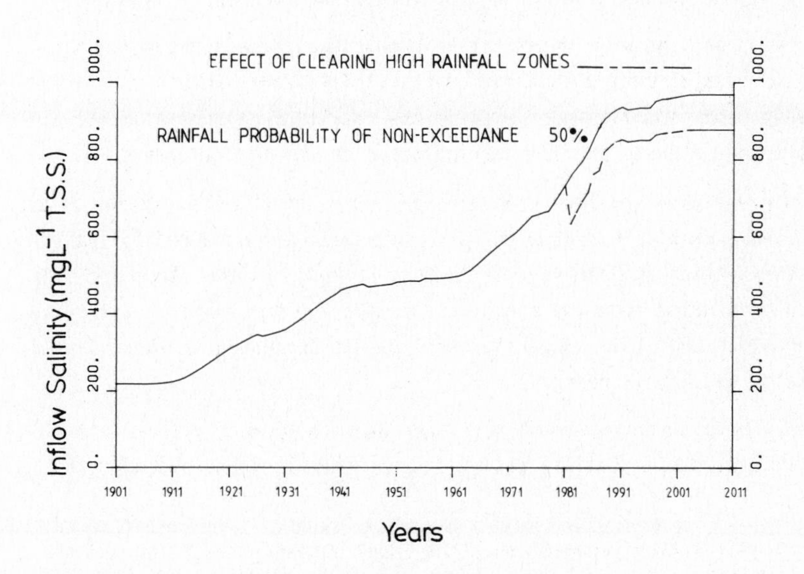

Fig. 12. Predictions of inflow salinities if 90% of the catchment with rainfall in excess of 1000 mm yr^{-1}, were cleared in 1980.

in a median rainfall year is increased by 25% while the salinity is only ultimately reduced by 10%. This figure is higher than the current predicted inflow salinity which suggests that increasing water yield in the high rainfall areas is insufficient to balance the effect of past clearing that has not yet affected the stream system.

Modelling of this type provides useful insights into the relative magnitude of the effects of past and possible future land management strategies on stream salinity. However it is only one of many inputs to land use planning and catchment management and results arising from it must always be considered in the light of the limitations and assumptions in the basic modelling.

7 CONCLUSIONS

Variations in stream salinity characteristic of a wide range of catchments in the south-west of Western Australia, have been described. Changes can occur over decades, between years, through seasons of the year and can occur within hours during flood periods. Approaches to predicting these changes will, of course, depend on the purpose of the prediction and the scale of the problem.

A regional model has been described which predicts the flow and salinity from a 2830 km^2 catchment (Wellington Reservoir Catchment) subject to a long clearing history and is designed to assist planning for long term catchment management.

The model is based on a general quantitative knowledge of the hydrologic characteristics of small catchments within different rainfall zones across the catchment and represents the compilation of much of the results from investigation and research into the salinity problem carried out over the last 10 years.

Data presented highlights the low salinities of the eastern Darling Range prior to clearing and the dramatic increases in salt load that have occurred following clearing. For the Wellington Reservoir catchment, some 25% of the flows and 21% of the salt load originated from the areas east of the 900 mm isohyet. With extensive clearing since 1900 this proportion of flow has increased to approximately 34% while the salt load has increased to 70%.

Only preliminary results of the model have been used to predict future salinities but its usefulness as a tool to assist in long term land use planning has been demonstrated.

Current historic simulations suggest that the model is over-predicting salinities in dry years and further research work on variations in groundwater discharge to streams from year to year is required.

Improvements could also be gained by the inclusion of results from more detailed groundwater modelling of the effects of different geometric patterns of clearing and aquifer dimensions on groundwater discharge. The adoption of daily or monthly rainfall data as a primary input will be necessary to improve predictions of salinity in individual flood and drought years.

However it is particularly the water and solute movement from hillside to streamline which requires both field investigation and modelling if regional models are to be improved and strategies to minimise saline groundwater discharge are to be developed.

8 ACKNOWLEDGEMENTS

The work presented in this paper represents the compilation of results of research and investigation programmes carried out by a number of State Government Departments and CSIRO over the last 10 years. The Water Resources Section of the Public Works Department has provided the basic hydrometric data for many of these programmes. The efforts of the section in collecting and processing the streamflow, rainfall and salinity data for this paper is greatly appreciated.

Mr. P.D.K. Collins developed the clearing history of Wellington Reservoir Catchment and Mr. R.J. Sharpless compiled the historic streamflow and salinity data.

9 REFERENCES

Batini, F.E. and Selkirk, A.B., 1978. Salinity sampling in the Helena Catchment, Western Australia. Res. Pap. 45, Forests Dept. of W.A.

Batini, F.E., Hatch, A.B. and Selkirk, A.B., 1977. Variations in level and salinity of perched and semi-confined groundwater tables, Hutt and Wellbucket experimental catchments. Res. Pap. 33. Forests Dept. of W.A.

Bettenay, E. and Mulcahy, M.J., 1972. Soil and landscape studies in Western Australia. (2) valley form and surface features of the south-west drainage division. J. Geol. Soc. of Aust., Vol. 18, Pt. 4, pp 359-369.

Collins, P.D.K., 1974. Murray River Basin Water Resources Survey. Tech. Note 45, Water Resour. Section, Public Works Dept. of W.A.

Conservation and Environment, 1980. Research into the effects of the woodchip industry on water resources in south western Australia. Bull. 81, Dept. of Conserv. and Environ., Western Australia, July 1980.

Hingston, F.J. and Gailitis, V., 1977. The geographic variation of salt precipitated over Western Australia. Aust. J. Soil Res., 14: 319-335.

Hurle, D.H. and Johnston, D.C., 1979. On the physical basis of salinity in the Darling Range of south western Australia. In Proc. Hydrol and Water Resour. Symp., Perth, 1979. Nat. Conf. Pub. No. 79/10, Inst. of Eng., Aust. pp 164-165.

Johnston, C.D., 1980. Salt content of soil profiles in bauxite mining areas of the Darling Range, Western Australia. Tech. Pap. No. 8, CSIRO Div. of Land Resour. Management, Perth.

Loh, I.C. and King, B., 1978. Annual rainfall characteristics of the Warren, Shannon and Donnelly River Basins. Tech. Rep. No. 78, Water Resour. Section, Public Works Dept., W.A.

Nulsen, R.A., 1978. Water movement through soil. J. of Agric., W.A., Vol. 19 No. 4 pp 106-107.

Peck, A.J., 1976. Estimating the effect of a land use change on stream salinity in south western Australia. In: Vansteenkiste, G.C. (Ed), System Simulation in Water Resources, North-Holland Publ. Company, pp 293-301.

Peck, A.J., 1978. Salinisation of non-irrigated soils: A Review. Aust. J. Soil Res., 16: 157-168.

Peck, A.J. and Hurle, D.H., 1973. Chloride balance of some farmed and forested catchments in south western Australia. Water Resour. Res. 9: 648-657.

Peck, A.J., Hewer, R.A. and Slessor, G.C., 1977. Simulation of the effects of dieback disease on river salinity. Tech. Pap. No. 3, CSIRO Div. of Land Resour. Management, Perth.

Peck A.J., Height, M.E., Hurle, D.H. and Yendle, P.A., 1979. Changes in groundwater systems after clearing for agriculture. In: Agriculture and the Environment in Western Australia. West. Aust. Inst. of Tech., October, 1979.

Public Works Department of Western Australia, 1976. Streamflow records of Western Australia to 1975.

Public Works Department of Western Australia, 1979. Clearing and stream salinity in the south-west of Western Australia. Doc. No. MDS 1/79.

Public Works Department of Western Australia, 1980. Tone River and Upper Kent River diversion study. Hydro. Studies Vol. 2. Appendix C, Doc. No. MDS 1/80.

Spencer, I., 1966. Darkan Early Days. West Arthur Shire.

Stokes, R.A., Stone, K.A. and Loh, I.C., 1980. Summary of soil salt storage characteristics in the northern Darling Range. Tech. Rep. No. 94, Water Resour. Section, Public Works Dept. of W.A.

Teakle, L.J.H., 1938. Soil salinity in Western Australia. J. of Agric., W.A. Vol. 15 pp 434-452.

Williamson, D. and Bettenay, E., 1979. Agricultural land use and its effect on catchment output of salt and water - evidence from southern Australia. In: Prog. Wat. Tech. 1979, Vol. 11, No. 6 pp 463-480. Pergamon Press, Great Britain.

Williamson, D.R. and Johnston, C.D., 1979. Effects of land use changes on salt and water transport. In: Agriculture and the Environment in Western Australia. West. Aust. Inst. of Tech., October, 1979.

Wood, W.E., 1924. Increases in salt in soils and streams following distribution of the native vegetation. J. of the Roy. Soc. of West. Aust. 10(7): pp 35-47.

THE INFLUENCE OF STREAM SALINITY ON RESERVOIR WATER QUALITY

J. IMBERGER
Department of Civil Engineering, University of Western Australia,
Nedlands, 6009, W.A.

ABSTRACT

Imberger, J., 1981. The influence of stream salinity on reservoir water quality. Agric. Water Manage., 1981.

The dynamic reservoir simulation model DYRESM is used to investigate the response of the Wellington Reservoir to changes in streamflow salinity and outflow strategies. Construction of a salinity diversion dam could be useful in substantially reducing reservoir salinities, but with the penalty of reduced water yield.

1 INTRODUCTION

The surface water resources of the populated south-western corner of Western Australia are drawn from a series of western flowing streams which rise in the plateau regions of the Darling Range, flow through increasingly defined valley systems, particularly near the western edge of the plateau (Darling Scarp), and out across the coastal plain to the Indian Ocean. Wellington Reservoir is typical of storages in the region, being located, as shown in Fig. 1, in the incised valley of the Collie River some 10 km from the scarp. However, its catchment differs from those of other reservoirs in the region, in that sections have been progressively cleared of native forest and converted to annual pastures for agricultural development. This land use change has resulted in substantial increases in the salinity of inflow to the reservoir. Loh and Hewer (1977) indicate that, prior to 1950, with only 5% of the catchment cleared, average annual inflow salinities were below 300 mg L^{-1}. Current estimates of the average annual inflow salinity, with some 23% of the catchment cleared, are 750 mg L^{-1} T.D.S., with a yearly increase between 1971 and 1975 of around 50 mg L^{-1} T.D.S.

It is generally accepted that the increased salinities are the result of the reduction of transpiration after forest clearing and the consequent increase in recharge to the groundwater system. The increased groundwater flows upset the existing salt balance by flushing large quantities of soluble salts, previously stored in the soil profile, to the stream system (Wood, 1924; Peck and Hurle, 1973). As groundwater systems respond over periods of tens of years (Peck et al., 1977) the full effect of recent clearing has yet to be felt. First order estimates made by Loh and Hewer (1977) indicate that if recent legislation limits catchment clearing to its current level, average annual inflow salinities of 1100 mg L^{-1} T.D.S. will ultimately develop, with dry year salinities as high as 1450 mg L^{-1} T.D.S. and peak salinities as high as 10 000 mg L^{-1}.

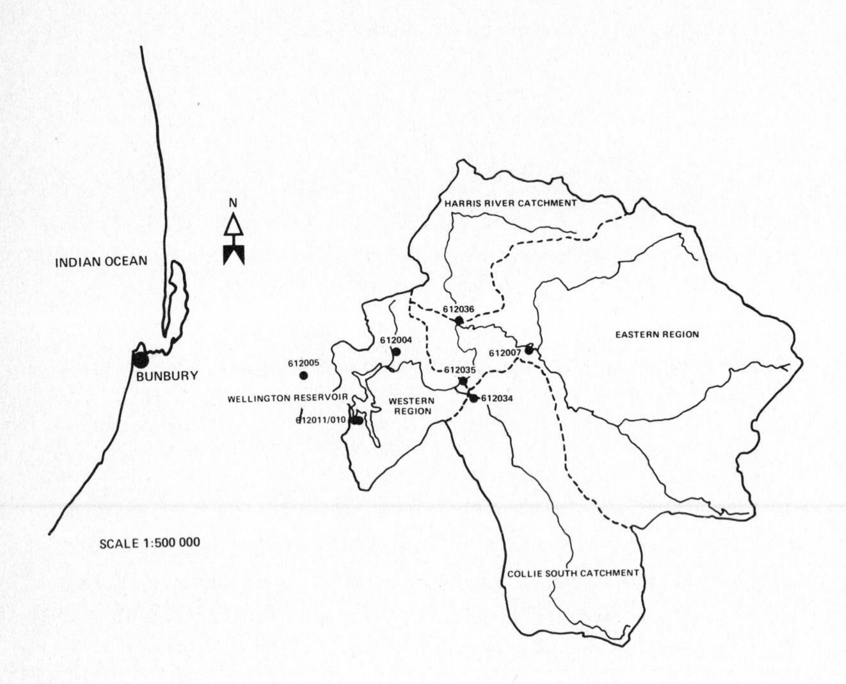

Fig. 1(a) - Wellington Reservoir. Map showing the reservoir and the major catchment regions. Numbers refer to gauging stations located on the streams.

Clearly, any improvement in the supply salinity by means of improved reservoir management will be increasingly beneficial as such high salinities develop, especially since long term solutions to the problem must involve improve catchment management and rehabilitation which take a long time to establish.

In the assessment of reservoir management strategies (Fischer et al., 1979) two questions arise. First, what is the effect of the streamflow salinity and the phase between the salinity and streamflow in determining the overall salt content of the reservoir. Second, what is the interplay between the streamflow salinity variations and the outflow volume and variability in determining the salt content of the reservoir. The dynamic reservoir simulation model DYRESM was used to simulate the response of the Wellington Reservoir to changing inflow salinities and outflow strategies in order to assess the answers to these two questions.

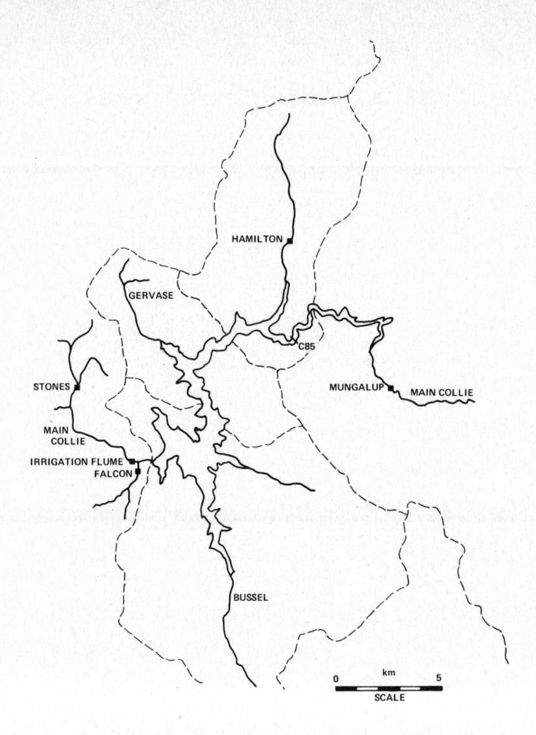

Fig. 1(b) - Wellington Reservoir. Enlargement showing the location of the dam wall and the Mungalup gauging station.

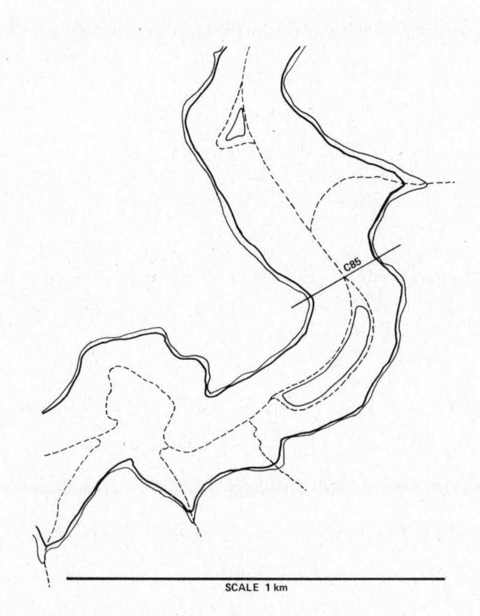

Fig. 1(c) - Wellington Reservoir. Site for the proposed salinity diversion dam.

2 STREAMFLOW SALINITY VARIATIONS

Much has been written about the salinity problem, its causes and the mechanisms responsible for the salinity increases observed over the last few decades in the south-west region of Western Australia. The purpose of this paper is to synthesize the effects of these increases on the salt content of the Wellington Reservoir, to establish the sensitivity of the result to variations in streamflow salinity and to explore techniques for the alleviation of the salt load.

The most developed model of the groundwater system active in the south-west region appears to be that proposed by Smith and Hebbert (1980), who modelled a two aquifer system as shown in Fig. 2. The water entering the higher areas is conducted to the deep groundwater via the preferred paths established by new and fossil root structures. Weak pressures are set up causing a constant upflow in the lower areas surrounding contributing creeks and streams. The upward flows negotiate a salt laden soil profile (see Johnston et al., 1980) introducing saline water to the streams in the winter. During periods of no flow this slow percolation continues leading to a gradual buildup, when combined with evaporation, of salt on and near the surface of the lower areas. This salt is stored ready for flushing into the streams during the first storms of the season.

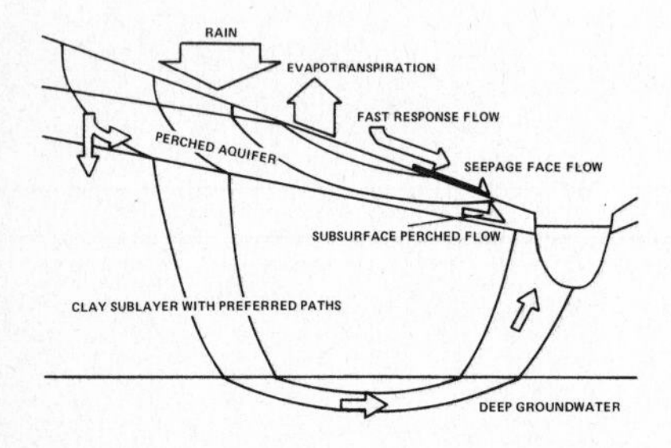

Fig. 2. (From Smith and Hebbert, 1980) - Schematic of the double aquifer concept. Water enters the stream directly via surface and subsurface flows in the perched aquifer and from below under weak artesian pressures.

During periods of rain the perched aquifer (Fig. 2) builds up, causing leaching of the salt accumulated by direct water ponding during the end of the previous winter and that accumulated by the upward percolation described above. In areas where the perched aquifer is saturated, direct runoff will occur leading to a very rapid response of the streams. The thickness of the perched aquifer obviously will vary greatly so that the contributing area will be a function of space as well as time. The response of the perched aquifer itself will be rapid once saturated (a few days). The salt stored on the surface and in the perched aquifer will thus be flushed out once the water has travelled a distance characterized by the width of the contributing area. Effective flushing requires a certain volume of continuous rain in order that the movement of water be continuous and not be interspersed by periods of strong evaporation.

Deep-rooted vegetation has two influences. First, it prevents the buildup of the upward flows into the contributing areas and secondly the vegetation drains much of the perched aquifer during the winter months.

This conceptually plausible model explained most of the features of the observed salinity and flow distribution, an example of which is shown in Fig. 3. The early rains in May and half of June would have progressively leached more and more of the upper salt laden areas. The increasing contribution from the rising perched aquifer leads to a general rise in salinity climaxing with the very large spike around the inflow at the end of June. From that time onward the salinity declined, indicating a general trend towards a flushed state. Each new flood naturally leads to a local spike as increases in flow encompass a bigger sphere of influence in the perched aquifer due to the growth in the contributing area.

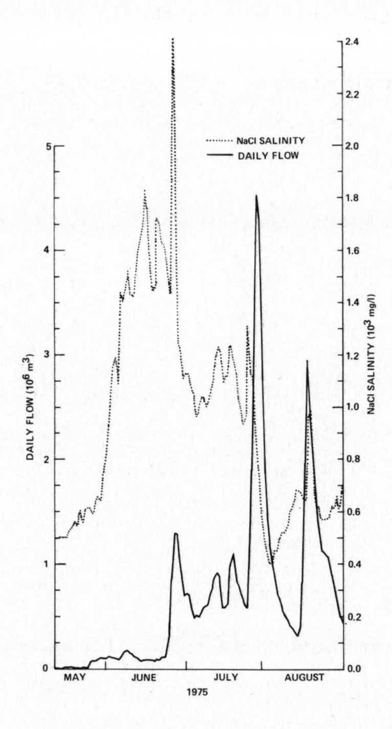

Fig. 3. Distribution of the salinity and flow in the Collie River at the Mungalup gauging station. The distribution typifies the rise in salinity during the early flows and the lowering of salinity as the upper surface areas are flushed.

Further, the model suggested by Smith and Hebbert (1980) is able to explain the very different salinity response recorded during the July flow and the high flow at the middle of August. A straight correlation of salinity and flow as suggested by Loh and Porter (1976) is obviously not warranted even if a correction for the time of year is included. A more plausible answer may be derived from the double aquifer model. The flow in July was preceded by perched aquifer flushing whereas the August flow was preceded by a period of low rainfall which had allowed a new salt load to accumulate in the contributing areas.

Smith and Hebbert (1980) have proposed this model, but it will take a few years to validate such a detailed model for the Collie Catchment and so no attempt was made to couple this model to DYRESM. Instead certain features of the model were used to change the recorded salinities to simulate reforestation and increased clearing in the following way. The essential features are that the peak salinities in the stream depend on the salt concentration in the contributing areas, the phasing of the flow in relation to the volume of the contributing areas and the percentage of salt rich area to fresh flowing catchment. By retaining the phasing as in 1975 to 1978, but increasing the percentage area contribution, increased salinity records may thus be generated. Further, by assuming that the nature of the clearing carried out in the last 15 years is similar to that done earlier it is reasonable to assume the flushing volumes contributing to the salinity have the same time responses. These assumptions would lead to a much increased peak salinity, but with about the same flushing time scale in relation to flow variability and the same "flushed" salinity. If this low state is taken as 500 mg L^{-1}, then a possible recipe for simulating an increased salinity would be :

$$S = \frac{S_e}{500} \cdot S_e \qquad (1)$$

where S_e is the existing salinity. This leads to a salinity record with an average salinity of about 1700 mg L^{-1} and a new peak salinity of 11 500 mg L^{-1} corresponding to expectations, outlined in the introduction. Further work is naturally required to verify this simple hypothesis expressed by equation (1), but for the purposes of the reservoir simulation it serves as a good first model.

Reduced salinities may again be expressed using the existing record by:

$$\begin{aligned} S &= 0.5\,(S_e - 500) + 350 \qquad & S_e > 500 \\ S &= 350 \qquad & S_e < 500 \end{aligned} \qquad (2)$$

This will be used to simulate conditions before the recent clearing took effect in 1960.

3 MODEL DESCRIPTION

The dynamic reservoir simulation model, DYRESM, is a one-dimensional numerical simulation model for the prediction of temperature and salinity in small to medium lakes. It is based on the assumption that in such lakes the thermal structure plays a dominant role and the isotherms are mostly horizontal planes. Deviations from this state of "rest" are allowed, but it is assumed that such deviations are sporadic or weak and may be treated as perturbations to the evolution of the one-dimensional structure.

These assumptions place certain restrictions on the applicability of the model. In particular the Wedderburn Number (see Thompson and Imberger, 1980) should be greater than one for the majority of the wind events, the inflow and outflow internal Froude numbers (see Imberger et al., 1976) should be less than one and the lake should be small or narrow enough for the influence of the earth's rotation to be negligible (see Fischer et al., 1979).

DYRESM was developed over the last 6 years in order to predict the salinity variations in the Wellington Reservoir, but in the meantime the model has found wide applicability as a base for more general water quality modelling. The development of the model is continuing and currently version 5 is the operational model. A full and detailed description of this version may be found in Imberger and Patterson (1980) and only a very brief summary will be given here.

The lake is divided into uniform horizontal slabs which form the computational building blocks of the model. The construction of the slabs is Lagrangian and they advect with the vertical velocity induced by the inflow and outflow. At each time step the model ensures that heat, salt, mass and energy are conserved for each slab and thus the reservoir as a whole. The vertical momentum equation reduces to the hydrostatic pressure assumption, since the assumption of one-dimensionality eliminates all motion except the slow vertical adjust required to accommodate in and outflows.

Once this slab structure has been established the meteorological inputs at the surface of the lake are calculated using bulk aerodynamic formulae. As shown in Fig. 4, these together with the daily inflow and outflow form the daily input data.

The basic time step of the model has been set at one day since only daily data is most commonly available. Certain assumptions are made regarding the distribution of these inputs over the 24-hour period and time steps down to $\frac{1}{4}$ hour are used where the processes to be simulated vary rapidly during the day itself. If more frequent data is available then this should be used.

The daily surface heat inputs are used by the model to calculate the temperature and salinity changes in the slab structure. The updated slab structure is then adjusted for mixed layer deepening and possible changes of the thermocline thickness due to the formation of shear instabilities. The calculation of the mixed layer deepening incorporates both deepening due to surface turbulence and turbulence generated by the shear instabilities at the base of the mixed layer.

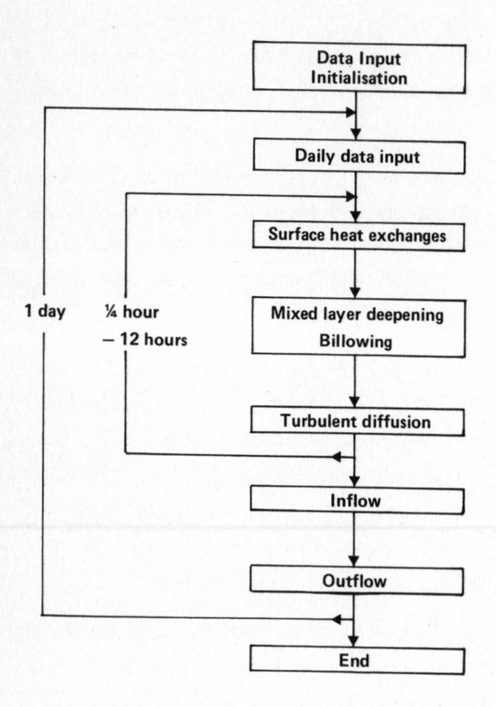

Fig. 4. Schematic of the programmed model DYRESM.

Once the new thermocline depth and thickness has been computed the model then calculates the net heat and salt transport from the bottom through the hypolimnion into the epilimnion. The basic algorithm by which this is carried out is an eddy diffusion parameterization. However, it was recognized very early in the development of DYRESM (see Imberger et al., 1978) that a constant diffusion coefficient was not applicable. Field experiments showed that the vertical transfer of heat and mass was higher during periods of high wind and inflow even though both of these processes did not directly contribute to the turbulence in the hypolimnion. It was postulated that both these disturbances induce basin scale internal oscillations which are damped at the boundary. The dissipation at the boundary in turn produces small scale turbulence capable of causing mixing. The active mixing in the boundary layer quickly leads to an adjustment of densities in the boundary layer and so induces horizontal intrusions which in turn cause an adjustment of the centre of gravity of the overall lake structure.

This concept has since then been postulated for the ocean (Armi, 1978) and recent experiments by Ivey (1980) have established rates of transfer in an idealized laboratory experiment. While the concept appears attractive, not enough is understood about the details of the individual processes to allow a process model to be constructed. Instead, a simple energy argument is used (Imberger et al., 1978).

It is assumed that a small, but constant fraction of the wind and inflow energies is used to generate mixing in the hypolimnion. This leads to a vertical diffusion coefficient :

$$\varepsilon_z = 0.048 \frac{H^2}{ST_M} , \tag{3}$$

where T_M is a time scale for mixing(equal to the potential energy of the density stratification of the lake divided by the power input of the wind and streams); and S is a normalized water column stability (=$H/\Delta\rho \; d\rho/dz$, where H is the depth of the lake; $\Delta\rho$ is the density difference between the bottom and the surface; and ρ is the density of the water at an elevation z).

This parameterization is successful (see Imberger and Patterson, 1980) provided the stability S is quite large. It obviously breaks down in the limit of a homogeneous water mass with an energy input. In this limit both S and T_M approach zero and equation (3) predicts an infinite diffusion coefficient.

In order to prevent this, an arbitrary cut off to equation (3) of $\varepsilon_z = 10^{-4} m^2 s^{-1}$ is provided in the model. Further work is required to rationalize this cut off especially since its value has quite a strong influence on the predicted salinity distribution. The mixing processes in the hypolimnion seem to require a two parameter model, equation (3) for strongly stratified water and an analogous formulation for a homogeneous hypolimnion where the mixing may be expected to be more evenly distributed throughout the water mass.

At the end of the diffusion routine, which is carried on the same time step as the mixed layer dynamics calculation, a new structure for a particular day has been calculated. This density structure is then used to route the inflowing water from the various contributing streams into the reservoir. The subroutine allows for turbulent entrainment and subsurface intrusions. Similarly, the outflow is calculated by the model using the structure left after the inflow has been added.

This routine is repeated for each day of the simulation.

4 MODEL PERFORMANCE

DYRESM has been tested on a number of different lakes, but its major evaluation and development has taken place with data from the Wellington Reservoir.

The seasonal variability of the various inputs to the reservoir over the period from the Julian day 133 in 1975 to day 365 in 1977 are shown in Fig.5. Depicted are the wind speeds, the short-wave solar radiation as computed from cloud cover records, the salinity and temperature of the inflowing water and the total rate of inflow from the Collie River which contributes approximately 85% of the total inflow. The remaining inflow is included in the simulations, but is not shown in Fig. 5.

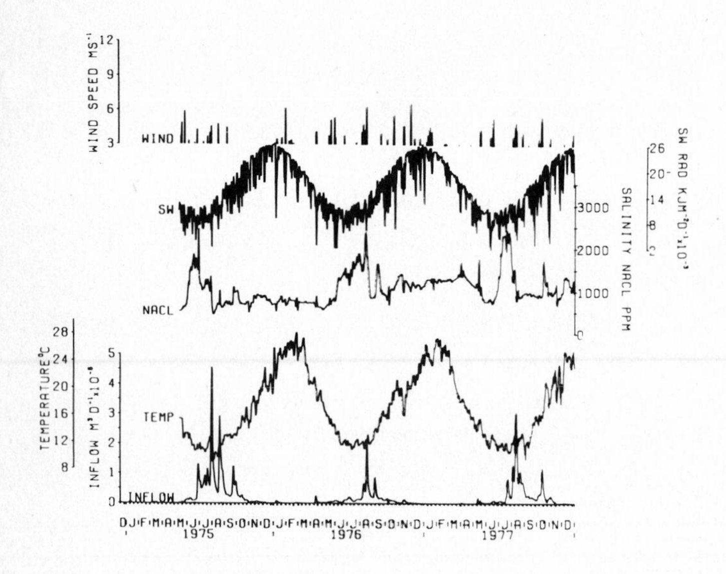

Fig. 5. Seasonal variability of wind speed, short-wave radiation, inflow salinity of the Collie River, inflow temperature and the inflow volume.

Figs. 6(a) and 6(b) show the field temperature and salinity structures length averaged along the Collie River Valley as a function of time over the period January 1975 to August 1978. The salinity data gathered between October 1977 and June 1978 is regarded as unreliable and is not shown. The broken lines in Fig. 6(a) indicate that no data was taken in the period covered and the thermal structure is interpreted from the structure before and after the period of interruption.

The yearly cycle is clearly evident in Figs. 5 and 6: the cold salty inflows lodge in the base of the homogeneous reservoir in the months of June, July and August; summer stratification builds up until December, when surface winds begin to mix the surface layers and a thermocline forms, protecting the waters below. In early winter, the air temperature falls and the winds increase, with the result that the reservoir is completely mixed before the following inflows arrive.

The marked difference in the thermocline structure between 1976 and 1977 was caused by a change in the withdrawal policy. In 1976 all the water was withdrawn from the offtakes at 15 m height, whereas in 1977 a large quantity of salty water was scoured through the offtake at the bottom of the dam wall. In addition about two thirds of the water for irrigation was taken from the bottom offtake.

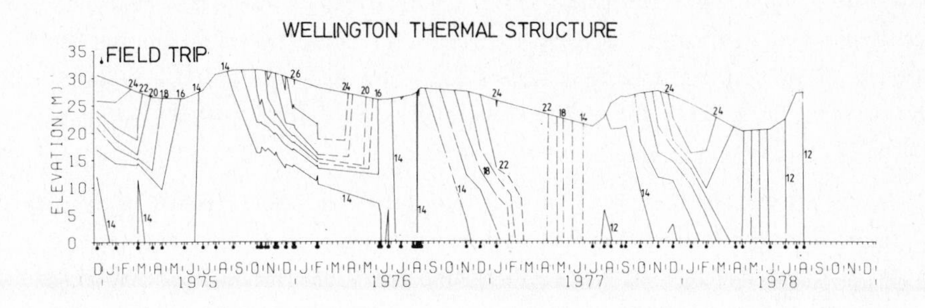

Fig. 6(a). Measured average reservoir temperature as a function of time for the years 1975 to 1978. (From Imberger and Patterson, 1980).

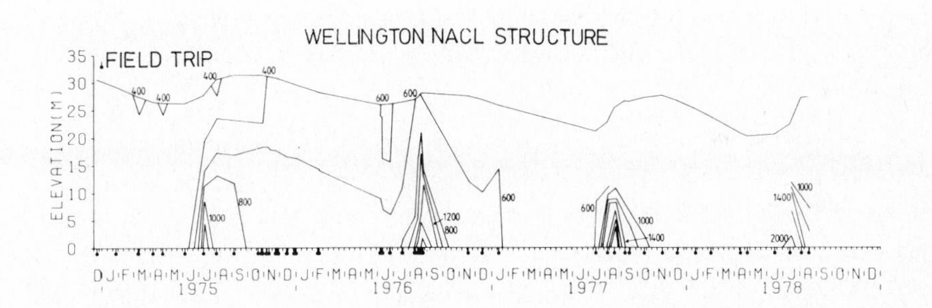

Fig. 6(b). Measured average reservoir salinity (mg L^{-1} NaCl) as a function of time for the years 1975 to 1978. (From Imberger and Patterson, 1980).

It is also clear from Fig. 6(a) that the temperature regime of the reservoir is determined by the inflows and the surface heating and cooling. The bottom temperature of the reservoir for most of the year is determined by the temperature of the coldest inflows, whereas the surface temperature is determined by the meteorological forcing.

There are seven constants which must be specified by the user before applying DYRESM. Of these only one is truly adjustable - the others are related to well identified physical processes and are determined from experimental or field data.

The constants are described below, together with experimentally determined values.

(i) C_D is the drag coefficient for inflowing streams. C_D was determined independently of DYRESM in a field study described by Hebbert et al. (1979). The value determined in that study, $C_D = 0.015$, is used here.

(ii) η_1 is an extinction coefficient for short-wave solar radiation penetrating the water. It relates the solar radiation received at the water surface, to that penetrating to a depth z. A single exponential decay formula was used as only limited field measurements were available. An average value of $\eta_1 = 0.35$ is taken, based on the fact that the Wellington is fairly clear in the summer months when surface heating is an important effect.

(iii) α_1 is a constant occurring in the expression for the diffusivity calculated for the deep hypolimnetic mixing. It basically represents the efficiency with which the power input from the surface wind and river inflows is converted to a gain in potential energy of the lake water due to vertical mixing. A 9.6% efficiency ($\alpha_1 = 0.048$) determined in earlier calibrations of DYRESM over the 100-day period from day 133 to 233 has proven satisfactory throughout. The cut-off value of $10^{-4} m^2 sec^{-1}$ was set by noting that this corresponds to the maximum value measured by a number of investigators (see Fischer et al., 1979).

(iv) C_K is the coefficient that describes the stirring efficiency of convective overturn. Experimental results summarized by Fischer et al. (1979) suggest an average value of $C_K = 0.125$.

(v) η, in combination with C_K as $C_K\eta^3$, is a coefficient measuring the stirring efficiency of the wind. The value given by Wu (1973) is adopted here, since it was shown by Spigel (1978) that in his experiments stirring effects dominated the entrainment, so that shear production, temporal effects and internal wave radiation losses were negligible. Wu's deepening law is $dh/dt = 0.23\ u_*/Ri^*$ giving $C_K\eta^3 = 0.23$, and thus $\eta = 1.23$.

(vi) C_S is the coefficient that describes the efficiency of shear production for entrainment. Values ranging from 0.2 to 1 are reported by Sherman et al. (1978). $C_S = 0.2$ was chosen as representing a good estimate for most experimental results. A value of 0.5 was used before the energy released by billowing was separately included as is done in version 5 of DYRESM.

(vii) C_T is associated with temporal, or unsteady, non-equilibrium effects due to changes in surface wind stress or surface cooling. C_T is constructed by the requirement that for a turbulent front entraining into a homogeneous fluid, $dh/dt = 0.3u_*$ (Zeman and Tennekes, 1977)giving a value of $C_T = C_K\eta/0.3 = 0.510$.

These coefficient values were used for a 962-day simulation of the lake dynamics starting with initial profile data on day 75133. The results from this simulation are shown in Fig. 7.

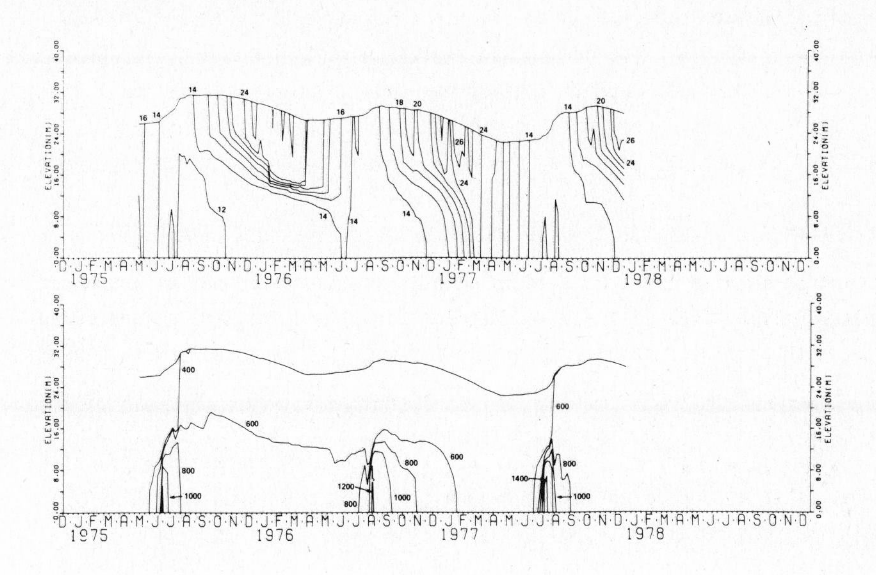

Fig. 7. Simulated average temperature and salinity for the period 1975 to 1978. (From Imberger and Patterson, 1980).

Comparison of Figs. 6(a) and 7 shows that the thermal structure is reproduced extremely well with all the stratifying and mixing regimes occurring at the correct times and of the correct magnitude. The slug of 12°C water predicted by the model in August to September 1975 does not appear in the field data. This slug is derived from the previous inflow and its presence may be due to errors in the inflow temperature data. In any case, the predicted temperature gradient in the bottom at this time is small, and the actual difference between field and predicted temperatures is less than 1°C.

The predicted and observed salinity variations also compare well with the most significant anomaly being the mismatch of the 600 mg L^{-1} line at the period of maximum inflow in 1976. The field data suggest somewhat more energetic surface mixing which may be due to small errors in the wind data. Additionally, the model does not reproduce the peak salinities of the inflows. This is due to the construction of the model layers in terms of layer volumes - hence the depth resolution in the upper part. The bottom three or four metres always appear as mixed, and any salinity slug occupying less than this will be mixed with the layer above. Thus there is no error in terms of the salt load, but one of display of the structure, and it was accepted in order to save computer time. If greater resolution were required, this could be achieved by specifying smaller minimum slab sizes.

Overall, DYRESM appears faithfully to reproduce even very severe changes in the reservoir structure caused by such diverse forcing as large saline inflows active scouring, strong wind deepening and winter convective cooling. Perhaps more importantly, the model correctly simulates two independent parameters, salinity and temperature to a resolution equal to that of the field programme. More accurate field data is required if more stringent tests are to be applied.

5 INFLUENCE OF SALINITY CHANGES AND POSSIBLE ALLEVIATION STRATEGIES

The simulation model DYRESM was used to investigate the influence of increased and decreased Collie River salinity and methods for alleviating the strong build-up of reservoir salinity for the case of increased Collie River salinity. The technique employed was to run the simulation for the three-year period using the measured inflow, outflow and meteorological data. This three-year period (1975 to 1978) represents a start of a severe drought period with the associated build-up of stream salinity and of course decreased flow. As mentioned in Section 2 it was felt that manipulating the magnitude of the salinity of this time series would be more rational than generating a model set since this data set preserves the phase relationship between the salinity and flow variations. The model then simulates the subtle interactions between the salinity structure, the outflow, mixing and inflow and allowed an assessment of the influence of any proposed changes. Obviously, many variations could be simulated but because of the cost involved only a set of extreme conditions will be described. It is felt that these results allow an assessment of the likely changes which may be expected.

5.1 Variations of Inflow Salinity

The simulation of the actual field data has already been discussed in Section 4. Two further cases were run. First, the Collie River salinity was decreased to that given by equation (2), the scour valve was turned off and flow with a salinity greater than 1300 mg L^{-1} NaCl was by-passed. The latter two strategies will be explained later, but they were incorporated in the one run in order to save computer time. The results are shown in Table 1 where it is shown that the average salinity is lowered from 650 mg L^{-1} to 480 mg L^{-1} and the dam over-flows. It must be remembered that leaving the bottom scour valve closed (see Fischer et al., 1979) has no effect in 1975/76 as it was essentially closed under normal operation. However, in 1976/77 much of the irrigation water (about two thirds) was withdrawn through the scour valve and keeping the valve closed thus leaves a surplus of water in the reservoir. However, the reduction of salinity is mainly due to the reduced inflow salinity as essentially no water was by-passed.

(salinity would have to be in excess of 2400 mg L^{-1}) and the scour valve salinity was always close to the average.

TABLE 1

Salinity of Inflow	Scour Valve Flow	Top Draw Flow	By-pass Flow	Day of Year							
				75133		76133		77133		77365	
				Vol*	Sal**	Vol	Sal	Vol	Sal	Vol	Sal
As measured	As operated	As operated	zero	11.4	308	12.1	450	8.1	530	13.0	650
Decreased	zero	As operated	Whenever S > 1300	11.2	308	12.0	354	14.6	413	Overflow	480
Increased	As operated	As operated	zero	11.2	308	12.0	804	7.6	956	14.3	1345
As measured	zero	As operated	zero	11.2	308	12.0	464	14.7	584	Overflow	641
Increased	zero	As operated	zero	11.2	308	12.4	810	14.7	1320	Overflow	1648
As measured	As operated	As operated	Whenever S > 1300	11.2	308	12.0	450	6.3	438	3.5	480
As measured	zero	As operated	Whenever S > 1300	11.2	308	12.2	430	12.9	444	Overflow	510
Increased	zero	As operated	Whenever S > 1300	11.2	308	9.7	371	8.9	357	11.4	367

* Volumes are in $10^7 m^3$

** Salinities are in p.p.m. NaCL (average for whole volume).

Second, the salinity of the Collie River inflow was increased as indicated by equation (1) and the reservoir was operated as was actually the case in the three-year period. The increase in the average reservoir salinity was quite dramatic reaching 1345 mg L^{-1} at the end of the simulation period. Such salinity levels would be quite unacceptable for all but a few selected crops. Further, the strategy of taking irrigation water from the scour valve becomes no longer possible as the salinities would rise to over 2000 mg L^{-1}. However, it does prevent an even greater increase of salinity buildup than would occur if all the irrigation flow were to be taken from the top offtake.

5.2 Variations of the Scour Valve Flow.

The effect of the scour valve flow is quite subtle and can easily be misleading. Table 1 shows the results from a simulation with inflow salinities, for the 1975 to 1978 period, but with the bottom scour valve completely closed during the whole period. As already mentioned this led to a strong reduction in the water available for irrigation in 1977/78 with a consequential overflowing of the reservoir. During the period 76133 to 77133 the scour policy had the obvious effect of reducing the average salinity from 584 to 530 a small but definite benefit. However, during the period from 77133 to 77365 scour policy actually had a detrimental influence since the water withdrawn in 1976 through the scour valve was of a lower salinity than the average salinity in this latter period. Had this water been wasted then the strategy would have had an adverse effect. It is therefore seen that a judgement must always be made whether the present year's poor quality water is likely to be of better quality than the best quality of the following year.

The reverse effect is noticed in the increased salinity case (see Table 1). Closing the bottom scour valve once again led to a full reservoir, but with an average salinity as high as 1648 mg L^{-1} at the end of the simulation period. Hence, little benefit would be gained by reducing the irrigation supply in anticipation of a better following year. Under such circumstances it would be better to irrigate generously in 1976 and bank one's profits in order to overcome the hardship of the following very salty inflows.

Fig. 8 shows the isohalines for this very extreme strategy. The higher salinities are reflected in the very much intensified structure, the increased peak salinities and the strong vertical salinity (and thus density) gradients. These gradients prevented mixing, leading one to suspect much increased efficiency of the scour policy (not shown in Table 1).

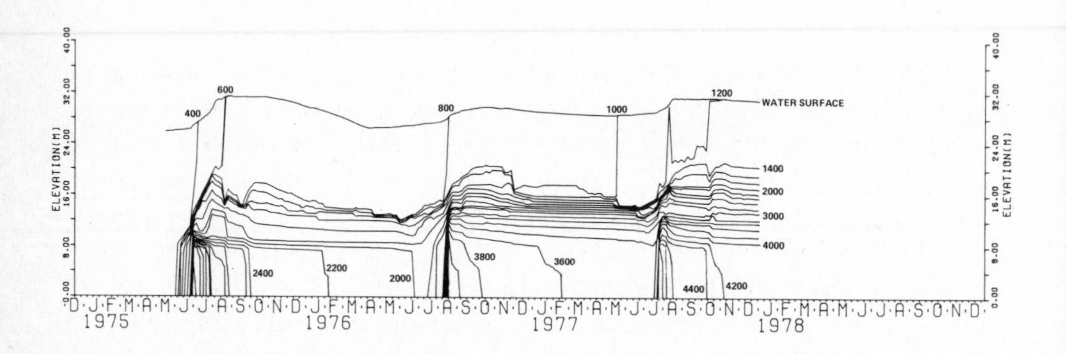

Fig. 8. Simulated salinity variations in 1975 to 1978 for increased streamflow salinity, no by-pass and no scour valve operation.

5.3 The Effect of By-passing High Salinity Water.

The full assessment of the benefits to be gained by by-passing the very high saline slug of Collie River water would involve a great many simulations. In this paper a cut off salinity of 1300 mg L^{-1} has arbitrarily been chosen to represent an acceptable upper limit. Introducing such a strategy of by-pass in addition to the scour practice of 1975 to 1978 reduced the average salinity from 650 mg L^{-1} to 480 mg L^{-1}, but the extra $1.8 \times 10^7 m^3$ water scoured would have reduced the water volume to a dangerously low level of $3.5 \times 10^7 m^3$ at the end of the simulation period and at the start of the 1978 irrigation season.

For this reason the by-pass policy was combined with a major reduction in the 1976 irrigation volume. This was achieved by closing the bottom scour valve cutting the irrigation flow in 1976/77 to about one third that actually delivered. The average salinity increased marginally, but the reservoir filled by the end of 1977. Obviously, a compromise between these two latter approaches would be most advantageous; a certain by-pass flow should be compensated for by a corresponding reduction in the irrigation volume.

The by-pass strategy was also evaluated for the case of increased salinities (see Table 1). In anticipation of a low volume it was decided again to close the scour valve. This resulted in a dramatic reduction in salinity from 1345 mg L^{-1} (1648 mg L^{-1} corresponding no scour case) to a very modest value of 367 mg L^{-1} at the end of the simulation period. The price paid for this reduction in salinity was a decrease in irrigation and an only partially full reservoir.

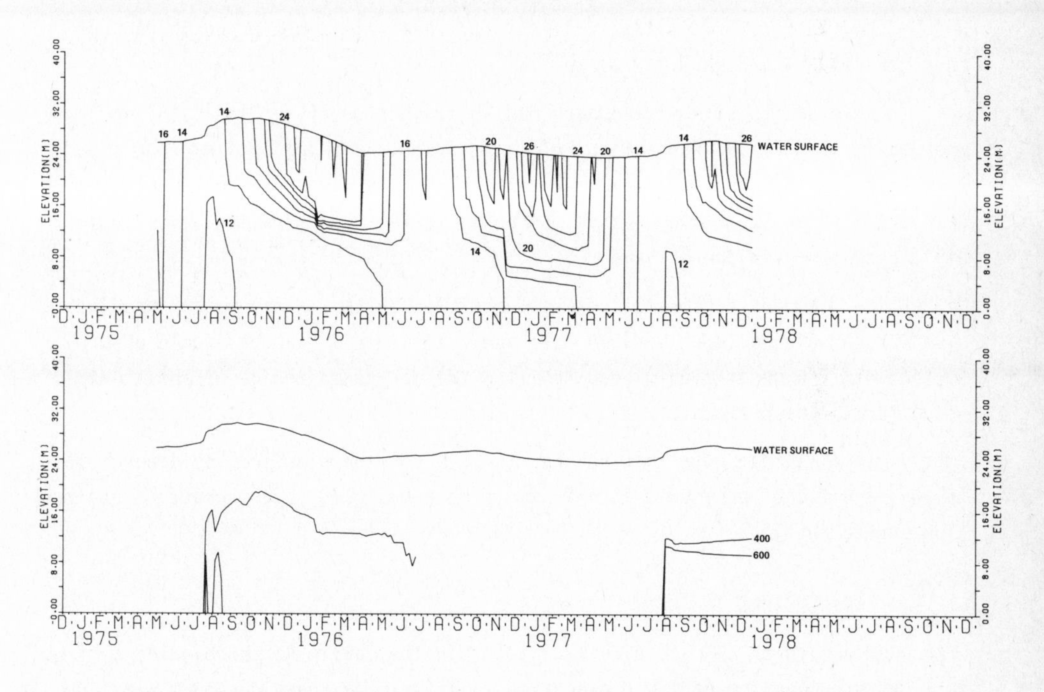

Fig. 9. Simulated temperature and salinity variation in 1975 to 1978 for increased streamflow salinity, by-pass for salinities greater than 1300 mg L^{-1} and scour valve operating as recorded.

The effectiveness of by-pass strategy depends greatly on variations of the Collie River salinity and most certainly equation (1) accentuates the benefits accrued by the by-pass strategy.

Fig. 9 shows the temperature and salinity structure resulting from such a policy. The reservoir still stratifies due to temperature gradients, but only weak salinity gradients remain.

5.4 Implementation of a By-pass Policy.

It was seen in Section 5.3 that a wise balance between reduction in irrigation and by-passing highly saline water could lead to very marked reduction in average reservoir salinities. The purpose of this paper was to demonstrate that such a flexibility may exist and should therefore be explored fully. However, an initial suggestion for a feasible diversion dam able to handle the by-pass water will now be made.

The choice of a cut off salinity of 1300 mg L^{-1} requires quite a large quantity of water to be by-passed yet the diversion dam should be as far upstream as possible in order to avoid contamination of the main reservoir. The location of the township of Collie further requires that the upstream level should not be raised above the existing high water marker.

It is proposed to site a secondary dam at station C85 (see Figs. 1(b) and (c)) where the river valley has a deep hole. This would require a 14 km long pipeline lying in the bed of the old river channel. The deep hole would allow a deep offtake making maximum use of the possible vertical stratification to inhibit drawdown over the offtake.

The pipe diameter required to prevent overflow of the saline water into the parent reservoir would be between 1.75 and 2 m. Such a diameter would ensure sufficient flow under a 25 m head to allow the upstream storage of about $4 \times 10^7 m^3$ to buffer the peak flows.

It is important to note that neither the dam nor the pipeline would be exposed to any significant pressures as both would be submerged. Light plastic construction should suffice.

6 CONCLUSIONS

An examination of the salinity variations in the Collie River revealed a structure dependent on the peak and base flow salinity, the phase during storms and the flushing time of the whole system. This variation was exploited in order to modify an existing salinity record to simulate possible future trends. It was shown that the prediction of the exact nature of the salinity variations is not important as the mixing in the reservoir acts as a strong filter. However, the overall sequences of the salinity variations are important. The simulation results revealed that without drastic action the salinity in the Wellington will soon become unacceptable. A by-pass strategy is suggested which could lead to dramatic reductions in salinity, but at the expense of a reduction in water for irrigation. This conclusion confirms the initial findings detailed in Fischer et al. (1979).

7 REFERENCES

Armi, L., 1978. Some evidence for boundary mixing in the deep ocean. J. Geophys. Res., 83: 1971-9.

Fischer, H.B., List, E.J., Koh, R.Y.C., Imberger, J. and Brooks, N.H., 1979. Mixing in inland and coastal waters. Academic Press, New York, 483 pp.

Hebbert, B., Imberger, J., Loh, I. and Patterson, J., 1979. Collie River underflow into the Wellington Reservoir. J. Hydraulics Div. ASCE, 105, No. HY5, 533-45.

Imberger, J., Thompson, R. and Fandry, C., 1976. Selective withdrawal from a finite rectangular tank. J. Fluid Mech., 78: 389-512.

Imberger, J., Patterson, J., Hebbert, B. and Loh, I., 1978. Dynamics of reservoir of medium size. J. Hydraulics Div. ASCE. 104, No. HY5, 725-43.

Imberger, J. and Patterson, J.C., 1980. A dynamic reservoir simulation model - DYRESM 5. Proc. Symp. on Predictive Abilities of Surface Water Flow and Transport Models, Berkeley, August, 1980, 75 pp.

Ivey, G., 1980. Boundary mixing in a stratified fluid in a rectangular tank. PhD Thesis, Univ. of California, Dept. of Civil Engineering, Berkeley.

Johnston, C.D., McArthur, W.M. and Peck, A.J., 1980. Distribution of soluble salts in soils of the Manjimup Woodchip Licence Area, Western Australia. CSIRO Aust. Div. Land Resources Manage. Tech. Pap. No. 5, pp. 1-29.

Loh, I., and Porter, J., 1976. Simulation of monthly flow and salt load inputs to Wellington Reservoir. Tech. Report No. 65. Water Resources Sec. Planning, Design and Investigation Branch, Public Works Department, Western Australia.

Loh, I.C. and Hewer, R.A., 1977. Salinity and flow simulation of a catchment reservoir system. Proc. Hydrol. Symp., Instn. Engrs., Aust., Brisbane.

Peck, A.J. and Hurle, D.H., 1973. Chloride balance of some farmed and forested catchments in south-west Australia. Water Resour. Res., 9: 648-57.

Peck, A.J., Hewer, R.A. and Slessar, G.C., 1977. Simulation of the effects of bauxite mining and dieback disease on river salinity. Tech. Rep. No. 5, CSIRO Aust. Div. Land Resources Manage.

Sherman, F.S., Imberger, J. and Corcos, G.M., 1978. Turbulence and mixing in stably stratified waters. Ann. Rev. Fluid Mech. 10: 267-288.

Smith, R.E. and Hebbert, R.H.B., 1980. Manuscript in preparation.

Spigel, R.H., 1978. Wind mixing in lakes. Ph.D. Thesis, Univ. of California, Berkeley.

Thompson, R.O.R.Y. and Imberger, J., 1980. Response of a numerical model of a stratified lake to wind stress. Proc. 2nd. Int. Symp. on Stratified Flows, Trondheim, June 1980.

Wood, W.E., 1924. Increase of salt in soil and streams following the destruction of the native vegetation. J. Roy. Soc. West. Aust. 10, 35-47.

Wu, J., 1973. Wind induced entrainment across a stable density interface. J. Fluid Mech. 61: 275-78.

Zeman, O. and Tennekes, H., 1977. Parameterisation of the turbulent energy budget at the top of the daytime atmospheric boundary layer. J. Atmos. Sci., 34: 111-23.

IMPACT OF WATER RESOURCE DEVELOPMENT ON SALINIZATION OF SEMI-ARID LANDS.

G.T. ORLOB
School of Civil Engineering,
University of California, Davis, Calif., U.S.A.

A. GHORBANZADEH
California Department of Water Resources,
Sacramento, Calif., U.S.A.

ABSTRACT

Orlob, G.T. and Ghorbanzadeh, A., 1981. Impact of water resource development on salinization of semi-arid lands. Agric. Water Manage., 1981.

Historical development of the water resources of California's San Joaquin Valley is described. Impacts of increased consumptive use of water and reallocation of available resources within the valley include reduction of flows in the lower reaches of the river system, and progressive deterioration of water quality. Salt accretions associated with the development of saline lands can be reduced by installation of drainage facilities. Preliminary assessments of the efficacy of tile drains using two-dimensional finite element models are presented.

1 INTRODUCTION

The San Joaquin Valley of California, shown in Fig. 1, is one of the most productive agricultural areas of the world. It has been subject to intensive development over the past 50 years, during which period all of its major rivers, with a combined natural runoff of more than 7×10^9 m^3 yr^{-1}, have been regulated for power production, water supply, flood control and irrigation. Consumptive use of water, principally by irrigated agriculture, has increased steadily, with the most dramatic changes occurring with the advent of the Central Valley Project (CVP) in the 1940's. This project provided for impoundment and extra-basin diversion of the major portion of the natural runoff of the Upper San Joaquin River. Water diverted to the Tulare Basin to the south was replaced, in part, by importation from the Sacramento - San Joaquin Delta through the Delta Mendota Canal.

The canal has a capability of supplying about 1.2×10^9 m^3 yr^{-1}, most of which has been allocated to development of lands along the semi-arid western side of the valley. These lands are generally saline, due to their origin as marine sediments and because they are situated in the rain shadow of California's coastal range where potential evaporation far exceeds annual precipitation.

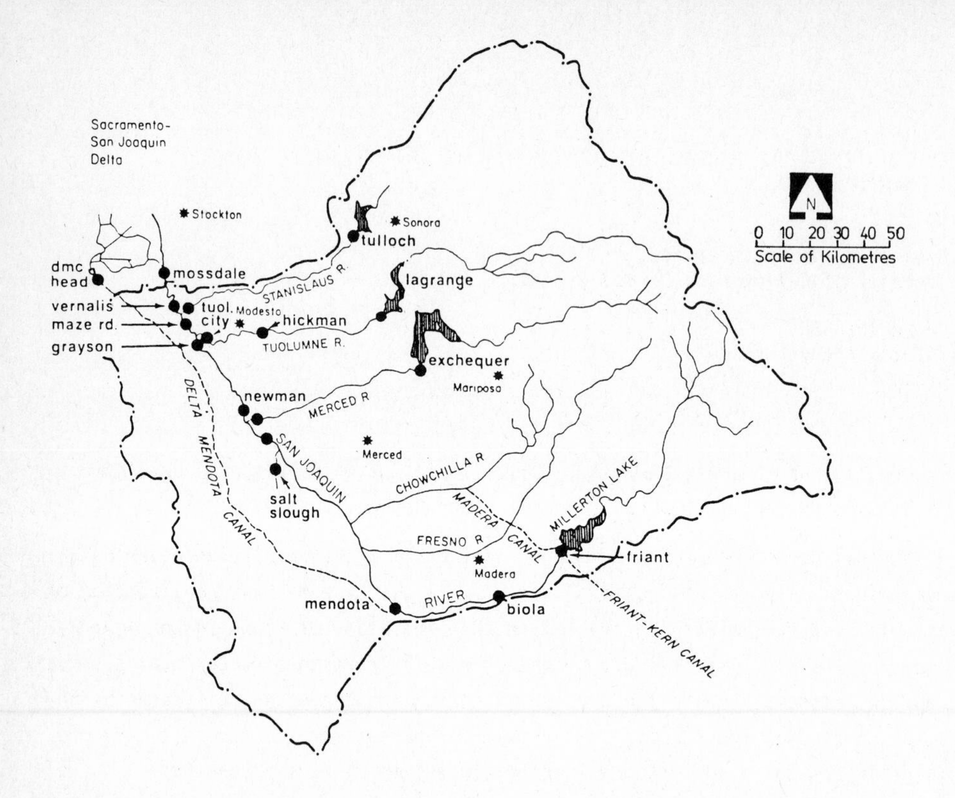

Fig. 1. San Joaquin Basin, California.

As a result of exploitation of the Upper San Joaquin River and the subsequent development of saline lands along the valley's western side, the main river system downstream of Mendota (see Fig. 1) has been deprived of the normal runoff which it received naturally, prior to the CVP and at the same it has inadvertently become the principal agricultural drainage course of the valley. The progressive diminution of runoff and deterioration in quality has seriously impacted agriculture in the northern portion of the valley, in particular, in the southern section of the Sacramento - San Joaquin Delta for which the San Joaquin River is the unique water supply.

2 CHANGES IN RUNOFF

2.1 Trends Due to Water Development

Changes that have occurred in the available water supply at the lower extremity of the San Joaquin Valley over the period 1930 through 1977 are depicted by the double mass diagram in Fig. 2. The figure compares the actual recorded cumulative annual runoff for the chronological period with the corresponding natural, or "unimpaired", runoff that occurs at the valley rim, above agricultural service areas.

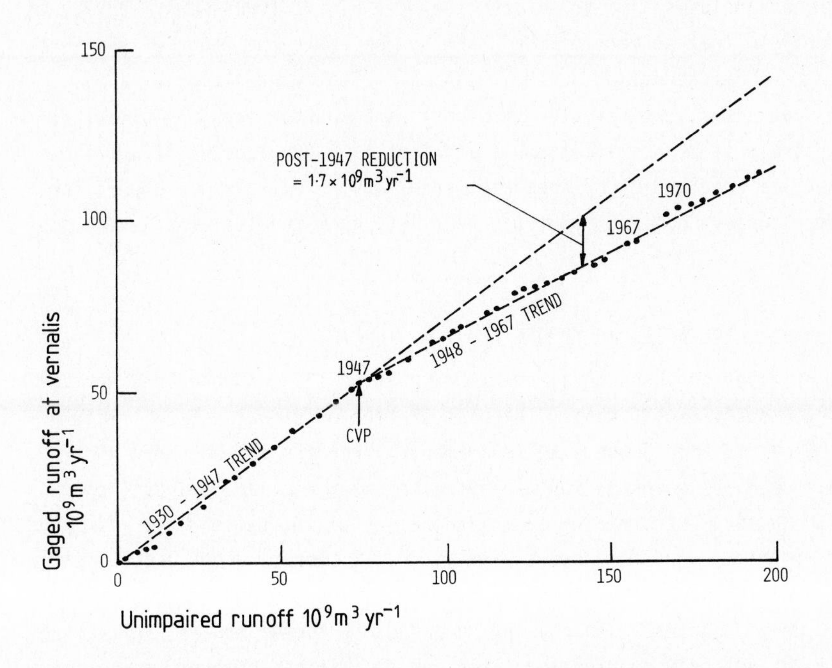

Fig. 2. Double mass diagram of runoff, San Joaquin River at Vernalis, 1930 to 1977.

It is noted that, except for minor hydrologic variations from year to year, the proportion of actual net runoff to unimpaired runoff - a ratio reflecting the depletion of available runoff by all downstream consumptive uses - remained more or less constant in the period 1930 through the mid-40's. However, subsequent to about 1947, the year when the Friant Reservoir on the San Joaquin River first came "on line", this proportion, indicated by a change in slope of the double mass plot, changed markedly. The difference in average slope between the segments 1930 to 1947 and 1948 to 1967* accounts for a net reduction in Vernalis flow of about 1.7 x 10^9 m^3 yr^{-1}. Adjusting for the slightly greater unimpaired runoff contributed during the earlier period (about 6%), the net reduction in flow due to upstream development is estimated to be about 1.6 x 10^9 m^3 yr^{-1}. Of this reduction about 70% occurs during the 6 month period, April through September, the principal agricultural season in the valley.

The implication of the mass curve analysis is that major water resource development, subsequent to the late 1940's, accounted for substantial reductions in

*Another major project initiated operation in 1967.

San Joaquin Valley runoff. Of course, not all of this occurred as a direct result of impoundment and exportation of the Upper San Joaquin River runoffs by the CVP. However, an analysis similar to that shown in Fig. 2 (WPRS-SDWA, 1980), but applied only to that segment of the river system above the mouth of the Merced River, indicated a net reduction during the April - September period of about 0.55×10^9 m^3 due almost entirely to extra-basin transport. The depletion that occurred downstream from this location is attributed to development of new lands in the CVP service area and other irrigated areas and to extra-basin export for municipal water supply, e.g., City of San Francisco from the Tuolumne River (see Fig. 1).

2.2 Seasonal Variation in Runoff

The impact of water resource development upstream on the San Joaquin River system is also to modify significantly the distribution of annual runoff. This effect may be seen in Fig. 3 where the mean monthly runoffs for each month of the hydrologic year are compared for two hydrologically similar periods*, one of which, 1930 to 1944, occurred prior to initiation of the CVP and the other, 1952 to 1966, subsequently. In this figure, the typical snowmelt runoff pattern of California's Sierra range is evident in the 1930 to 1944 period, the major watershed yield occurring naturally in the spring and early summer months. After the implementation of upstream storage and water use facilities, however, the pattern is substantially modified; the snowmelt runoff is captured, transferred out-of-basin or diverted to consumptive use. Flows have been generally reduced in the river system during the spring, summer and early fall periods. Overall, a net reduction of about 1.6×10^9 m^3 yr^{-1} occurred. It is significant to note that the greatest proportionate reduction, pre- to post-project, occurred in the month of July, a factor of specific concern in relation to water quality management.

3 CHANGES IN QUALITY

3.1 Trends Due to Water Development

Changes in water quality that have occurred over the 50-year period since 1930 are attributable to many factors, the most prominent of which appear to be reduction in diluting flows of natural runoff and the increase in salt accretions to the natural drainage courses due to irrigation drainage. In the latter instance, the effects are compounded beyond the normal problems of irrigation tail water drainage by development and irrigation of new lands, large acreages of which, in

---

*Runoffs were adjusted for difference in average annual unimpaired flows above water use areas; the earlier period experienced about 6% greater natural inflow.

the case of the San Joaquin Valley, are of historically saline soils. These are located largely on the western side of the valley within the service area of the Delta-Mendota Canal.

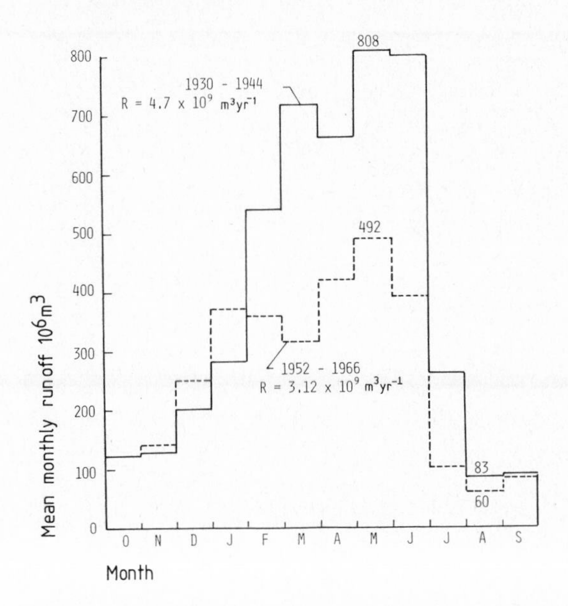

Fig. 3. Seasonal runoff of the San Joaquin River at Vernalis for two similar hydrological periods, before and after start-up of the Central Valley Project. The runoffs were adjusted for a 6% difference in unimpaired flow above water use areas.

The historic changes that have taken place in the quality of water at Vernalis at the lower extremity of the valley are exemplified by the seasonal variations in mean monthly total dissolved solids, for each decade since the 1930's, as illustrated in Fig. 4. It is observed that an upward trend occurred between the 1940's and 1950's, especially in the spring and summer months. This trend accelerated subsequently so that by the 1960's average TDS levels at Vernalis were more than double those of the pre-project period, 1930 through 1949. A marked increase occurred in even the winter and early spring months, a fact that suggests an imbalance in salt in the system, the salts accumulating in the soil profile during the irrigation season being returned to the river during the succeeding periods of higher runoff. This pattern apparently has persisted into the 1970's, with some further exacerbation in the winter and spring periods.

3.2 Extreme Values

It is obvious that while the trend of mean values is indicative of the direction and rate of change in quality, it does not characterize the seriousness of

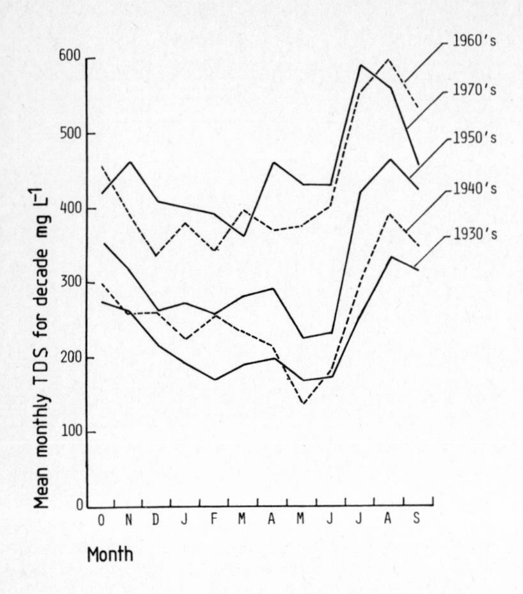

Fig. 4. Mean monthly TDS at Vernalis by decades, 1930 to 1979.

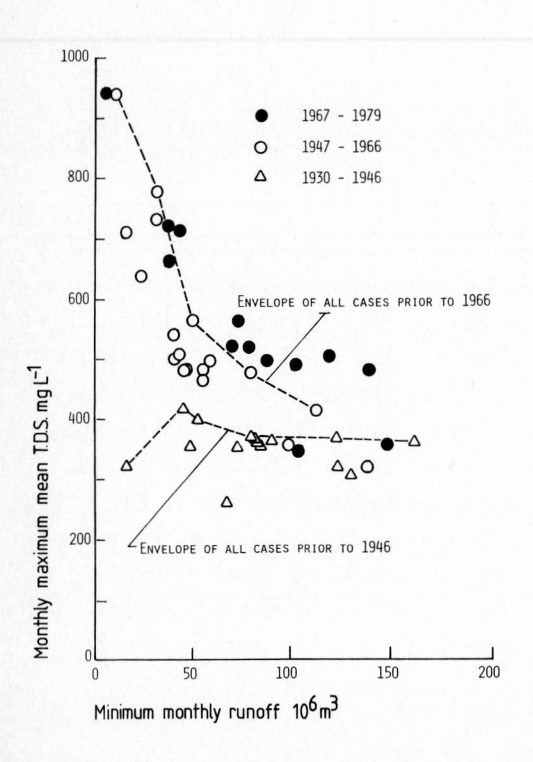

Fig. 5. Water quality and flow extremes at Vernalis, 1930 to 1979.

of degradation from the agriculturist's viewpoint. This is illustrated more dramatically by extreme mean monthly values of TDS in relation to actual runoff at Vernalis, as shown in Fig. 5. In this example, the maximum value of the monthly mean TDS for each year of record is plotted against the corresponding monthly runoff. The result is a reflection not only of the TDS extremes that could maximize impact on agricultural production, but also of the changes that have occurred historically. Two envelope curves are shown, one for conditions prior to 1946, i.e., prior to the CVP, and another for conditions prior to 1966, i.e., including the influence of the CVP but excluding subsequent water resource development.

The generally high quality of water available during the pre-1946 period is contrasted to the degradation exhibited in later years, even at the same levels of runoff. Also, general chronological ordering of individual data points in the region above the lower envelope suggests, once again, the progressive deterioration of conditions in the lower San Joaquin basin. It is noted that in the 1970's the maximum TDS values tend to be appreciably higher in the years of greater runoff, say those with a mean monthly runoff of greater than about 70 x 10^6 m^3. This is taken as further evidence of release of salts from storage in years succeeding very dry years, with the further implication of a progressive accumulation of salt within the basin.

4 IMPACT OF WATER RESOURCE DEVELOPMENT

The combined impact of reduction in runoff in the San Joaquin River and the gradual deterioration in quality at Vernalis is reflected in the ratio of TDS concentration and runoff, an impact factor defined as:

$$IF = \frac{\bar{C}}{\bar{Q}} \qquad (1)$$

where $\bar{C}$ = mean concentration of total dissolved solids, mg L^{-1}, and $\bar{Q}$ = mean runoff, m^3 x 10^6.

For purposes of indicating the relative changes in both quality and runoff in pre- and post-project periods, values of IF for each month for the decades of the 1950's, 1960's and 1970's are normalized against those for the period 1930 to 1949 as depicted in Fig. 6. Here the consequences of upstream development are more dramatically evident. The relative combined impact rises rapidly during the post-project years, especially in the early part of the irrigation season, April through July, where by the 1970's the impact reached a level about 9 times that of the pre-project period, 1930 to 1949. It will be recalled that this is the period most affected by reductions in flow (see Fig. 3), but also, perhaps more significantly, it is the period when crops are likely to be in the most sensitive stages of growth and most susceptible to damage due to insufficiencies in supply or poor quality.

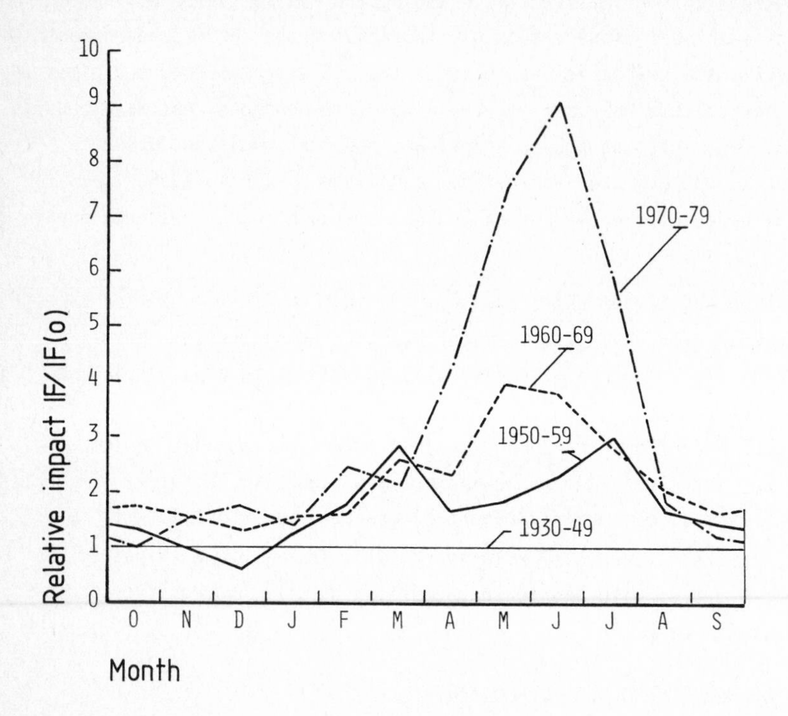

Fig. 6. Impact of upstream development relative to mean of period 1930 to 1949, San Joaquin River near Vernalis.

5 IDENTIFYING SOURCES OF SALT ACCRETION

5.1 Quality Variations Along the River System

To identify the major sources of salt accretion to the main stem of the San Joaquin River System, it was necessary to examine the spatial variations in quality, i.e., to obtain a synoptic view of the river's quality. This was accomplished by means of complete chemical analysis of samples collected at various key sampling locations (see Fig. 1) during periods of maximal quality degradation, i.e., during the mid-irrigation season of dry or below normal years. Fig. 7 is an example of this diagnostic method, showing the distribution of principal anions in the San Joaquin River and its tributaries from Mendota to Vernalis.

The figure depicts the considerable deterioration that occurs in the upstream portion of the river system, from 80 km to 200 km. In this reach, drainage return flows, typified by Salt Slough at 100 km, cause substantial degradation of the flow entering above Mendota (M) from Friant Reservoir (F) and that supplied to the local users by the Delta-Mendota Canal (DMC). It is especially significant to note that the quality of the main river is closely identified with this drainage, as may be seen in the proportions of the various anions. Sulfate, in particular, is prominent in drainage waters from the westside of the valley while it appears only in very low concentrations in the Sierra streams entering the San Joaquin from the eastside. Noncarbonate hardness and boron are other characteristics of the westside soils and their drainage, that serve to identify the agricultural lands between 80 km and 200 km as the primary sources of salt accretion to the San Joaquin (WPRS-SDWA, 1980).

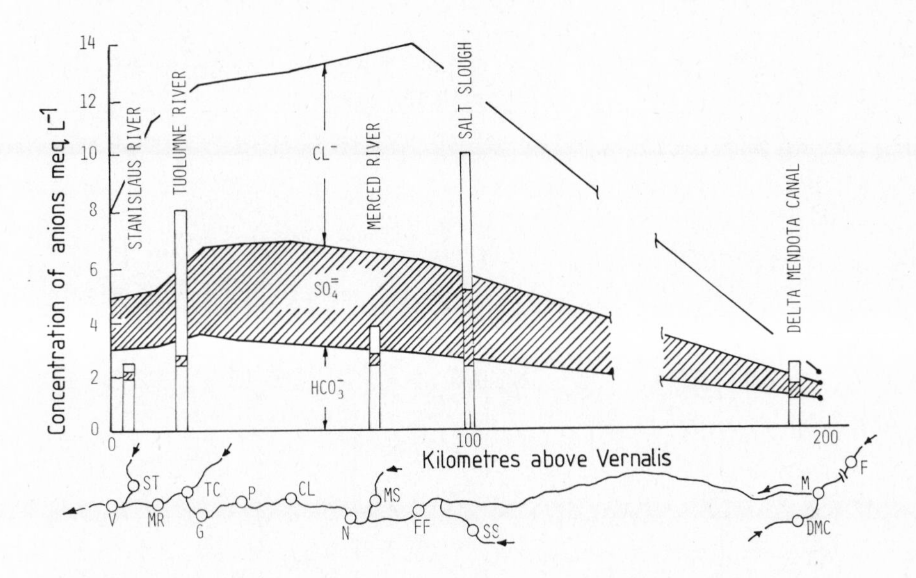

Fig. 7. Concentrations of principal anions in the San Joaquin River and its major tributaries. Period: 3-9 May 1966.

5.2 Preliminary Salt Balances

Given both runoff and water quality at a sufficient number of locations along the river, it is possible to identify more accurately the major sources of salt accretion and to quantify them roughly in relation to the total basin salt balance. This has been done for a dry year of record, 1960/61, for each of the major quality constituents that characterize San Joaquin River waters (WPRS-SDWA, 1980). Two examples, for sulfates and boron, are shown in Fig. 8.

In the upper section of the figure it is seen that virtually all of the sulfate that reaches Vernalis originates with drainage upstream of the mouth of the Merced River (75 km); very little is contributed to the high quality eastside Sierra streams. Overall, the basin has a negative balance for this year

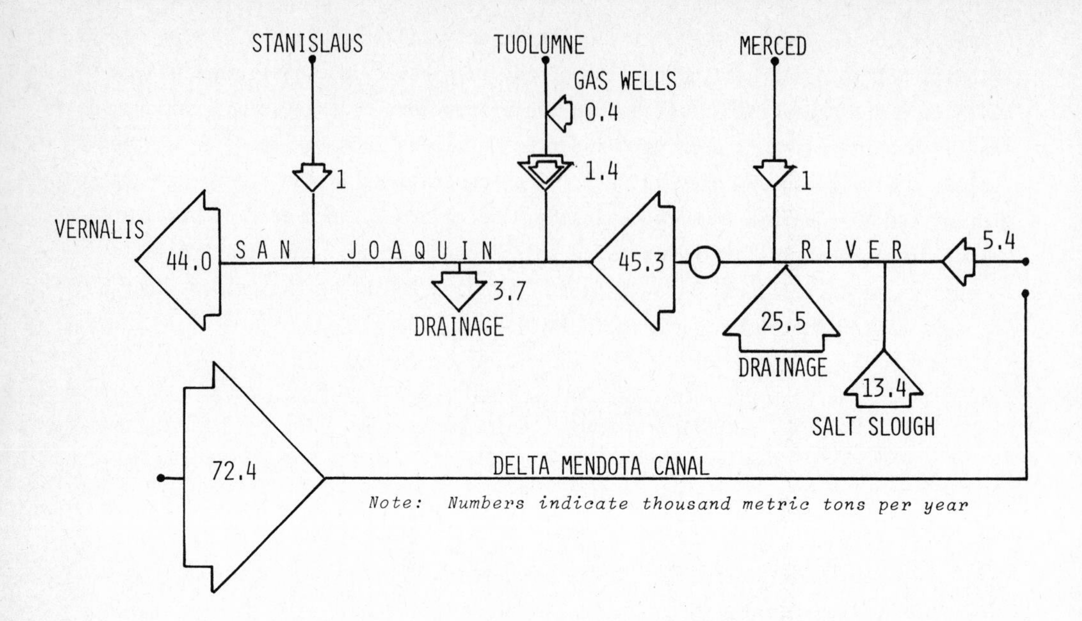

A. SULFATES

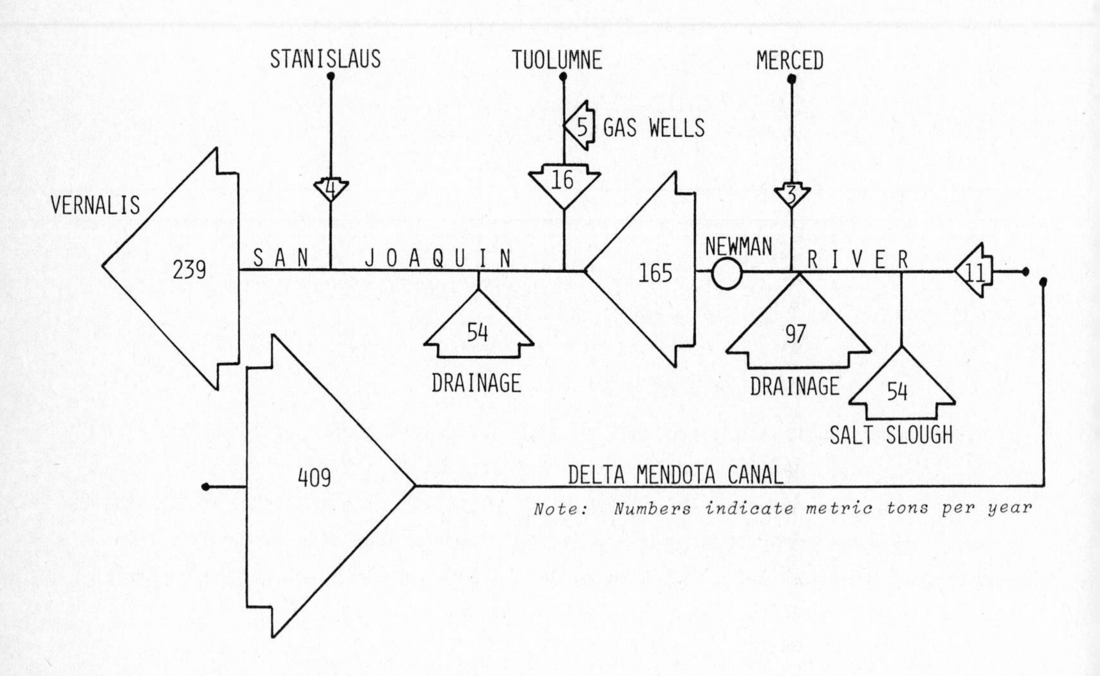

B. BORON

Fig. 8. Salt balance for San Joaquin River System, 1960/61, sulfates and boron.

since a greater tonnage is introduced by the Delta-Mendota Canal than leaves through the San Joaquin River at Vernalis. It will be noted that the San Joaquin River itself is a major contributor of salts to the canal because of their close proximity in the Sacramento-San Joaquin Delta. Even though the average quality of the canal water is much better than that of the river, the canal's greater flow (about 1.2×10^9 m^3 yr^{-1} compared to the river's runoff of about 0.5×10^9 m^3 yr^{-1} in 1960/61) resulted in a negative sulfate balance for 1960/61.

A similar balance picture results for boron, as shown in the lower portion of Fig. 8. The major boron contribution, about 70% of the total at Vernalis, originates above the Merced River (75 km); very little from eastside streams, although localized sources, some abandoned gas wells on the Tuolumne River and drainage between the Tuolumne and Stanislaus Rivers add about 30% of the total. Once again, importation by the DMC exceeds losses through the San Joaquin River.

The general result of salt balance studies has been to identify the Central Valley Project service area above the mouth of the Merced as a major source of salt accretion to the river. The proportion of the total salt load at Vernalis contributed by this area varies with the hydrologic conditions in dry years (lower quartile in unimpaired annual runoff). As much as 70% originates in this area, while in wet years (upper quartile) the proportion drops to about one half, due mainly to the larger inflows from eastside streams.

5.3 Trends in Salt Load

While data are not sufficient for a year-by-year salt balance computation, a comparison of pre-project to post-project conditions (1930 to 1944 vs 1952 to 1966) indicates that the salt load carried by the San Joaquin River at Vernalis is steadily increasing. It is estimated that the average annual increase in tonnage between these periods has been approximately 130×10^3 metric tons, corresponding to about 55% of pre-CVP levels (WPRS-SDWA, 1980). Of this, about 93×10^3 metric tons is considered to have originated within the CVP service area. Taken together with the changes that have occurred in runoff and quality during the past 30 years or so, this evidence points strongly toward the need to rectify the basin's salt balance. An important step in this direction will be the installation of drainage facilities to convey out of the basin accretions from discrete sources. This will, no doubt, include tile drainage systems, the efficacy of which is still a subject of investigation.

6 PRELIMINARY MODEL STUDIES

6.1 Technical Approach

It is desired to estimate the concentration and salt load carried by tile drains, once installed in saline soils of the westside of the San Joaquin Valley. The processes of transport between the semi-saturated and saturated porous media profile and the ultimate drainage conveyance facility are essentially dynamic, following the intermittent and seasonally varying irrigation activity. Thus, to simulate these processes it is necessary to have access to, or to develop, a suitable mathematical model capable of representing unsteady two-dimensional flow and dispersion in an anisotropic porous medium that may at the same time be both saturated and unsaturated.

The most promising approach to modeling such flows and the salt transport in porous media appears, in the opinion of the authors, to be one based on the Galerkin-type Finite Element Method (FEM). Details of the classic formulation of the finite element method are given in Zienkiewicz (1971), Norrie and deVries (1973) and Pinder and Gray (1977).

Galerkin's method, which has been gaining favor among investigators recently, offers an alternative way of formulating a problem for finite element solution without using variational principles, the more traditional approach. Pinder and Frind (1972) have chosen the Galerkin formulation in an investigation of two-dimensional flow of groundwater in a confined aquifer. Pinder (1973) subsequently extended this technique for simulation of groundwater contamination on Long Island, New York. Neuman et al. (1974) used Galerkin's finite element method to simulate flow in saturated - unsaturated soils, considering water uptake by plants. Ghorbanzadeh (1980) employed the method to analyze nonsteady, two-dimensional tile drainage in a saturated - unsaturated soil profile supplied both from above, by rainfall or excess irrigation, and from below through an artesian aquifer at very high permeability. This model forms the basis for the preliminary investigation of tile drainage techniques described here. A brief description of one case study is presented to indicate at least one promising avenue for future research on salinity problems.

6.2 Governing Partial Differential Equations

The simulation of solute transport in porous media entails setting up two sets of partial differential equations; one which describes the transient flow in a unified saturated - unsaturated porous medium and another which characterizes the movement of the solute by advective and dispersive transport.

The equation of motion describing transient flow in a unified saturated - unsaturated porous medium is:

$$[c(h) + \frac{\Theta}{n}S_s] \frac{\partial h}{\partial t} = \frac{\partial}{\partial x_i} [K_r(h) K^s_{ij} \frac{\partial h}{\partial x_j} + K_r(h) K^s_{i3}] + G_v \qquad (2)$$

where h = pressure head (negative in unsaturated flow and positive in saturated flow); Θ = soil moisture content; n = porosity; $c(h) = \frac{d\Theta}{dn}$, specific water capacity (zero in the saturated zone); S_s = specific storage (negligible in the unsaturated zone); t = time; $x_i (i=1,2,3)$ = spatial coordinates; $K_r(h) = K(h)/K^s_{ij}$ = relative hydraulic conductivity, $(0 \leq K_r \leq 1)$; K^s_{ij} = a saturated hydraulic conductivity tensor; G_v = source (+) or sink (-).

The convection - dispersion equation for transport of a conservative substance in a porous medium is:

$$\frac{\partial C}{\partial t} = - \frac{\partial}{\partial x_i} (q_i C) + \frac{\partial}{\partial x_i} (D_{ij} \frac{\partial C}{\partial x_j}) + q_c \qquad (3)$$

$$i,j = 1,2,3$$

where C = mass concentration of a conservative substance in solution; q_i = seepage velocity; D_{ij} = coefficient of hydrodynamic dispersion; q_c = source (+) or sink (-) of substance C.

Hydrodynamic dispersion is considered a function of seepage velocity (Bear, 1972), described by the relationship:

$$D_{ij} = \alpha_2 \, q \, S_{ij} + (\alpha_1 - \alpha_2) \, q_i \, q_j / q \qquad (4)$$

$$i,j = 1,2,3$$

where α_1 = longitudinal dispersivity coefficient; α_2 = lateral dispersivity coefficient; q = seepage velocity; S_{ij} = Kronecker delta = $\{ \begin{matrix} 1 \text{ if } i = j \\ 0 \text{ if } i \neq j. \end{matrix}$

Solution of these equations for a specified mesh is attained by the finite element method, given the necessary initial and boundary conditions. The flow regime is first defined to obtain seepage velocities and the configuration of the saturated zone as functions of time and space. These quantities are then used to determine D_{ij} and to solve the convection - dispersion equation (3). Details concerning initial and boundary conditions and solution technique are given by Ghorbanzadeh (1980). An illustrative case is presented here.

6.3 Finite Element Array

The problem chosen for illustration is that of a tile drainage system in a heterogenous soil profile 3 m in depth in which 10 cm drainage tiles are located 20 m on center. Drains are located 2 m below the surface and the normal water table is situated 0.5 m below the surface. For modeling purposes the domain, 3 m x 10 m, from the drain and the centerline between drains is represented as a system of 116 rectangular elements, as shown in Fig. 9. Element sizes are varied arbitrarily in accordance with the detail required to describe adequately the anticipated flow and quality regimes. For example, smaller elements are positioned in the vicinity of the drain where higher velocities and steeper gradients are expected.

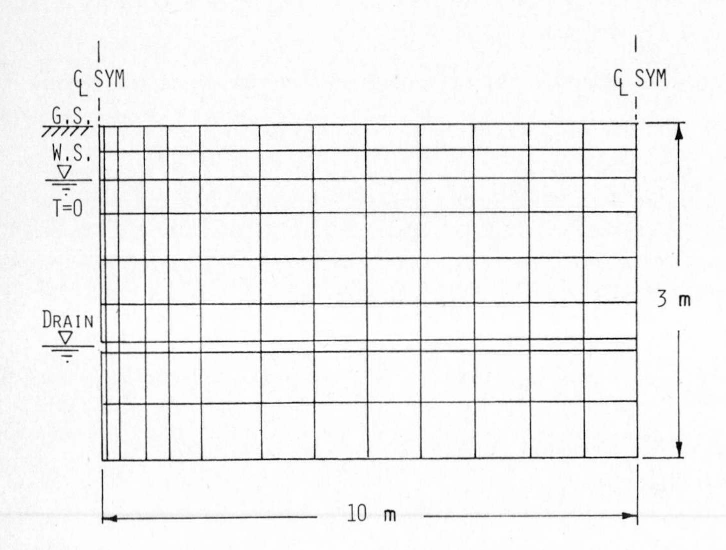

Fig. 9. Finite element mesh for flow region.

6.4 Specification of the Problem

Since the flow regime includes an unsaturated zone it is necessary to define both moisture content Θ and relative conductivity K_r as functions of soil moisture tension. For this example experimentally determined relationships for these quantities, shown in Fig. 10, were employed.

Boundary and initial conditions for the case study are presented in Fig. 11. Irrigation water application (corresponding to net drainage into the soil profile) is specified at 0.012 m per unit time.* This variation is considered representative of actual profiles of poorly drained lands in the westside of the San Joaquin Valley. At the location of the drain $C/C_o = 2.25$ initially. Water reaching the free water surface by percolation through the unsaturated zone is assumed to have a quality represented by $C/C_o = 1.0$.

---

*Temporal variations are expressed in relative terms, i.e., T/T^* where T^* is an appropriate standard unit, say one day in this case. The initial water surface is situated at a depth of 0.5 m and the drain is assumed to be flooded at $T/T^* = 0$. Initial quality conditions (also expressed in relative terms as C/C_o) range from 1.0 at the soil surface, to 10.0 at a depth of 0.5 m to 1.5 at the lower impermeable boundary, 3 m deep.

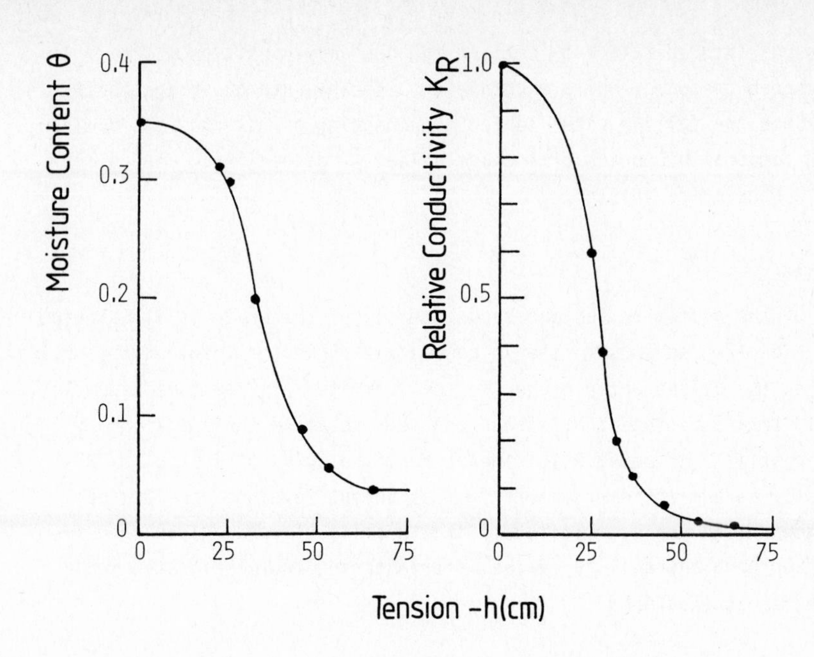

Fig. 10. Hydraulic properties of the porous medium.
(θ_s = 0.352; K_s = 0.57 m/T*)

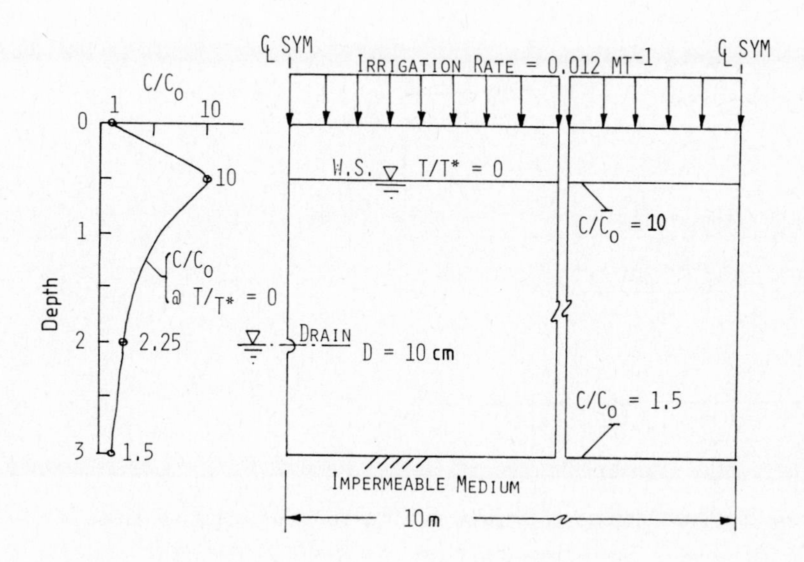

Fig. 11. Initial and boundary conditions for simulation of flow and salt transport.

The system is put into operation by setting the boundary condition at the drain to atmospheric pressure, while maintaining a steady rate of irrigation.* The model describes the falling water table, the changing efflux from the drain, the spatial and temporal histories of seepage velocity, dispersivity and concentration of the conservative solute.

6.5 Results

The response of the system to instantaneous opening of the drain is illustrated graphically in Fig. 12. An abrupt rise in concentration at the level of the drain is experienced as the saline water higher in the saturated profile (see Fig. 11) is drawn rapidly toward the point of egress. A peak relative concentration of 5.13 is achieved at T/T* of about 0.4. Thereafter, the curve of C/C_0 vs T/T* declines gradually as a mixture of waters from different levels in the drainage horizon moves toward the drain. Ultimately, the quality of water discharged from the drain will approach the quality of the water entering the unsaturated soil horizon from the irrigated field.

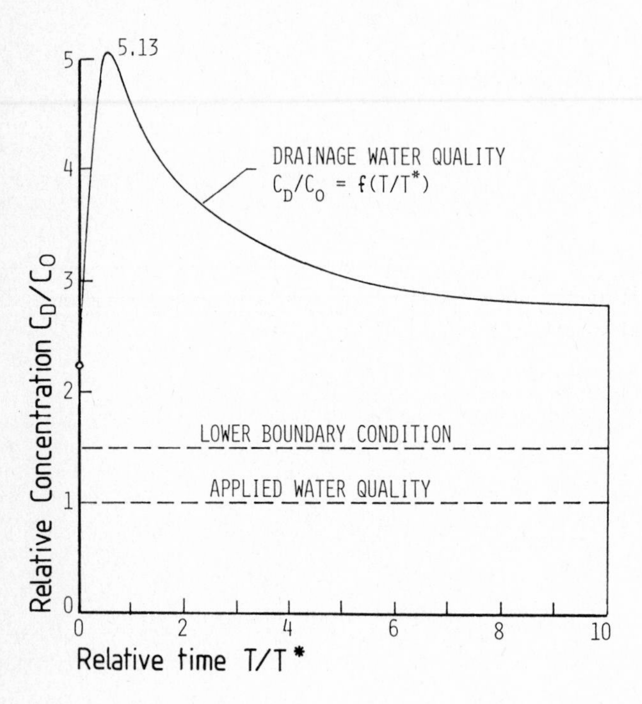

Fig. 12. Relative concentration at the drain as a function of relative time after opening drain.

---

*However, the rate of application may be unsteady or intermittent, if desired.

At the midpoint between drains, as shown in Fig. 13, the concentration of solutes along the vertical axis is modified gradually by the downward movement of applied water at $C/C_0 = 1.0$. The upper saline zone is depleted of salts and the maximal level of salinity is gradually moved downward with time. Since this location is a boundary of symmetry with zero flux through the vertical boundary, reductions in salinity due to the superior quality of applied water are first experienced here. As seepage flows accelerate horizontally toward the drain, this zone of improved quality will expand, until ultimately the salinity of the entire soil horizon will be dominated by the quality of water entering the unsaturated soil horizon from above.

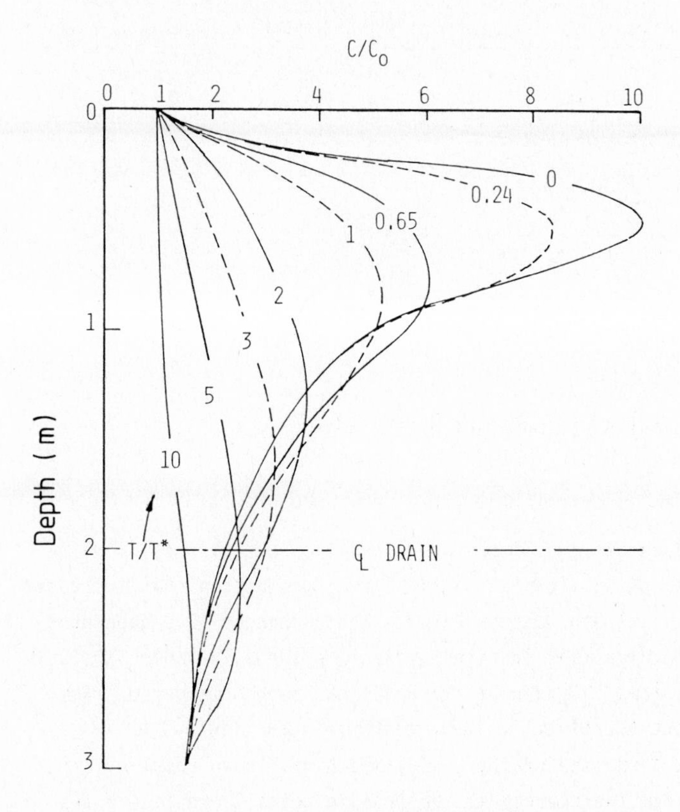

Fig. 13. Relative concentration as a function of depth and time at midpoint between drains.

Fig. 14 shows the distribution of salinity along the horizontal plane at the level of the drain, for various values of T/T*. It is noted that the pattern during transport is bimodal, with a peak at the drain and one at the centerline of the field between drains. The depression between these peaks is a result of the diluting effect of the applied water at $C/C_0 = 1.0$. Ultimately the entire profile will reach the level of the applied water as the saline water is flushed toward the drain. While the simulation was not carried beyond T/T* = 24, it is estimated, from these preliminary results, that the drain concentration would decline to C/C_0 = 1.5 in a relative time probably in excess of 100.

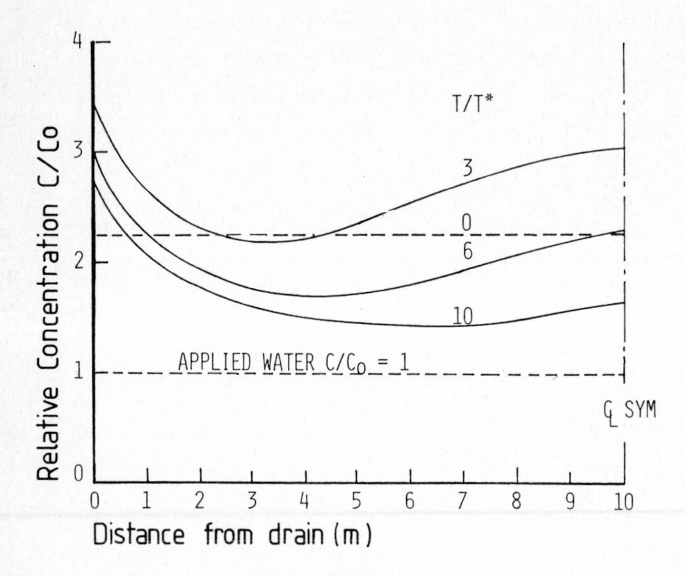

Fig. 14. Concentration profile at the level of the drain.

7 CONCLUSIONS

Development of water resources in the San Joaquin Valley of California, while allowing for the expansion of California's agricultural production, has increased the salt burden carried by the San Joaquin River. These changes have been progressive with time, even accelerating in recent years, to the point where agricultural productivity in the lower portion of the basin has been endangered. The principal sources of saline accretion to the system are identified with saline lands along the western side of the valley which have been supplied largely with water imported from the Sacramento-San Joaquin Delta to replace extra-basin exports. Solution of drainage problems in this area includes consideration of tile drains to improve productivity of saline soils and to convey drainage water to suitable locations where impact on downstream users will be minimized. A model of salinity transport in a drained field has been developed and applied in preliminary studies of the efficacy of tile drains. Research is continuing on methods of alleviating salinization of agricultural lands in the San Joaquin Valley.

8 REFERENCES

Bear, J., 1972. Dynamics of fluids in porous media. American Elsevier Publ. Co. Inc., New York.
Ghorbanzadeh, A., 1980. Non-steady, two-dimensional tile drainage of saturated - unsaturated artesian lands analyzed by finite element method. Ph.D. Thesis, Civil Engrg., Univ. of California, Davis, 268 pp.
Neuman, S.P., Feddes, R.A. and Bresler, E., 1974. Finite element simulation of flow in saturated - unsaturated soils considering water uptake by plants, Technion, Haifa, Israel, Proj. No. A10-SWC-77.
Norrie, D.H. and deVries, G., 1973. The finite element method. Academic Press, New York.
Pinder, G.F., 1973. A Galerkin finite element simulation of groundwater contamination on Long Island, New York. Water Resour. Res. 9: 6.
Pinder, G.F. and Frind, E.O., 1972. Application of Galerkin's procedure to aquifer analysis. Water Resour. Res., 8: 108-120.
Pinder, G.F. and Gray, W.G., 1977. Finite element simulation in surface and subsurface hydrology. Academic Press, New York.
WPRS-SDWA, 1980. Report on the effects of the CVP upon the Southern Delta water supply, Sacramento-San Joaquin River Delta, California. Water Power Resour. Serv., Depart. of the Interior and South Delta Water Agency, Limited Distribution, 179 pp. plus appendices.
Zienkiewicz, O.D., 1971. The finite element method in Engineering Science. McGraw-Hill, London.

DRYLAND CROPPING STRATEGIES FOR EFFICIENT WATER-USE TO CONTROL SALINE SEEPS IN THE NORTHERN GREAT PLAINS, U.S.A.

A.L. BLACK,
Northern Great Plains Research Centre,
U.S.D.A., SEA-AR, Mandan, North Dakota.

P.L. BROWN,
A.D. HALVORSON,
Northern Great Plains Soil and Water Research Centre,
U.S.D.A., SEA-AR, Bozeman, Montana.

F.H. SIDDOWAY,
Northern Plains Soil and Water Research Centre,
U.S.D.A., SEA-AR , Sidney, Montana.

ABSTRACT

Black, A.L., Brown, P.L., Halvorson, A.D. and Siddoway, F.H., 1981. Dryland cropping strategies for efficient water-use to control saline seeps in the Northern Great Plains, U.S.A. Agric. Water Manage., 1981.

Investigations of the characteristics, hydrology, and causes of most saline seeps in the Northern Great Plains of the United States and Canada show that the same basic principles are involved in their development regardless of geographic location. Precipitation received in excess of the soil-root zone storage capacity, primarily during fallow or idle periods between crops, is the source of the water. One method of controlling saline seeps involves the establishment of a perennial deep-rooted crop, such as alfalfa (*Medicago sativa L.*), on the recharge area for enough time to use existing and anticipated available-water supplies to maximum possible soil depth. Another method is to establish intensive, flexible cropping systems using adaptable crops in combination with proper soil, water and crop management practices to improve crop production-water use relationships sufficiently to reduce or eliminate the need for summer fallow. The key to developing more efficient cropping systems is to improve existing, and develop new, soil water conservation techniques to increase soil water storage during non-crop periods, thereby increasing the opportunity of developing cropping systems that use water more efficiently than conventional crop-fallow systems. This paper identifies and discusses the soil, water, and crop management strategies needed for the control of saline seeps. These strategies are based on the specific water requirements and rooting depths of crops, soil water conservation and storage, crop residue management, disease and weed control, and proper fertilization.

1 INTRODUCTION

Saline seeps in the semiarid Northern Great Plains of the United States (Montana, North Dakota, and north-western South Dakota) and the southern regions of the prairie provinces of Canada (Alberta, Saskatchewan, and Manitoba) are caused by a combination of geological, climatic, and cultural conditions (Doering and Sandoval, 1975; Halvorson and Black, 1974; Van der Pluym, 1978). Approximately 800 000 ha of dryland in the Northern Great Plains region is now severely affected by saline seeps (Van der Pluym, 1978). Investigations of the characteristics, hydrology, and causes of most saline seeps reveal that the same basic processes are involved in their development regardless of geographic location (Halvorson and Black, 1974; Doering and Sandoval, 1975 and 1976; Brown et al.,1976; Ballantyne, 1963; Greenlee et al.,1968). Precipitation that penetrates below the root zone during fallow or idle periods between crops is the primary source of water that causes most saline seeps. This water moves through salt-laden substrata to impermeable or slowly permeable layers and forms a perched water table. It then flows laterally, downslope from the recharge area and resurfaces wherever soil substrata conditions force the water table near the soil surface at a lower topographic position. As the saline water surfaces, whether by hydraulic pressure or capillary movement, it evaporates and leaves a salt deposit. The salts are primarily sodium, calcium, and magnesium sulfates and nitrates.

The efficiency for storing precipitation in the soil by summer fallowing in the semiarid Northern Great Plains has improved considerably in recent years (Greb et al.,1970; Black, 1973a; Van der Pluym, 1978; Bauer, 1968). However, 60 to 80% of the precipitation still evaporates, runs off, or penetrates below the root zone. Limiting the amount of water moving below the root zone is the key to halting the development and expansion of saline seeps. The advantages and disadvantages of using alternate crop-fallow have been documented (Haas et al.,1973; Black et al., 1973). The need for more intensive and flexible cropping systems to use precipitation and available water supplies more efficiently than conventional crop-fallow systems has also been recognised (Haas and Willis, 1962; Halvorson and Black, 1974).

The objective of this paper is to identify and discuss the strategies for soil, water, and crop management that are needed to increase the efficiency of precipitation-use. These strategies are based on specific crop water requirements and rooting depths, soil water conservation and storage, crop residue management, disease and weed control, and proper fertilization.

2 CLIMATE-SOIL WATER RELATIONS

In the Northern Great Plains, annual precipitation ranges from 250 mm in the western and north-central regions to 500 mm in the eastern and north-eastern regions. About 70% is received during the April through August growing season. Although the September through March precipitation accounts for only 30% of the annual precipitation, it frequently accounts for 60%, or more, of the total soil water stored between crops or during the fallow period.

In a spring wheat (*Triticum aestivum* L.)-fallow cropping system, Haas and Willis (1962) showed that during the 1915 to 1954 period, 63% of the total water stored in the soil during a 21-month fallow period was stored during the first 8-month overwinter period from harvest to April 20, and 84% for a 10-month period from harvest to July 1. During the 1957 to 1970 period, Black et al. (1973) showed that 83% of the total water stored during the 21-month fallow period was stored during the first 9-month segment. These data show that if an additional 20 to 30 mm of water could be stored during the first overwinter period, fallowing would not be necessary.

3 PRINCIPLES OF SALINE SEEP CONTROL

The control of saline seep areas involves the efficient utilization of available soil water supplies in the recharge area before water penetrates below the root zone. Establishing a perennial crop like alfalfa on the recharge area to use existing available water supplies to the maximum depth possible is one of the most dependable and fastest means of stopping the source of water from migrating to the seep-discharge area (Brown and Cleary, 1978; Halvorson and Reule, 1976 and 1980). However, Halvorson and Reule (1980) showed that an alfalfa crop must be established on a major portion of the recharge area to provide hydrologic control of the seep discharge. Once water movement below the root zone was controlled, the water table receded, the soil surface in the seep area dried, and soil salinity decreased. The salinized area can then be managed to decrease soil salinity in the root zone.

In contrast, a buffer strip of alfalfa occupying about 20% of another recharge area on the upslope side of a different saline seep did not provide hydrologic control (Halvorson and Reule, 1980). A companion greenhouse study showed that alfalfa removed less and less water from a soil column as in situ soil salinity increased. Therefore, the concept of establishing a narrow buffer strip of alfalfa in the recharge area to intercept and use water from a saline water table failed because; (i) the alfalfa's capacity to use water decreased as soil salinity increased and (ii) the alfalfa established on a minor portion of the recharge area could not utilize enough of the surplus water. To control saline seeps, the cropping system employed must significantly reduce the total volume of water entering the groundwater system below the root zone over a major portion of the recharge area.

Delineation of the total recharge area contributing to a specific seep is difficult. Deep soil sampling by coring or augering with hydraulic equipment on a field grid system is one reliable but costly and time consuming method. Another method developed by Halvorson and Rhoades (1974) is to measure in situ soil electrical conductivity versus soil depth using a four-probe Wenner electrode configuration in the field. Figures constructed for soil electrical conductivity versus electrode spacing (soil depth) revealed characteristically different curve types for nonsaline productive areas, for potential saline seep areas, and for the seep area itself. In addition, the four-electrode technique can be used for field mapping to identify potential seep and recharge areas (Halvorson and Rhoades, 1976).

Establishing a flexible intensive cropping system using adaptable grain, forage and oilseed crops with judicious use of fallow is also an effective method for controlling seeps (Black and Siddoway, 1976; Halvorson and Black, 1974; Dubbs, 1976a; Brown and Miller, 1978). Flexible cropping involves sowing a crop in years when stored soil water and expected growing season precipitation are sufficient to produce a reasonable crop yield. Fallowing is employed whenever soil water and expected precipitation are not sufficient to produce a reasonable crop yield. To assess the soil water supply on the recharge area at planting time for a given crop, a farmer could use the hand-operated soil moisture probe method (P. Brown, personal communication, 1980). However, the success of the flexible cropping method depends directly upon many other soil and crop-management factors, like weed control and soil fertility. A computerized management guide and a users' manual have been developed to help farmers in deciding the best cropping and soil management options for wheat, barley (*Hordeum sativum*, Jess.), oats (*Avena sativa* L.) and safflower (*Carthamus tinctorius*) based on stored soil water and expected growing season precipitation (Kresge and Halvorson, 1979).

Because most areas susceptible to saline seeps have relatively shallow phreatic water, observation wells can be installed at strategic locations on the recharge area using perforated plastic pipe (5 to 10 cm diameter) to monitor changes in the phreatic water surface. These wells should be deep enough to reach the slowly permeable parent material but no deeper than 5 or 6 m. The depth to water should be recorded at regular intervals. If the perched water table is at least 2 m deep and receding over time, then the current cropping system is using most or all of the precipitation. If the water table continues to rise, then a different cropping practice is needed to use available water supplies more efficiently.

4 WATER CONSERVATION

A key factor in developing efficient cropping systems is increasing soil water storage during noncrop periods. Such systems should provide adequate water for more frequent cropping than does the conventional crop-fallow system. Two of the best methods for increasing soil water supplies are : (1) control of fall weeds and volunteer grain after harvest and (2) leaving standing stubble to trap snow.

In the Northern Great Plains and Canadian Prairies, snow is an important component of the annual precipitation. Therefore, snow management through the use of standing stubble from the previous crop (Willis and Haas, 1969; Willis and Frank, 1975; Black and Siddoway, 1977; Smika and Whitfield, 1966), annual or perennial barriers spaced at various intervals across a field (Greb and Black, 1961; Greb and Black, 1971; Black and Siddoway, 1976), and properly designed and managed field windbreaks (George et al., 1963; Staple and Lehane, 1955) can augment overwinter soil water storage considerably compared with land having no vegetative materials present to trap snow. Such soil water conservation techniques greatly increase the opportunity to crop more frequently and use available water supplies more efficiently than with conventional crop-fallow.

The amount of additional soil water stored through the use of snow management using the tall wheatgrass (*Agropyron elongatum*) barrier system on a Williams loam soil (Black and Siddoway, 1976 and unpublished data) is illustrated in Table 1. The tall wheatgrass barriers were spaced 15 m apart and yearly growth maintained a barrier height of about 1.2 m. Soil water recharge for the continuously cropped strips averaged 78 mm of water to a 1.5 m depth within the grass barrier system the first winter (9-month period) as compared with 53 mm on an undisturbed stubble field, an increase of 47%. Outside the barrier system, only 90 mm of soil water was stored during the entire 21-month fallow period. The 12-year average storage efficiency with the grass barrier system as compared with standing stubble without barriers was somewhat lower than reported previously (Black and Siddoway, 1971; Black and Siddoway, 1976) for two reasons. Overwinter precipitation was below average for 3 of the last 4 years; and beginning in 1972 safflower was used in the annual cropping system, which extended the crop water use period through August and September, thus reducing the time period between crops from 9 to 7 months. Nevertheless, during the entire 1965 to 1979 study period (Black and Siddoway, 1971; 1976) there was only 1 year when available soil water in the upper 1.2 m soil depth at seeding time within the annually-cropped grass-barrier system was less than 76 mm. Recropping is not recommended whenever plant-available soil water stored at planting time is much less than 76 mm (Black and Ford, 1976).

TABLE I

Twelve-year average soil water gains and storage efficiences, with and without tall wheatgrass barriers for annual crop and spring wheat-fallow systems, (Black and Siddoway, 1976 and unpublished data).

Cropping Sequence	Precipi-tation	With Grass Barrier		Without Grass Barriers	
		Soil Water Gain	Storage Efficiency*	Soil Water Gain	Storage Efficiency
	mm	mm	%	mm	%
Annual Cropping:					
First Winter (9 months)	163	78	47.9	53	32.5
Spring wheat-fallow:					
†First Winter (9 months)	172	78	45.3	53	32.5
Summer Fallow (5 months)	230	19	8.3	20	8.6
Second Winter (7 months)	140	10	7.1	17	12.1
Total (21 months)	542	107	20.1	90	16.6

* Storage efficiency is the percentage of precipitation stored in the soil profile for each period.

† First Winter = August to April
Summer Fallow = April to October
Second Winter = October to May

Precipitation-use efficiency of different cropping systems is expressed as a percentage of the total precipitation received from harvest of one crop to harvest of the next crop (Table 2). Precipitation at harvest to date of fall sampling was added to the precipitation available to the next crop. Year to year variations in the periods between harvest dates varied from 11 to 13 months and from 23 to 24 months in the annual and crop-fallow cropping systems, respectively. For this reason, the average precipitation available per crop does not equal the annual or twice the annual average precipitation for the 12-year study period.

Crops grown under annual cropping systems used an average of 75 to 81% of the precipitation received between crop harvest, whereas conventional spring wheat-fallow used only 40% (Table 2). Therefore, the average amount of unused water between crops, a portion of which may contribute to saline seeps, averaged 473 mm for spring wheat-fallow and only 72 to 98 mm for annual cropping. During the spring-wheat growth period, nearly all precipitation was used. However, during the noncrop periods, much of the precipitation is lost to evaporation, runoff, or deep percolation in the wheat-fallow system.

Similarly, at Brandon, Manitoba (Ferguson, 1962), evapotranspiration accounted for only 38% of total rainfall in a spring wheat-fallow rotation. With continuous spring wheat, evapotranspiration accounted for 78% of total rainfall available to the crop. Total water unaccounted for in evapotranspiration was 417 mm for the wheat-fallow system and only 75 mm for continuous spring wheat. These studies illustrate the relative magnitude of the potential for losing water below the root zone in a spring wheat-fallow system as compared with annual cropping.

Annual grain yields with 34 kg of N applied to each crop averaged 49 to 71% greater for annual cropping systems, than for conventional spring-wheat fallow

(Table 2). The winter wheat-fallow and the 3 year crop rotation of spring wheat-winter wheat-fallow yielded 17 and 33% more on an annual basis than conventional spring wheat-fallow. Yield data for a higher N rate (67 kg N ha^{-1}) are not shown in Table 2 because the average annual yields did not differ from the 34 kg N rate over the 12-year period. In only 1 year out of 12 did the 67 kg ha^{-1} rate of N increase grain yields (about 1.3-fold) over the 34 kg ha^{-1} rate, and that occurred only in the annual cropping system in 1976 when two growing seasons in succession (1975 and 76) were exceptionally wet. Fertilization with either 34 or 67 kg of N ha^{-1} increased average annual production for annual cropping, spring wheat-winter wheat-fallow, and crop-fallow systems about 60, 40 and 20% respectively.

Without N-fertilization, the amount of grain produced per unit of available water per cropping system was about 2.5 kg ha^{-1} mm^{-1} of water, independent of cropping frequency within the rotation (Table 2). Increases in the production efficiency per unit of available precipitation due to N-fertilization were about 20% for crop-fallow, 42% for spring wheat-winter wheat-fallow and 32 to 84% for annual cropping compared with the same production-water use efficiencies without N.

The importance of establishing proper fertilization management practices for various dryland crop rotations has been documented by other investigators (Dubbs, 1976a; Black and Ford, 1976; Brown, 1971; Black et al.,1973; Bergman et al.,1979; Choriki et al.,1971; Schneider et al.,1980). These studies all showed that the need for adequate levels of available soil P is a prerequisite to realization of maximum N-fertilization responses of crops grown on fallow. As cropping frequency increases, the need for additional N-fertilizer increases proportionately, and P-fertilizer responses become dependent upon first satisfying the crop needs for N. However, the full benefit of additional N-fertilizer is not realized in crop rotations that involve only one crop which tend to increase specific weed problems. As an example, over a 9-year period, winter wheat yields averaged 2150 kg ha^{-1} in a winter wheat-fallow rotation and 2630 kg ha^{-1} in a winter wheat-barley-fallow rotation (Dubbs, 1976b). Including spring barley in the cropping sequence helped control winter annual weeds, principally downy brome (*Bromis tectorum*). In another comparison (Dubbs, 1976b), spring barley yields in the winter wheat-barley-fallow rotation increased 66% with 45 kg ha^{-1} of N applied while in the continuous barley cropping sequence yields increased only 34% with the same rate of N applied because summer annual grassy weed populations like green foxtail grass (*Seteria viridis*) and wild oat (*Avena fatua*) were a problem in the continuous barley rotation.

TABLE 2

Average precipitation-use efficiency as influenced by cropping system within the tall wheatgrass barriers over a 12-year period, (Black and Siddoway, 1976 and unpublished data).

Cropping System	Number Crops yr^{-1}	Total Precip. per Crop	* Total Water-use per Crop	†Water-Use Eff. per Cropping System	Annual Grain Yield		Grain Yield per unit of Precip.	
					No N	34 kg N ha^{-1} per crop	No N	34 kg N ha^{-1} per crop
		mm	mm	%	-----kg ha^{-1}-----		-----kg ha^{-1} mm^{-1}-----	
Annual Cropping								
1. 6 WWh, B, Saf, B, WWh, Saf, B	1.00	396	322	81.3	1328	1794	3.4	4.5
2. ‡5 SpW, Saf, B, WWh, B, WWh, B, WWh	1.00	394	296	75.1	993	1822	2.5	4.6
3 4 SpW, Saf, B, WWh, Saf, SpW, B, WWh, B	1.00	390	318	81.5	969	1590	2.5	4.1
Spring Wheat-Winter Wheat-Fallow	.66	569	333	58.5	997	1416	2.6	3.7
Winter Wheat-Fallow	.50	788	404	51.3	1019	1247	2.6	3.1
Spring Wheat-Fallow	.50	786	313	39.8	853	1065	2.2	2.7

*Total water use per crop is based on soil water use plus precipitation received from seeding to harvest.

†Water-use efficiency per cropping system is expressed as a percentage of the total precipitation received for each crop from harvest of one crop to harvest of the next.

‡Symbols: SpW = Spring Wheat, WWh = Winter Wheat, Saf = Safflower, and B = Spring Barley

5 CROPPING SYSTEMS STRATEGIES

After choosing among crops adapted to a given region, cropping decisions should be based on available soil water present at time of seeding and expected precipitation during the growing season. Available soil water can be estimated by knowing the depth of moist soil plus having a general knowledge of soil texture (Black and Ford, 1976). However, plant available water stored in the soil profile is not uniformly available to each crop because of differences among crops in rooting depths and water extraction potentials (Black and Ford, 1976; Brown and Cleary, 1978; Brown et al., 1976). Roots of safflower and alfalfa can extract soil water to a lower soil water content and from greater soil depths than small grain crops (Black and Brown, unpublished data). In our studies, safflower and alfalfa extracted 30 to 40 mm more water from the 30 to 120 cm soil depth than did cereal grains. Some water was also used from depths greater than 120 cm because of increased rooting depth. Therefore, differences among crops in their efficiency to deplete or utilize available soil water supplies becomes a factor in monitoring soil water use and recharge cycles in various cropping systems.

Based on the crop water-use data obtained on a deep, well-drained glacial till soil near Ft. Benton, Montana (Table 3), safflower used more soil water to greater soil depths in one year than any other annual crop. Alfalfa used only slightly less water the first year than safflower or sweetclover (*Melilotus officinalis*), but its ability to use growing season precipitation plus existing-soil water supplies from progressively deeper soil depths in successive years

TABLE 3

Rooting depth and soil water use by annual dryland crops (Brown, Cleary and Miller, 1976; Black, unpublished data).

Crop	Ft. Benton Mt Rooting Depth	Ft. Benton Mt Soil Water Use	Culbertson, Mt Rooting Depth	Culbertson, Mt Soil Water Use
	m	mm	m	mm
Safflower	2.2	249	2.1	229
Sunflower	2.0	206	-	-
Winter Wheat	1.8	200	1.6	190
Rapeseed	1.5	170	-	-
Spring Wheat	-	-	1.2	152
Barley	1.4	190	1.1	135
Corn	1.2	94	-	-

TABLE 4

Rooting depths and cumulative soil water depletion for some perennial crops and biennial sweetclover. (Brown, Cleary and Miller, 1976; Brown, unpublished data).

Year	Vernal Alfalfa Rooting Depth	Vernal Alfalfa Soil Water Use	Eski Sanfoin Rooting Depth	Eski Sanfoin Soil Water Use	Russian Wildrye Rooting Depth	Russian Wildrye Soil Water Use	Biennial Sweet Clover Rooting Depth	Biennial Sweet Clover Soil Water Use
Site A	m	mm	m	mm	m	mm	m	mm
1971	2.1	178	1.5	150	-	-	-	-
1972	3.0	305	2.7	262	-	-	a 1.8	276
1973	4.3	513	3.4	455	2.1	318	b 2.4	353
1974	5.5	666	4.0	561	2.7	409	-	-
1975	5.8	785	4.0	564	3.0	475	b 2.7	403
1976	6.1	787	4.3	594	3.0	475	-	-

a. First year sweet clover
b. Second year sweet clover

Site B	Kane Alfalfa		Beaver Alfalfa		Ladak 65 Alfalfa		Rambler Alfalfa	
1976*	-	-	-	-	-	-	-	-
1977	3.7	384	4.0	521	4.5	568	2.7	351
1978	3.7	414	5.8	721	4.9	582	3.0	373
1979	3.7	505	6.1	846	4.9	673	5.2	482
Mean Yield kg ha^{-1}	6700		5470		5360		5840	

*1976 - Establishment year.

marks alfalfa as the best crop for initial use to gain hydrologic control on recharge areas (Table 4). Although less effective than alfalfa, Sanfoin (*Onobrychis viciifolia* Scop. 'Eski') and Russian wildrye (*Elymus junceus* Fish.) depleted soil water to a depth of 4 and 3 m, respectively, after 3 years. Biennial sweet clover used more soil water than safflower the first year, but total soil water use was about equal to alfalfa when both were compared at the end of the second year. Following alfalfa, sweet clover and safflower, the ranking of annual crops in decreasing order of rooting depth and soil water use is sunflower (*Helianthus annuus* L.), winter wheat, rapeseed (*Brassica napus* L.), spring wheat, barley, and corn (*Zea mays* L.) (Table 3).

Within each crop, varietal differences in rooting depth and water use are important, as shown for alfalfa in Table 4. Some alfalfa varieties attain maximum rooting depth and soil water depletion in 3 years, others may require 4 to 5 years. Therefore, land owners should choose alfalfa varieties compatible with their soil profile characteristics and the amount and depth of soil water to be extracted. The subsoil characteristics which determine rooting depth include texture, depth to impermeable soil layers, and abrupt changes in subsoil materials. However, once the effective rooting depth is known for a given soil or field, a cropping system can be designed to manage the soil water use and soil water recharge cycles of the soil profile.

To illustrate how a cropping system may be developed, consider two broad climatic regions for the Northern Great Plains based on average annual precipitation of 250 to 380 and of 380 to 500 mm. Assume that the subsoils are moist nearly to field capacity initially and that the effective rooting depth is unrestricted to depths of 1.5 m, 1.5 to 3.0 m, or 3.0 to 6.0 m. Cropping systems to match these climatic and soil profile conditions can now be designed similar to those suggested in Table 5. It is important that cropping sequences be established on individual fields of a farm in different years and not on all fields over the entire farm so that all fields do not follow the same crop sequence. This pattern reduces the risk of crop losses due to crop-specific diseases and insect pests. It also eliminates the possibility of having all the crop-land in fallow, providing no source of income in any one year.

To hydrologically control seeps as fast as possible alfalfa should be the first choice for recharge areas that have no root-restricting soil layers to a depth of 3 to 6 m. Soils having a root-restricting layer in the 1.5 to 3 m depth could be cropped with either alfalfa or biennial sweet clover. In either case, these legumes may be seeded alone or with a companion small grain crop of oats, barley, or a semidwarf spring wheat. If the legumes are cut for hay and only the crowns and roots returned to the soil, there is very little net gain in soil N in semiarid dryland areas from growing a legume. However, if either biennial sweet clover or alfalfa is plowed under at the bud stage, there is a net gain of about 100 or 80 kg ha^{-1} of N, respectively, with about half becoming available in time for the next crop. As fertilizer costs continue to increase, farmers have an additional incentive to produce some of their own N while at the same time obtaining hydrologic control of water contributing to seep development.

Plowing legumes under in the early summer or late fall is expensive because of fuel costs, and the soil is exposed to the hazards of wind and water erosion. Legumes can be chemically treated to stop plant growth at the desired growth stage and left unharvested to protect the soil and trap snow during the winter. This improves both soil water and N supplies, thereby increasing the probability of a successful crop after a deep-rooted legume.

Once the soil water is depleted from the root zone, fallow becomes an optional practice within the cropping sequence (Table 5). However, spring seeded crops should not be planted when stored soil water in the 0 to 1.2 m soil depth is less than 76 mm. (Brown and Miller, 1978; Black and Ford, 1976; Kresge and Halvorson, 1979). One exception is sowing winter wheat into standing spring grain stubble. In this instance, 30 to 50 mm of soil water should be available in the upper 30 cm soil depth, including adequate seedbed-depth soil water to obtain a satisfactory stand. Over-winter soil water recharge of the upper 1.2 m of soil following a small grain crop in the Northern Great Plains depends on the snow trapping effectiveness of the standing stubble (Black and Siddoway, 1977).

The decision making logic for what crop should succeed another requires a knowledge of how effectively and to what depth the previous crop depleted the available soil water supply. For example, crops should be grown that sequentially have greater rooting depths in short-term rotations until the depth of removal exceeds recharge during noncrop periods. One of the reasons the spring wheat-winter wheat-fallow rotation has proved successful is because the rotation follows this logic (Black and Siddoway, 1976; Halvorson et al., 1976; Black and Siddoway, 1977; Schneider et al., 1980). Spring wheat typically uses soil water to to a depth of about 1.2 m. Winter wheat is then no-till seeded into standing spring wheat stubble, and the stubble then traps snow to provide soil water recharge of the upper 1.2 m soil depth. Winter wheat uses available water to a depth of 1.5 to 2.0 m. The probability of replenishing available soil water supplies in one overwinter period to a depth of 1.5 to 2.0 m after a winter wheat crop is relatively low. Therefore, fallow logically follows a deep-rooted crop like alfalfa, safflower, or winter wheat rather than a relatively shallow-rooted small grain crop or corn.

TABLE 5

Types of cropping strategies needed, based on effective soil root zone depth, crop rooting, and precipitation.

Precipitation mm	Cropping Strategies for Three Effective Depths of Soil Root Zone				
	<1.5 m	1.5 to 3.0 m		3.0 to 6.0 m	
250 to 380	Winter Wheat Barley Winter Wheat Fallow[†] or Spring Wheat Winter Wheat Barley Fallow[†]	*Alfalfa (3 to 4 years) Fallow[†] and/or Barley Fallow[†] Wheat[‡] Safflower Barley Fallow[†]	and/or Winter Wheat Barley Fallow[†] or Spring Wheat Winter Wheat Fallow[†]	Alfalfa (4 to 5 years) Fallow[†] and/or Wheat[‡] Safflower Barley Fallow[†]	and/or Spring Wheat Winter Wheat Fallow[†]
380 to 500	Spring Wheat Oilseed (Sunflower) Barley or Oats Fallow[†] or Barley Spring Wheat Sunflower Barley or Oats Fallow[†]	*Alfalfa (3 to 4 years) Fallow[†] and/or Sunflower Spring Wheat Barley or Oats Fallow[†] Spring Wheat Sunflower		Alfalfa (4 to 5 years) Fallow[†] and/or Spring Wheat Sunflower Barley or Spring Wheat Fallow[†] Spring Wheat Sunflower	

*Sweet clover could be substituted for alfalfa (2 years).
[†]Fallow is optional
[‡]Spring or Winter Wheat.

6 WEED CONTROL

For weed control, oilseed crops can be included in the cropping sequences (Table 5); for example, safflower in regions with less than 380 mm of precipitation and sunflower in regions with more than 380 mm of precipitation. In the Canadian Prairie Provinces, a rapeseed crop is used in spring grain cropping systems for weed control. With the recent advent of several excellent pre-emergence herbicides for use with these oilseed crops, and their particular effectiveness on annual grass weed types, weed populations can be significantly reduced. Annual green foxtail, barnyard grass (*Echinochloa crusgalli*) and to some extent wild oats can now be controlled with these herbicides. The supplemental use of appropriate herbicides to control broadleaf weeds and wild oats during cereal grain cropping provides a well-balanced weed control program.

In our experience, wild oats have been the most persistent and hard to control weeds. We found that wild oat populations could be controlled more effectively with current available herbicides if the land had a history of shallow tillage - no more than 7.5 cm - than if some tillage operations had been performed deeper. Deeper tillage mixes wild oat seeds throughout the soil to the depth of tillage. Thus, several generations of wild oats germinate over a period of 3 to 6 weeks each spring as appropriate soil temperatures are reached at successive depths.

There are valid reasons why specific crops should not follow the same crop or certain other crops in a cropping sequence. The first rule is that the same crop, and especially the same variety, should not be grown on the same land each year because this tends to promote specific weed, disease, and insect populations (Black and Siddoway, 1976; Bergman et al., 1979). A rigid cropping system like spring wheat-fallow or winter wheat-fallow encourages specific weed populations involving summer annual-and winter annual-grassy weed types, respectively (Black and Siddoway, 1976; Dubbs, 1976a; Choriki et al., 1971). Breaking these weed cycles simply involves a winter wheat-barley-fallow rotation (Dubbs, 1976a) or a spring wheat-winter wheat-fallow rotation (Black and Siddoway, 1976). However, to complete the weed control program and also disrupt cereal grain disease cycles, an oilseed crop should be inserted into the cereal grain cropping sequence every third or fourth crop year. Oilseed crops should not be grown on the same land more often than every third or fourth year because of the increased danger of disease and insect damage (Bergman et al., 1979; NDSU, 1975).

After 3 to 5 years of alfalfa, fallowing may be necessary before another crop can be grown. However, if there is 76 mm, or more, of stored water in the profile, then those shallow rooted crops that respond well to biologically fixed N, such as spring barley, sunflower, or corn, should be grown. Fallowing after safflower or sunflower should be avoided if possible, because these crops do not provide enough crop residue to protect the soil from erosion, particularly during the second fall and winter period of fallow.

In continuous cropping systems, winter wheat should not be grown after safflower or sunflower because these oilseed crops are normally harvested later than the recommended seeding date for winter wheat. In addition, most cereal grain crops are sensitive to possible residues of the herbicides used on oilseed crops, but the level of sensitivity varies. The relative sensitivity of barley, spring wheat, or oats seeded directly into a trifluralin-treated soil layer showed respective stand establishments of 90, 50, and 5% (Bergman et al., 1979). Therefore, spring barley is probably the best choice of crops following an oilseed crop (Table 5). However, for this same reason, safflower or sunflower should not be grown after spring barley because barley is resistant to the herbicide and thereby escapes control as a volunteer weed.

7 RESIDUE MANAGEMENT

Overwinter soil water storage increased proportionately to stubble height overwinter (Willis and Haas, 1969; Smika and Whitfield, 1966; Black and Siddoway, 1977; Willis and Frank, 1975) and to the quantity of residue maintained on the soil surface during idle periods between crops (Greb et al., 1970; Black, 1973a). Quantity of crop residue returned to the soil proportionately improves specific chemical and physical soil properties (Black, 1973b; Ferguson, 1967), such as increased soil organic matter, soil nitrogen, soil carbon, available phosphorus, exchangeable potassium, nitrogen mineralization, and dry aggregate soil structure. These changes, plus an associated decrease in soil bulk density in the 0 to 7.6 cm depth, improve soil fertility and physical structure. Therefore, the quantity of crop residue produced and returned to the soil becomes an important factor in maintaining soil productivity.

With or without N-fertilization, wheat grown on fallow produced 30 to 40% more straw than wheat grown in annual cropping sequences (Black and Siddoway, 1976). With 34 kg ha^{-1} of N applied, winter wheat after fallow produced a 12-year average of 2490 and 4150 kg ha^{-1} of grain and straw, respectively. Winter wheat grown after spring wheat in the spring wheat-winter wheat-fallow rotation produced 2030 and 2900 kg ha^{-1} of grain and straw, respectively. In the latter case, fallow increased grain yields 23% and straw yields 43%. However, the total amount of straw returned to the soil was about 800 kg $ha^{-1}yr^{-1}$ greater in an annual cropping system.

The relatively large quantities of straw present from winter wheat grown on fallow in years of above average precipitation causes some difficulty in preparing a seedbed for the next crop. Management of large amounts of crop residues is a major concern in any cropping system involving more than one crop every two years. Burning crop residue to facilitate seeding operations is not the answer. However, fewer residue management problems are encountered in annual cropping systems because less crop residue is present, and seedbeds thus require little or no tillage. Both quantity and height of stubble remaining in the field can be controlled by carefully selecting crop varieties, adjusting the combine platform height, and chopping, bunching, or bailing a selected amount of straw for removal from the field. Recent developments and acceptance of short-strawed semidwarf varieties of spring wheat with high yield potentials has made recropping and no-till seeding of other crops feasible (Black and Siddoway, 1977; Halvorson et al., 1976). Conventional no-till drills for seeding crops in undisturbed crop stubble are being developed. To obtain the best stand while not aggravating weed problems, the types of devices needed for placing seed in undisturbed stubble, involve in-line mountings of a plain coulter, double disk opener, and a V-shaped steel press wheel (Krall et al., 1979). Krall et al.(1979) reported that minimum disturbance of the soil during no-till slot planting reduces weed populations considerably as compared to seeding with openers that moved varying amounts of soil between the rows.

8 CONCLUSIONS

Controlling saline seep development requires a thorough evaluation of the necessity of summer fallow in existing and potential cropping systems. The practice of summer fallowing a given field every other year restricts a farmer to a fixed cropping system with limited flexibility for adjusting cropping patterns to fit available water supplies. The selection of alternative cropping strategies must be based on a knowledge of the total supply of water available at any given time, the specific water requirements and rooting depth of various crops adaptable to the area, and expected growing season precipitation. Summer fallow must be used judiciously instead of as a regular practice.

Paramount in importance is a knowledge of the quantity and depth of available water stored within the effective rooting zone in the recharge area to be cropped, including a knowledge of the rooting depth to some restricting or impermeable soil layers. The depth to any existing shallow water table should be known, as well as the salinity level of such water, which can affect rooting depth and water use by crops. As salinity increases, rooting depth and water use decrease proportionately.

Once the quantity and depth of available soil water are known, the crop best suited to a given supply of water can be selected. On deep soils, alfalfa can remove more soil water than any other adaptable crop in a 3 to 5 year period to depths of 6 m if no root restricting factors are present. Biennial sweet clover ranks second in soil water use in a 2-year time period. Of the annual crops, safflower uses the most soil water to the greatest soil depth followed in order by sunflower, winter wheat, rapeseed, spring wheat, spring barley, and corn.

After available soil water is depleted to the maximum depth possible for a given soil, a flexible cropping system can be developed. The cropping system can include cereal grains, oilseed crops, and limited use of fallow: the crop choice should be based on stored soil water at planting time plus expected growing season precipitation. If, with time, enough water should escape below the root zone of relatively shallow-rooted crops to become a potential source of water for regenerating a saline seep area, then a deep rooted crop would have to be grown again, depending on the depth, amount, and positional availability of the water present. However, soil water supplies in the upper soil profile must be adequate to establish a deep-rooted crop and to support root growth to reach the deeper soil depths before soil water can be depleted . Saline seep areas can be hydrologically controlled in a relatively short time (2 to 5 years), provided a major portion of the contributing recharge area is cropped with the suggested cropping strategy in mind.

The physical and economic success of any cropping strategy depends strongly on other important soil-management and water-management practices. Methods of increasing water conservation during noncrop periods must include the control of volunteer cereal grains and weeds in the fall plus provisions for snow management through proper use of crop residue and vegetative barriers in the winter.

Leaving crop stubble standing or establishing the tall wheatgrass barrier system, or both, to trap and hold snow augments available water supplies. Optimizing the efficient use of available soil water supplies and growing season precipitation requires careful variety selection, a good weed program, optimum fertilization, and timely farming operations.

9 REFERENCES

Ballantyne, S.K.,1963. Recent accumulation of salts in the soil of southeastern Saskatchewan. Can. J. Soil Sci. 48: 43-48.

Bauer, A., 1968. For dryland wheat production: Evaluation of fallow to increase water storage. North Dakota Agric. Exp. Sta. Farm Res. 25(5): 6-9.

Bergman, J.W., Hartman, G.P., Black, A.L., Brown, P.L., and Riveland, N.R., 1979. Safflower production guidelines. Mont. Agric. Exp. Sta., Bozeman, MT, Capsule Info. Series, No. 8 (Revised). July 1979.

Black, A.L., 1973a. Crop residue, soil water, and soil fertility related to spring wheat production and quality after fallow. Soil Sci. Soc. Am. Proc. 37: 754-758.

Black, A.L., 1973b. Soil property changes associated with crop residue management in a wheat-fallow rotation. Soil Sci. Soc. Am. Proc. 37: 943-946.

Black, A.L., and Ford, R.H., 1976. Available water and soil fertility relationships for annual cropping systems. pp. 286-290, *In* Proc. Regional Saline Seep Control Symposium, Mont. State Univ., Bozeman, MT, Coop. Ext. Serv. Bull.1132.

Black, A.L. and Siddoway, F.H., 1971. Tall wheatgrass barriers for soil erosion control and water conservation. J. Soil Water Conserv. 26: 107-111.

Black, A.L. and Siddoway, F.H., 1976. Dryland cropping sequences within a tall wheatgrass barrier system. J. Soil Water Conserv. 31: 101-105.

Black, A.L. and Siddoway, F.H., 1977. Winter wheat recropping on dryland as affected by stubble height and nitrogen fertilization. Soil Sci. Soc. Am. J. 41: 1186-1190.

Black, A.L., Siddoway, F.H. and Brown, P.L., 1973. Summer fallow in the Northern Great Plains (winter wheat). pp. 36-50. *In* Summer Fallow in the Western United States. Conserv. Res. Rept. No. 17, U.S. Dept. of Agric., Washington, D.C.

Brown, P.L., 1971. Water use and soil water depletion by dryland winter wheat as affected by nitrogen fertilization. Agron. J. 63: 43-46.

Brown, P.L. and Cleary, E., 1978. Water use and rooting depths of crops for saline seep control. pp. 7.1 to 7.7. *In* Dryland Saline Seep Control; Proceedings meeting of sub-commission on salt-affected soils at 11th. Int. Soil Sci. Soc. Congr., Edmonton, Alberta, Canada, June 1978.

Brown, P.L., Cleary, E.C. and Miller, M.R., 1976. Water use and rooting depths of crops for saline seep control. pp. 125-136. *In* Proc. Reg. Saline Seep Control Symposium, Mont. State Univ., Bozeman, MT, Coop. Ext. Serv. Bull.1132.

Brown, P.L. and Miller M.R., 1978. Soil and crop management practices to control saline seeps in the U.S. Northern Plains. pp. 7-9 to 7-15, *In* Dryland Saline Seep Control; Proceedings meeting of subcommission on salt-affected soils at 11th. Int. Soil Sci. Soc. Congr., Edmonton, Alberta, Canada, June 1978.

Choriki, R.T., Smith, C.M., Brown, P.L. and Dubbs, A.L., 1971. Evaluation of dryland cropping systems for central and northcentral Montana. *In* Proc. Saline Seep-Fallow Workshop, Feb. 22-23, 1971. Highwood Alkali Control Association, Highwood, MT.

Doering, E.J. and Sandoval, F.M., 1975. Saline-seep development on upland sites in the Northern Great Plains. USDA, ARS, NC-32.

Doering, E.J. and Sandoval, F.M., 1976. Hydrology of saline seeps in the Northern Great Plains. Trans. ASAE 19: 856-861, 865.

Dubbs, A.L., 1976a. Dryland cropping systems in central Montana. pp. 145-153. *In* Proc. Reg. Saline Seep Control Symposium, Mont. State Univ., Coop Ext. Service, Bozeman, MT, Bull.1132.

Dubbs, A.L., 1976b. Fertilization of dryland crop rotations. pp. 293-294, *In* Proc. Reg. Saline Seep Control Symposium, Mont. State Univ. Bozeman, MT, Coop. Ext. Serv. Bull.1132.

Ferguson, W.S., 1962. Effect of intensity of cropping on the efficiency of water use. Can. J. Soil Sci. 43: 156-165.

Ferguson, W.S., 1967. Effect of repeated applications of straw on grain yields and on soil properties. Can. J. Soil Sci. 47: 117-121.

George, E.J., Broberg, D. and Worthington, E.L., 1963. Influence of various types of field windbreaks on reducing wind velocities and depositing snow. J. Forestry 61: 345-349.

Greb, B.W. and Black, A.L., 1961. New strip cropping pattern saves moisture for dryland crops. Crops and Soils 13:23.

Greb, B.W. and Black, A.L., 1971. Vegetative barriers and artificial fences for managing snow in the Central and Northern Plains. Symposium on Snow and Ice in Relation to Wildlife and Recreation. Ames, Iowa. Feb. 11-12. pp. 96-111.

Greg, B.W., Smika, D.E. and Black, A.L., 1970. Water conservation with stubble mulch fallow. J. Soil Water Conserv. 25: 58-62.

Greenlee, G.M., Pauluk, S. and Bowser, W.E., 1968. Occurrences of soil salinity in drylands of southwestern Alberta. Can. J. Soil Sci. 48: 65-75.

Haas, H.J. and Willis, W.O., 1962. Moisture storage and use by dryland spring wheat cropping systems. Soil Sci. Soc. Am. Proc. 26: 506-509.

Haas, H.J., Willis, W.O. and Bond, J.J., 1973. Summer fallow in the Northern Great Plains (Spring Wheat). pp. 12-35. *In* Summer Fallow in the Western United States. USDA Conserv. Res. Rept. No. 17 U.S. Dept. Agric., Washington, D.C.

Halvorson, A.D. and Black, A.L., 1974. Saline-seep development in dryland soils of Northeastern Montana. J. Soil Conserv. 29: 77-81.

Halvorson, A.D., Black, A.L., Sobolik, F.and Riveland, N., 1976. Proper management--Key to successful winter wheat recropping in Northern Great Plains. North Dakota Agric. Exp. Sta. Farm Res. 33(4): 3-9.

Halvorson, A.D. and Reule, C.A., 1976. Controlling saline seeps by intensive cropping of recharge areas. pp. 115-124. *In* Proc. Reg. Saline Seep Control Symposium. Montana State Univ. Coop. Ext. Service, Bozeman, MT, Bull. 1132.

Halvorson, A.D. and Reule, C.A., 1980. Alfalfa for hydrologic control of saline seeps. Soil Sci. Soc. Am. Proc. J. 44: 370-374.

Halvorson, A.D. and Rhoades, J.D., 1974. Assessing soil salinity in identifying potential saline-seep areas with field soil resistance measurements. Soil Sci. Soc. Am. Proc. 38: 576-581.

Halvorson, A.D. and Rhoades, J.D., 1976. Field mapping soil conductivity to delineate dryland saline seeps with four electrode technique. Soil Sci. Soc. Am. J. 40: 571-575.

Krall, J., Dubbs, A. and Larson, W., 1979. No-till drills for recropping. Montana Agric. Exp. Sta. Bull. 716.

Kresge, P.O. and Halvorson, A.D., 1979. FLEXCROP USER'S MANUAL. Computer-assisted dryland crop management. Coop. Ext. Serv., Montana State Univ., Bozeman, MT. Bull. 1214.

North Dakota State University, 1975. Sunflowers, production, pests and marketing. Cobia, D. and Zimmer, D. (Editors). Ext. Bull. 25, June 1975.

Schneider, R.P., Johnson, B.E. and Sobolik, F., 1980. Saline seep management: Is continuous cropping an alternative? North Dakota Agric. Exp. Sta. Farm Res. 37(5): 29-31.

Smika, D.E. and Whitfield, C.J., 1966. Effect of standing wheat stubble on storage of winter precipitation. J. Soil Water Conserv. 21: 138-141.

Staple, W.J. and Lehane, J.J., 1955. The influence of field shelterbelts on wind velocity, evaporation, soil moisture, and crop yield. Can. J. Agric. Sci. 35: 440-453.

Van der Pluym, H.S.A., 1978. Extent, causes and control of dryland saline seepage in the Northern Great Plains Region of North America. pp. 1-48 to 1-58. *In* Dryland Saline Seep Control, Proceedings Mtg. of the Subcommission on salt-affected soils, 11th. Int. Soil Sci. Soc. Congr., Edmonton, Alberta, Canada, June 1978.

MANAGEMENT OF SOIL WATER BUDGETS OF RECHARGE AREAS FOR CONTROL OF SALINITY IN SOUTH-WESTERN AUSTRALIA.

R.H. SEDGLEY,
Department of Agronomy, University of Western Australia,
Nedlands, W.A., 6009.

R.E. SMITH,
U.S.D.A. S.E.A.,
P.O. Box E, Fort Collins, 80522, Colorado, U.S.A.

D. TENNANT,
Department of Agriculture,
South Perth, W.A., 6151.

ABSTRACT

Sedgley, R.H., Smith, R.E. and Tennant, D., 1981. Management of soil water budgets of recharge areas for control of salinity in south-western Australia. Agric. Water Manage., 1981.

Replacement of deeper-rooted indigenous flora with relatively shallow-rooted agricultural species typically causes small but consequential changes in the water balance in south-western Australia. Such changes, estimated as from 23 - 65 mm yr^{-1} out of annual rainfalls of from 400 - 1100 mm yr^{-1}, have been estimated to cause significant increases in streamflow salinities.

One method to help control salinization of land and stream water in agricultural areas is to manage the crop to utilize water so as to minimize recharge to the salt storage region. In studying the potential for recharge control in wheat belt agriculture, water profiles under wheat (*Triticum aestivum*) and subterranean clover (*Trifolium subterraneum L.*) were compared at a 385 mm yr^{-1} area. Mostly because of the shallower rooting, the annual recharge under clover was almost twice that of wheat.

In another demonstration, a numerical simulation of the same two species was performed in a high rainfall (680 mm yr^{-1}) zone. Similar differences were obtained, though not as dramatic, with the model providing insight into relations between recharge and rain intensity patterns.

Suggestions are made for several strategies to minimize recharge during and after the growing season, by better management of existing species, by the introduction of deeper-rooted species, and by use of crops which are more in phase with the availability of water.

1 INTRODUCTION

The increased average annual recharge responsible for salinization of land and streams over much of south-western Australia is generally small - of the order of 23 to 65 mm yr^{-1} (Peck and Hurle, 1973) - relative to annual rates of potential evaporation (> 1500 mm yr^{-1}). Peck (1977) concluded from this that the prevention or reclamation of dryland salinity in the region by the most generally satisfactory method - land management practices designed to maintain or restore catchment water balances to those of the pristine environment - was physically feasible.

To develop new management strategies which are also economic, is a formidable challenge. In the mediterranean environment of Western Australia endemic summer drought and infertile subsoils mitigate against the most direct solution to the problem, viz: exploitation of water by deep-rooted perennial crops. So far attempts to introduce such crops have not succeeded and a return to the indigenous vegetation in areas where agriculture is practiced is not acceptable, because of its low economic value.

Water lost from the root zone to recharge represents a loss of production. Dry matter increase is quantitatively related to the use of water by crops (de Wit, 1958). Each millimetre of water lost to recharge in the region can be equated with a loss of 10 kg ha^{-1} of grain (Rickert, K.G. and Sedgley, R.H., unpublished data). Excessive recharge can also be related to losses of production and increased costs due to excessive leaching of nutrients and an increase in rate of soil acidification in the root zone.

In this paper we briefly outline theory relating to analysis and simulation of soil water budgets in recharge areas. Ways in which their components may be affected by the interaction between weather patterns and changes caused by clearing of the original deep-rooted flora are summarized. Limited local data from the cereal belt are presented to indicate the scope for minimizing recharge by agronomic approaches. Recharge is simulated as a function of rainfall pattern in a first step towards a systematic analysis of the control of salinization by management of recharge. Emphasis is on the hydrological characteristics of agricultural species.

2 ANALYSIS AND SIMULATION OF THE WATER BALANCE IN RECHARGE AREAS

Evidence for the substantial disparity of rooting depth (and associated hydrological differences) between indigenous forest or woodland (Specht and Rayson, 1957, and Kimber, 1974) and introduced agricultural species (Ozanne et al., 1965) suggests that clearing will generally be accompanied by some increase in drainage beyond the root zone. This is supported by the estimates by Peck and Hurle (1973) of additional recharge under agricultural species in the higher rainfall areas (receiving > 500 mm yr^{-1}) of south-western Australia following clearing and of Holmes and Colville (1968) for crop and grassland in South Australia.

In attempting to restore water balances to those of the pristine state the problem of reducing recharge can be separated into two parts:

(a) in situ reduction of recharge by relatively shallow-rooted agricultural species;

(b) extraction of recharge below the root zone of agricultural species by deep-rooted native or other tree species.

The need for strategic location of deep-rooted species for maximum extraction may involve lateral drainage of recharge away from cropland or pasture to adjacent belts of deep-rooted species where it is extracted.

Recharge in situ can be described by one-dimensional theory of soil water flow, whereas two-dimensional drainage theory is required for a simple description where lateral drainage is involved.

2.1 The Water Balance

For a given depth of soil in the recharge phase any change in water stored over a period of time - the water balance - in the root zone is given in units of depth of water by :

$$\Delta S = P - E_t - U \tag{1a}$$

where ΔS = the change in storage (mm); P = rainfall (mm); E_t = evapotranspiration (mm);and U = recharge or drainage below the rooting zones (mm).

Over a number of years in the pristine environment ΔS is assumed to average zero on an annual basis, indicating an equilibrium water balance. Average P is assumed not to change, but E_t and U are expected to change in varying degrees as a result of clearing. E_t can be analyzed in terms of components, E_{pl}, E_s and E_i: plant soil and interception evaporation respectively. Runoff and surface ponding are assumed negligible for the permeable soils that constitute the recharge areas.

2.2 Evapotranspiration

Evapotranspiration (E_t) can be determined in a number of different ways (see Shuttleworth, 1979), including the energy balance - Bowen ratio approach, eddy correlation method, lysimetry, aerodynamic method and combination methods of which that originally due to Penman (1948) is widely applied. For purposes of modelling effects of land use changes on evapotranspiration the combination approach is the most useful because it incorporates biophysical parameters subject to such changes (see Feddes et al., 1978).

The combination equation as modified by Monteith (1964) and others is given as:

$$LE_t = \frac{\Delta(R_n - G) + \rho C_p \{e_s[T(z)] - e(z)\}/r_a}{\Delta + \gamma(1 + r_c/r_a)} \tag{2a}$$

where LE_t = evapotranspiration (Wm^{-2}); LE_t (Wm^{-2}) x 0.352×10^{-1} = E_t (mmd^{-1});
L = latent heat of vaporization ($J\ kg^{-1}$);
R_n = net radiation (Wm^{-2}),
= $R_s(1 - \alpha) + R_N^L$ where R_s is the incoming total global short wave radiation; α is the albedo or short wave reflection coefficient;and R_N^L is the net long wave radiation between the surface and its environment;
G = soil heat or storage flux density (Wm^{-2}); ρ = density of air ($kg\ m^{-3}$); Δ = slope of the saturation vapour pressure curve for water ($mb\ {}^oC^{-1}$); C_p = specific heat of air ($J\ kg^{-1}\ {}^oC^{-1}$); γ = psychrometric constant ($mb\ {}^oC^{-1}$); $e_s(T(z))$ = saturation vapour pressure of water at temperature T at height z above the ground surface (mb); e(z) = vapour pressure at height z;
r_a = aerodynamic resistance under neutral conditions of stability ($s\ m^{-1}$),
= $\{\ln(\frac{z - d}{z_o})\}^2 / k^2u$, in which z_o is the aerodynamic roughness parameter of the vegetation; d is the zero-plane displacement; k is von Karman's constant; and u is the horizontal wind velocity; z_o and d functions of vegetation height (m);
r_c = surface or canopy resistance ($s\ m^{-1}$).

Land clearing may significantly affect LE_t through albedo effects on net radiation, the effect of vegetation height and roughness on r_a and through the effect of vegetation canopy geometry and leaf anatomy on r_c. r_c can often be given as a function of stomatal resistance (r_s) and leaf area index (L); for cereals Monteith et al. (1965) found $r_c \simeq r_s/2L$. This may not apply to less homogeneous surfaces where a relationship between r_c and ground cover is more appropriate (Feddes, 1971).

When r_c in equation (2a) goes to zero, as for a completely wetted surface - e.g., open water or just after rain - we write the expression as:

$$LE_o = \frac{\Delta(R_n - G) + \rho C_p \{e_s[T(z)] - e(z)\}/r_a}{\Delta + \gamma} \tag{2b}$$

where LE_o (Wm^{-2}) is the potential evaporation of the surface characterized radiometrically by the albedo, and aerodynamically by r_a for a range of wind speeds. Other symbols are as before.

2.3 Simulation Of Recharge Through the Root Zone

To simulate the movement of water into and out of the root zone of a plant, we use the basic equation for unsaturated flow in porous media. We require information on root depth, extraction pattern in response to water availability, soil hydraulic characteristics, and rainfall rate pattern. The extent of information required for each process is a function of a sensitivity of our results to the respective process, which in turn depends on the particular objective of our simulation. The model outlined below employs assumptions made with regard to the objective of predicting upland recharge. One of the values of simulation is to assemble interactive processes and verify our assumptions regarding sensitivity to the parameters involved.

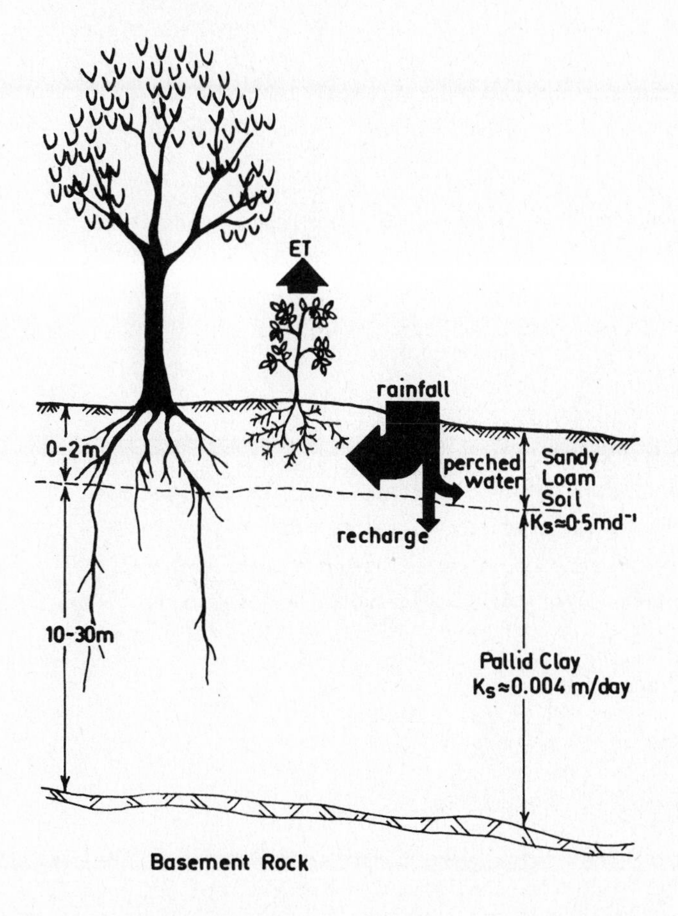

Fig. 1. Characteristic soil profile of recharge areas in south-western Australia.

For the soil mantle typical of Western Australian agricultural areas, we may assume the soil profile to be roughly typified as in Fig. 1. A relatively porous soil overlies a comparatively tight clay layer of great depth, within which flow occurs mostly in occasional cracks and root channels. A saturated layer, or perched aquifer, is often formed at the interface during the rainfall season. Root depth for annual crops will vary through the year.

The original forest vegetation is able to take water from the upper soil when it is available, and the trees are rooted into the clay for a water source during the dry periods (Carbon et al., 1980). Rainfall rates are very rarely greater than the saturated hydraulic conductivity K_s, of the upper soil so that only rarely is there overland flow from rainfall excess, by the classical hydrologic response often known as the Horton mechanism. The simulation model encompasses this process but it is not described here.

Our treatment of water flow through the soil depends on assumptions related to both the unsaturated flow mathematics and the particular properties of the system outlined above. The general one-dimensional equation is (Philip, 1969):

$$\frac{\partial \theta}{\partial t} - \frac{\partial}{\partial z} \{D(\theta)\frac{\partial \theta}{\partial z} - K(\theta)\} = 0 \tag{3}$$

where θ = soil water content by volume ($m^3\ m^{-3}$); t = time (s); z = distance below soil surface (m); $K(\theta)$ = hydraulic conductivity (ms^{-1}); and $D = K(\theta)/\frac{d\theta}{d\psi}$ is soil water diffusivity ($m^2 s^{-1}$).

The second or Darcy term in equation (3) is alternatively $\partial V/\partial z$, where:

$$V = -D(\theta)\frac{\partial \theta}{\partial z} + K(\theta) \tag{4}$$

This is a flux rate term composed of a diffusive and gravitational component. In our system θ is characteristically below θ_0 (saturated θ), and D varies most at high values of θ. Furthermore, our time scale is large with respect to the time scale of diffusive velocities. Therefore we assume a water movement in which each change in soil flux is an individually routed nonlinear wave. Thus $D(\theta)\partial\theta/\partial z$ is some function of θ, i.e. $D(\theta)\partial\theta/\partial z = \xi(\theta)$, so that $V = V(\theta)$ alone, and we may write, adding a loss term to equation (3):

$$\frac{\partial \theta}{\partial t} + \frac{dV}{d\theta}\frac{\partial \theta}{\partial z} = -\Delta(\theta) \tag{5}$$

Where Δ is local rate of evaporative extraction.

At larger times, such as necessary for water to move from the soil surface to below the root zone, gravitational flow dominates the wave movement, so that we assume that in equation (4) $dK/d\theta >> d\xi/d\theta$. Equation (5) may in any case be solved by characteristic methods, with each wave routed through the soil by tracing the advancing characteristics (see Henderson and Wooding, 1964).

Another assumption which adds efficiency to the simulation procedure is the mathematical expression of unsaturated soil hydraulic characteristics. For this simulation the relation of Brooks and Corey (1964) is modified. The original relations relate effective saturation, S_e, to other properties as follows:

$$S_e = \frac{\theta - \theta_r}{\theta_o - \theta_r} = \begin{cases} \left(\frac{P_b}{\Psi}\right)^{\lambda}, & \Psi > P_b \\ 1, & \Psi \leqslant P_b \end{cases} \tag{6a}$$

$$K = S_e^{\frac{2 + 3\lambda}{\lambda}} \tag{6b}$$

where Ψ is capillary potential (m) and λ, θ_r and P_b are parameters.

In this simulation, the original handicap of a singularity at $\Psi = P_b$ is removed by expressing :

$$\{S_e - (\frac{P_b}{\Psi})^{\lambda}\}(S_e - 1) = C_c \tag{6c}$$

where C_c is a curvative coefficient.

Expression (6b) and (6c) are used to describe soil properties, including the wave characteristic velocity, v_c.

From equations (6a) and (6b), we have :

$$v_c \equiv dK/d\theta = \varepsilon\, S_e^{\varepsilon - 1}\, dS_e/d\theta \tag{7}$$

$$= \frac{\varepsilon}{(\theta_o - \theta_r)} \left|\frac{\theta - \theta_r}{\theta_o - \theta_r}\right|^{\varepsilon - 1} \tag{8}$$

where $\varepsilon = \frac{2 + 3\lambda}{\lambda}$ (9)

A rainfall pulse at the rate r at the surface where $r \ll K_s$ results in a square wave moving in the z-direction, of magnitude θ, given by equations (6a) and (6b), solved for θ (K = r). Use of rain patterns composed of a series of square waves, as commonly obtained from weather records, results in "shock" waves of θ moving downward in response to the gravitational potential gradient. The leading edge with water content jump from θ_1 to θ_u moves with velocity $\{K(\theta_u) - K(\theta_1)\}/\{\theta_u - \theta_1\}$. The various points on the trailing wave, where θ changes from θ_u to θ_1, move with velocities v_c.

Since values of ε typically range from approximately 3 to 5 (Brooks and Corey, 1964), v_c increases rather rapidly with θ, and the trailing waves elongate and overtake leading waves. After overtaking, the wave attenuates and loses net velocity. Treatment of a random sequence of waves, as in rainfall, requires careful routing and accounting as much overtaking occurs. The initial velocity of a shock is related directly to the pulse rate of the rain, and thus the initial routing of rainfall is dependent on an accurate record. The wave model of soil water flow under appropriate conditions may be verified by comparison with numerical solution of equation (3). More detailed discussion is beyond the scope of the present work.

At the lower boundary, the kinematic wave model must assume no upward effect from the lower boundary. This is not an unreasonable assumption when (a) our concern is with what penetrates below the root zone, and (b) the capillary "fringe" is small as in the case of sand. In any case, the vertical flux is only minimally affected by water table proximity beyond a depth equal to 2 or 3 times the "bubbling pressure" P_b.

This simulation allows us to treat a long sequence of rainfall events with relatively large time steps, so long as our assumptions hold approximately.

2.4 Approximation of Plant Evapotranspiration (E_t).

In connection with simulating the water fluxes resulting from a seasonal pattern of rainfall, the loss of water through the root zone may be approximated by a simplified model of root growth, plant water demand, and plant response to water stress. The model described herein is currently still being developed and improved, and our present plant model uses a simplified plant transpiration model from Feddes et al. (1978), which assumes that at any given rooting depth water is taken at potential rate up to a limiting value of total root potential H_c, and declines linearly for increasing potentials beyond H_c up to 15 bars. H is defined as $z - \Psi$.

Plant potential water use is related to leaf area index L or LAI by assuming plant demand increases with L up to a limit, depending on species (Fig. 3a). L is described by a growth curve relating L to time. We assume rooting depth increases linearly with L. Potential demand of the plant surface (E_o) is related to meteorological factors by assuming E_o to be related to pan evaporation, E_{pan} simply as:

$$E_o = 0.51\ E_{pan} + 0.49 \qquad (10)$$

which has been obtained from E_o-pan correlations (Sedgley, 1979). Pan evaporation data is obtained from local records, as is half hourly rainfall data. Evaporation is assumed to be negligible during periods when rain is falling and soil evaporation is not considered.

The plant model estimates a value $\Delta\theta$ (θ), and over a given time period, Δt, evaporation is calculated as :

$$Et\ (\Delta T) = \int_0^{\Delta t} \int_0^{z_t} \Delta(\theta)\ dz\ dt \tag{11}$$

2.5 Estimation of Recharge from Water Content Profiles

For analysis of field water content profiles, recharge - drainage below the root system - is described by the velocity term :

$$U = \left[-D(\theta)\frac{\partial\theta}{\partial z} + K(\theta)\right]_{z = L} \tag{12a}$$

which in a gravity dominated system, may be approximated by :

$$U = [K(\theta)]_{z = L} \tag{12b}$$

where L is the depth of the rooting zone.

For computation equations (1a) and (12b) are combined and written as :

$$\Delta S = P - \Delta t\ \bar{K}(\theta)_{z = L} - E_t \tag{1b}$$

where ΔS, P and E_t are the changes in storage, rainfall and evaporation during the time interval Δt. $\bar{K}(\theta)_{z = L}$ is the time averaged hydraulic conductivity at depth $z = L$.

2.6 Interaction Between Adjacent Areas of Deep and Shallow Rooted Vegetation.

Peck (1975) analysed the situation where relatively small areas of phreatophytic vegetation are used to transpire recharge from crops and pastures where they replace the original deep-rooted vegetation higher in the catchment.

From simple water balance equations it can be shown that the ratio of areas of phreatophytes A_p(ha) and agricultural species A_g(ha) for total transpiration of unused water is given by :

$$A_p/A_g = \Delta U/(E_p - E_t) \tag{13}$$

where ΔU = the difference between groundwater recharge under the crop or pasture and the forest vegetation (mm yr^{-1}); E_p = the evaporation rate from the phreatophytic vegetation (mm yr^{-1}) regarded as approximately equal to the potential evaporation rate E_o given by equation (2b); E_t = the evaporation rate from the crop (mm yr^{-1}) given by equation (2a).

In the majority of areas in the Darling Ranges, Peck estimated that a ratio of A_p/A_g of 1:10 would be needed to maintain the original water balance of the area. This simplified approach assumes that recharge from the root zone of crops and pastures moves laterally through the soil until it is intercepted and extracted by deep-rooted vegetation sown in belts across the flow paths.

Assuming a porous uniform soil overlying a horizontal slightly permeable clay layer, for phreatic vegetation planted in parallel layers, water is assumed to flow to the roots as groundwater flows to a drainage ditch. For this situation Peck calculated the maximum spacing l between belts of phreatophytes using drainage theory (Van Schilfgaarde, 1957):

$$l = 2\{(H_o^2 - h_o^2)K/U\}^{\frac{1}{2}} \quad (14)$$

where l = the distance between drainage ditches (or belts of phreatophytic vegetation) (m); H_o = maximum height of the water table above the impermeable layer midway between ditches (m); h_o = height of the water table above the basement in the ditches (or below the phreatophytes) (m); K = the hydraulic conductivity of the soil (m d^{-1}); U = the average rate of recharge of water beneath the pasture or crop (m d^{-1}).

3 HYDROLOGICAL CONSEQUENCES OF REMOVING DEEP-ROOTED VEGETATION

Any successful approach to salinity control directed towards restoring pristine water balances must involve a detailed understanding of how such balances were achieved, and to what extent they are changed by clearing. This will be dealt with in another paper and mention here will be restricted to aspects immediately relevant to management of soil water budgets in recharge areas.

3.1 Interception

Changes in the interception component of evapotranspiration (E_i) due to differences in canopy properties are possible. Effects of interception on E_i may be large over small time scales, but the effect of a large E_i component on annual E_t may still be only small. In a study of the effects of mediterranean vegetation on the water balance, Shachori et al. (1967) concluded that although total annual interception was 19 mm the additional annual evaporation or net

4 REDUCTION OF RECHARGE BY AGRICULTURAL SPECIES

4.1 Current Distribution of Species

Agriculture in the high rainfall areas (> 500 mm yr^{-1}) is primarily wool and meat production and dairying based on subterranean clover (*Trifolium subterraneam* L.) and associated annuals. As rainfall recedes towards the east the pasture phase of the rotation shortens and makes way for cereals, mainly wheat. These dominate the drier limits (270 - 320 mm yr^{-1}) of the cereal belt. This trend in pasture production stems partly from the inability of the subterranean clover to regenerate under the lighter more variable rainfall pattern. The outstanding virtue of subterranean clover is its ability to continue growing and to set and bury seed under intense grazing.

Quinlivan and Francis (1976) estimated that of 7 000 000 ha of sown pasture in the region all but 500 000 are in subterranean clover. A major proportion of these pastures occupy 'light land'* which includes the recharge areas for aquifers responsible for land and stream salinization throughout the region (Smith, 1962; Bettenay et al., 1964). Subterranean clover is well adapted to 'light land' with its slight to moderately acid soils - a condition indicative of leaching and hence of recharge.

4.2 Evaluation of Recharge Under Different Species

Water lost by drainage below the root zone represents inefficiency in the use of water. We shall attempt by water balance simulation to evaluate the efficiency of water use by existing agricultural species and to assess the scope for increasing the efficiency of water use in the region by selection from the wider range of species available.

4.2.1 Field trial at Wongan Hills

Data were obtained from a field trial conducted on a deep loamy sand at Wongan Hills (Lat. S 30° 52', Long. E 116° 43') in the cereal belt (mean rainfall 384 mm yr^{-1}) in 1966 (Tennant, unpublished data). Species treatments were replicated three times in a randomized block design. Treatments sown on 20/6/66 comprised *Trifolium subterraneum* L. (subterranean clover), *Trifolium hirtum* (rose-clover), *Triticum aestivum* L. (spring wheat cv. Gamenya), *Secale cereale* (cereal rye), bare fallow, and wheat undersown with subterranean clover and wheat undersown with rose clover.

---

* Light land comprises sands of lateric origin occupying several million hectares (Mulcahy, 1973) of land found in the upland parts of landscapes (sand plain) and on the pediments in south-western Australia. Many have sandy clay subsoils at varying depth.

interception loss was much smaller and therefore negligible. Further data on net interception loss are needed for specific sites.

3.2 Surface Cover

The duration and extent of vegetative surface cover largely determines transpiration (E_{pl}). Where vegetation is uniformly distributed over the surface as for cereals and pasture, surface cover is a function of leaf area index (L) - the ratio of green surface area of vegetation to area of ground covered and can be accounted for in the combination equation (2a) by r_c. For widely spaced tree or shrub plantings L may have no simple relationship with ground cover and R_n may also be significantly modified.

In mediterranean environments the change from perennial to annual leaf canopy is the most significant change to normally accompany clearing. In such environments duration of surface cover is highly corellated with depth of root system. The problem with shallow rooted annual species is how to compensate for their shorter growing season by producing high leaf area indices for longer periods in order to minimize drainage during winter and to achieve soil water deficits comparable to those under perennials.

The answer must come from plants achieving full canopy development soon after the first rains and in sustaining green canopies for longer into spring/summer when potential evaporation rates (E_o) are relatively high. In environments receiving 300 - 1100 mm yr^{-1} and with E_o 1500 mm yr^{-1} indigenous species growing under low fertility ration water by maintaining lower leaf area indices throughout the year than found in fully developed agricultural crops. During winter when rates of E_o are low E_t may not differ greatly between introduced and native species.

3.3 Root Systems

Control of surface cover as indicated above will depend on finding a suitable combination of canopy and root system characteristics.

The most important species root system parameters include rate of development and depth of rooting zone, density profile of roots, and diameter of water conducting vessels (see Passioura, 1972 and Carbon et al., 1979). Other characteristics however may modify the expression of these in a given environment such as tolerance to adverse soil physical, chemical and biological conditions, e.g. mechanical resistance, poor aeration, soil pathogens, aluminium and manganese toxicities in acid sub-soils and nutrient deficiencies.

Soil water content profiles were sampled gravimetrically from 9/8/66 at approximately 2 week intervals until 31/10/66 and again on 9/1/67 at depths of 0.05,

0.10, 0.15, 0.30, 0.45, 0.60, 0.75, 0.90, 1.10, 1.20 and 1.85 m. Recharge below the root zone was estimated for each sampling interval using equations (1a) and (1b). Hydraulic conductivities used were estimated by the method of Marshall (1958) from field moisture characteristic curves (Rickert, K.G., Sedgley, R.H. and Hamblin, A.P., unpublished data). Daily rainfall was available for the site.

4.2.2 Water content profiles

Water content profiles at selected times are shown in Fig. 2 for wheat, subterranean clover and bare treatments. The first profile sampling on Fig. 2 on 9/8/66, two months after sowing when surface cover would be complete, represents a fully recharged root zone. The second profile is at maturity and the third in midsummer at least a month after harvest. In the interval prior to maturity the subdued pattern of water extraction under clover at depth contrasts strongly with that of wheat indicating substantial differences in recharge during the growing season. Continuing drainage of water after maturity was greatest under clover, indicating that water not used by subterranean clover during the growing season was substantially lost as recharge over summer and therefore not available in the following season.

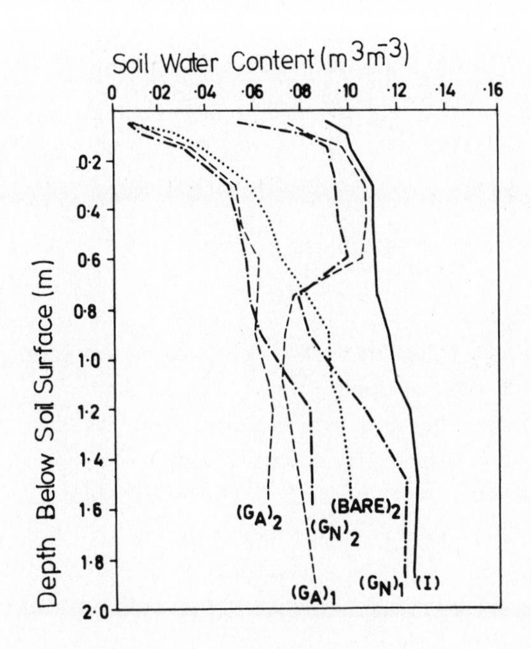

Fig. 2. Water content profiles for deep loamy sand at Wongan Hills during the season 1966/67 for wheat, subterranean clover and bare treatments: (G_A---) wheat cv. *Gamenya*; (G_N-.-.) subterranean clover, *Geraldton* strain; (BARE····) bare unsown plot; (I——) initial profile common to all treatments on 9/8/66. Subscripts 1 and 2 refer to sampling dates 31/10/66 and 9/1/67 respectively.

At the end of the season (9/1/67) the soil water deficit under wheat was greater than under clover, indicating that drainage below the root zone was likely to start earlier in the next winter after clover than after wheat.

4.2.3 Water balance and recharge

Water balances (shown in Table 1) were constructed for the period 9/8/66 to 31/10/66 by calculating recharge from the fortnightly water contents at 1.85 m using equation (12b) and ΔS was determined graphically from initial and final soil water content profiles using the mean profile of all treatments for the initial profile (maximum standard error of water contents = ± 0.003). Equation (1b) was solved for E_t.

TABLE 1

Water balance for a loamy sand over depth 0 - 1.85 m at Wongan Hills under different species for the period 9/8/66 to 31/10/66.

	Bare	Subterranean clover	Rose clover	Wheat	Wheat/ subterranean clover	Wheat/ rose clover	Cereal rye
P (mm)	103	103	103	103	103	103	103
ΔS (mm)	-	-30	-25	-57	-56	-58	-61
U (mm)	20	17	21	9	15	15	7
E_t (mm)	-	116	107	151	144	146	157
100 $E_t/P-\Delta S$ (%)		71	65	92	88	89	96

P = rainfall; ΔS = change in storage; U = drainage below 1.85 m;
E_t = evapotranspiration; 100 $E_t/P-\Delta S$ = water use efficiency.

Interpretation of recharge requires qualification. The water balance neglects the wettest part of the season when the root zone was initially recharged and some drainage should have occurred. Averaging $K(\theta)$ over two weekly periods may also under-estimate recharge because of insensitivity to drainage soon after heavy rain unless this occurs close to the sampling date. Estimated recharge therefore represents less than the annual figure for this soil.

Williamson (1978), on a highly permeable sand at Meckering in the cereal belt receiving similar rainfall, estimated annual recharge in 1966 at about 50 mm under annual pasture. Recharge, but not water use, was similar under the two clover and the wheat undersown with clover treatments. Water use in the mixed stands was similar to that in wheat alone, but recharge from the mixed stands was comparable with that for the clovers alone. Inspection of two weekly profiles indicated that in the mixed stands extraction was slower than for wheat alone and that this allowed greater loss to recharge below the rooting zone. Final soil water deficits were similar.

The above data indicate therefore that analysis of recharge should include drainage below the root zone during the current season and the carry over effect of the size of the soil water deficit on drainage in the following season.

4.2.4 Water use efficiency

Water use efficiency for the period 7/8/66 to 30/10/66 was calculated as the percentage of total water available - defined as the sum of the rainfall plus water extracted over the period down to 1.85 m for cereal rye - that could be attributed to evapotranspiration E_t. Clearly the presence of clover reduces the efficiency of water use, and this must involve loss of production and increased recharge. More complete data are required for calculating efficiency on an annual basis.

4.2.5 Implications

Data from Wongan Hills indicate that recharge on 'light land' will be greater under clover than wheat. The extent to which this is modified by depth to the clay B-horizon should be investigated. Current trends in chemical weed control designed to maintain pure clover stands will intensify recharge by eliminating deeper-rooted species normally found in annual pastures.

Attempts to extend clovers to drier parts of the cereal belt where regeneration is difficult seem likely to hasten salinization.

Alternative approaches which have prospects for minimizing recharge are:

(i) Narrowing of rotations in higher rainfall areas to include longer periods under deep-rooted agricultural species. This would essentially require better adapted and higher yielding wheats for these areas.

Part of this adaptation may involve root system modification as suggested by Passioura (1972), but with the purpose of increasing water intake from depth rather than rationing uptake of stored water (Walter and Barley, 1974).

(ii) Use of mixed stands of wheat and legume species. Early development of transpiring surface by the legume could be used to exploit soil water otherwise lost to recharge because of normally slower canopy development by wheat early in the season. Undersown legume species would encourage better utilization of stubbles by livestock and accumulation of nitrogen for wheat in the following season, thereby eliminating the pure pasture phase. Wheat cultivars incorporating the uniculm habit (Atsmon and Jacobs, 1977) may be suited for this role, and a deep-rooted legume species such as *Ornithopus* (serradella) may find a role in this situation where it is no longer vulnerable during seed set, as in normal pasture.

(iii) Use of deep-rooted grain legumes such as lupins (*Lupinus* sp.). *Lupinus* is widely recognised as a deep-rooted species. Water use, productivity and nitrogen fixation should be assessed. As for mixed wheat-legume stands new better adapted cultivars will have to be developed.

(iv) Use of winter-active strains of *Medicago sativa* (lucerne, alfalfa). Winter-active lucernes are new to Australia and must be assumed to have great potential under dryland conditions with summer drought and stored water at depth. In principle, they can extend the winter growing period into summer, survive the dry summer months and rapidly recover with the opening rains.

Previously only a single genotype (Hunter River) was available in Australia and this was a winter dormant, summer active type well suited to irrigation but unadapted to the severe summer drought of south-western Australia. Mixed stands of lucerne and clover have already been managed successfully in other parts of Australia (Wolfe and Southwood, 1980) and may be used to combine the best production features of both species and maximize water use. The above alternatives call for a review of the basis for developing new cultivars in the region. The need for genotypes with specific attributes offers scope for applying the ideotype concept of plant breeding as outlined by Donald (1968, 1980).

5 SIMULATION OF RECHARGE THROUGH ROOTING ZONE

5.1 Simulation of a Plant Season

The model described in Section 2.3 and 2.4, provides a means of routing rainfall from the surface through the root zone, and extracting water by a plant in response to the available water, in a pattern which approximates the distribution in the field. Use of the approximate routing procedure allows us to treat a long and detailed rainfall record with a relative thoroughness prohibited by the complexities and time increment constraints imposed by numerical solution of the more rigorous equation (3).

A brief demonstration of the application of the model is provided in Fig. 4 and Table 2. The rainfall year chosen (arbitrarily) is a relatively wet one, and not "typical" of the wheat growing region. Two crops are compared: wheat and sub-clover, with simulation begun 2 weeks before the modelled germination date.

The two LAI growth patterns used are illustrated in Fig. 3a., showing relatively small but distinct differences in water demand, mostly at the end of the growing period. In drier years this would be more significant than the relatively wet year 1974 whose simulation results are shown in Table 2. The soil used, Collgar loamy sand (see Fig. 3b) had the following parameters, determined from field data: P_b = 10 cm, λ = 1.0, C_c = 0.05, S_r = 0.28, S_o = 0.9, ϕ (porosity) = 0.32.

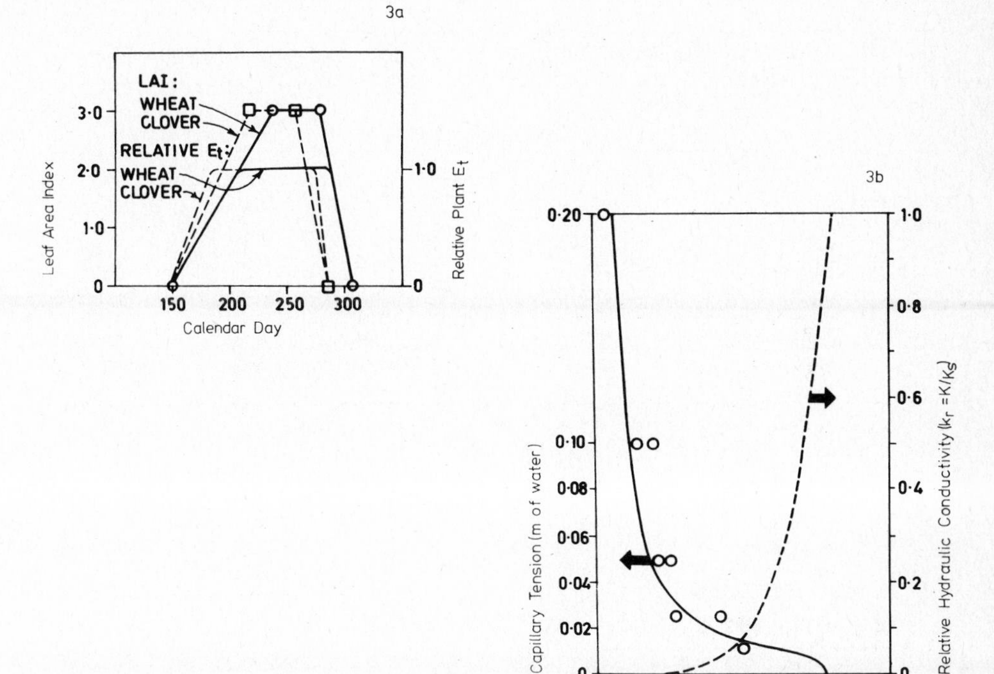

Fig. 3a. Leaf area index (L) growth patterns for wheat and subterranean clover assumed in the demonstration simulation, and relative plant evaporation rate E_t/E_0 for the same two crops. Root depth is assumed to reach its peak coincidental with L, but plant evaporation peaks earlier.

Fig. 3b. Hydraulic properties of Collgar loamy sand. Lines shown are fitted to equations (6b) and (6c).

5.2 Results

Table 2 shows the effect on water balance of the differences in root depth and plant growth period. The difference between the total E_t for the two crops is not dramatic, and so the overall water balance figures are not dramatically different. The wheat uses some 88% of its potential evapotranspiration, and the clover 81%, indicating relatively little stress at any time through the season with the rainfall extending late into the season. The wet year also caused a great depth of recharge, but the deeper roots of the wheat caused extraction of more water, deeper and later into the season, so that total recharge is reduced by 35 mm. We expect the difference would be more dramatic in dry years.

TABLE 2

Simulated comparison of two crops for drainage management.

Time (calendar days)	Crop simulated	Water* in 2 m of soil (mm)	Accumulated Rain* (mm)	Accumulated drainage (mm)	Accumulated E_t (mm)	Root depth (m)
150 (germination)		288	82	13	0	0
240	wheat	248	412	216	80	1.85
	clover	223		224	87	0.93
300	wheat	228	481	257	183	1.85
	clover	230		270	141	0.93
360	wheat	222	507	288	184	0†
	clover	141		323	141	0

* Simulation starts at day 136; these are not annual values.
† Plants assumed to die at time when LAI returns to zero.

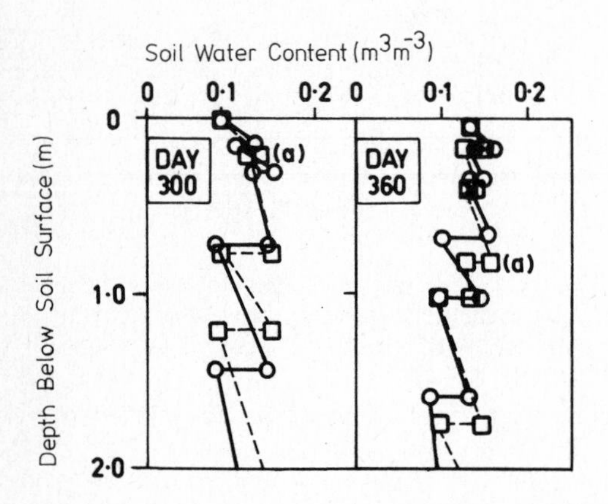

Fig. 4. Simulated water content profiles under wheat and subterranean clover at days 300 and 360, after most root growth has ceased; symbols as for Fig. 3a. The relatively uniform water contents result from the wet rainy season during which the plants were rarely stressed. Wave (a) has moved only some 400 mm in the 60 day interval.

Fig. 4 illustrates the simulated water profile at day 300 and 360. Soil under the wheat is slightly drier, especially at depth, than under the clover. Routing of rain pulse "waves" is illustrated by the translation of the wave marked (a) in the 60 day period, which has been augmented by additional rains in the interval. The subtle differences in water content are more easily recognised in differential wave velocities.

6 CONCLUSIONS

Additional recharge caused by replacing indigenous deep-rooted species is considered responsible for serious salinization of land and streams in south-western Australia. Estimated average additional annual recharge is small compared to annual potential evaporation rates. As a result of this there is considerable scope for controlling salinity through management of water budgets in recharge areas.

The objective in controlling salinity by better management of recharge areas is to increase evapotranspiration. This largely involves increasing the duration and extent of plant cover. Deeper root systems, longer growing seasons and perennial habit make this possible.

Root systems of agricultural species form a distinctly shallower rooted group by comparison with deep-rooted indigenous and exotic tree species. Within the range of agricultural species there is a wide range of root system characteristics, including depth, awaiting exploitation for controlling recharge. Subterranean clover strains are grown across almost the entire south-western region of Australia, and is one of the shallowest rooted agricultural species. On 'light land' subterranean clover may often permit substantially higher annual recharge, with consequent effects on salinity, than cereals.

Flexibility is needed in choosing management strategies at the beginning of each season. Although average additional annual recharge is small variations in environmental and plant factors from year to year means that recharge will also fluctuate. A flexible response to such variations is needed if recharge is to be controlled.

Management of water budgets in recharge areas involves two processes:

(a) minimizing recharge below root zones of agricultural species by careful selection and cultivation of deeper-rooted agricultural types,

(b) interception and extraction of recharge below the root zone of agricultural species by strategic placement of deep-rooted indigenous or exotic tree species.

Specific suggestions for modifying water budgets in recharge areas include:

(a) better management of subterranean clover pastures to maximize E_t by exploiting the planophile leaf geometry, which enable clovers to rapidly cover the ground surface,

(b) replacement of subterranean clover pasture leys with a grain legume phase using lupins (*Lupinus* sp.),

(c) replacement of subterranean clover pasture leys with cereal-legume mixtures,

(d) development of better adapted and higher yielding wheat cultivars for closer rotations and reduction of the period under annual pasture in higher rainfall areas,

(e) development of cereal cultivars and clover strains with root systems better able to extract water at depth (through lower hydraulic root resistances),

(f) testing and development of strains of lucerne (*Medicago sativa*) with aphid and disease resistance and markedly higher winter production than previously available strains.

The above requires a broad agronomic approach paying particular attention to definition of recharge areas, rainfall patterns and soil profile characteristics, and of crop ideotypes to meet specific breeding objectives.

In combination with forestry the above practices should make possible a more discriminating approach to the imposition of clearing bans, for the purpose of protecting land and streams from salinization.

Research into the possibilities raised here may need the support of simulation models to guide field testing and studies. To benefit from the simulations illustrated here more accurate models, of root and plant performance as water users, for various species will be required. The hydraulic properties of the soils common to recharge areas in south-western Australia provide an opportunity for relatively accurate simulation without resort to excessively complex models.

There should be greater awareness in the cereal belt of the consequences of excessive recharge from the root zone of subterranean clover pastures and of the scope for reducing such recharge by agronomic management.

7 REFERENCES

Atsmon, D. and Jacobs, E., 1977. A newly bred 'Gigas' form of bread wheat (*Triticum aestivum* L.): morphological features and thermoperiodic responses. Crop Sci., 17: 31-35.

Bettenay, E., Blackmore, A.V. and Hingston, F.J., 1964. Aspects of the hydrologic cycle and related salinity in the Belka Valley, Western Australia. Aust. J. Soil Res., 2: 187-210.

Brooks, R.H. and Corey, A.T., 1964. Hydraulic properties of porous media. Hydrology Paper No.3, Colorado State Univ., Fort Collins, Colorado.

Carbon, B.A., Bartle, G.A., Murray, A.M. and Macpherson, D.K., 1980. The distribution of root length, and the limits to flow of soil water to roots in a dry sclerophyll forest. Forest Sci., 2(4), in press.

Donald, C.M., 1968. The breeding of crop ideotypes. Euphytica, 17: 385-403.

Donald, C.M., 1981. A wheat ideotype and its application in Australia. In: Wheat Science Today and Tomorrow. Symp. in honour of Otto Frankel's birthday. CSIRO Div. of Plant Ind. Held at Canberra, A.C.T., Australia. To be published by Cambridge Univ. Press.
Feddes, R.A., Kowalik, P.J. and Zaradny, H., 1978. Simulation of field water use and crop yield. Simulation monographs series. Pudoc, Wageningen.
Feddes, R.A., 1971. Water, heat and crop growth. Meded., Landbouwhogesch. Wageningen, Nederland. 71 (12).
Henderson, F.M. and Wooding, R.A., 1964. Overland flow and groundwater flow from limited rainfall of finite duration. J. Geophys. Res., 69: 1531-1534.
Holmes, J.W. and Colville, J.S., 1968. On the water balance of grassland and forest. Trans. 9th Int. Congr. Soil Sci., Adelaide. Vol. 1, 39-46.
Kimber, P.C., 1974. The root system of jarrah (*Eucalyptus marginata*). Res. Paper No. 10, Forests Dept., Western Australia.
Malcolm, C.V. and Stoneman, T.C., 1976. Salt encroachment - the 1974 saltland survey. J. Agric., Western Australia, 17: 42-9.
Marshall, T.J., 1958. A relation between permeability and size distribution of pores. J. Soil Sci., 10: 79-82.
Monteith, J.L., 1964. Evaporation and environment. In: The State and Movement of Water in Living Organisms. 19th Symp. of Soc. of Experimental Biology, 205.
Monteith, J.L., Szeicz, G. and Waggoner, P.E., 1965. Measurement and control of stomatal resistance in the field. J. Appl. Ecology, 2: 345-55.
Mulcahy, M.J., 1973. Landforms and soils of southwestern Australia. J. Roy. Soc., Western Australia, 56: 17-22.
Ozanne, P.G., Asher, C.J. and Kirton, D.J., 1965. Root distribution in a deep sand and its relationship to the uptake of added potassium by pasture plants. Aust. J. Agric. Res., 16: 785-800.
Passioura, J.B., 1972. The effect of root geometry on the yield of wheat growing on stored water. Aust. J. Agric. Res., 23: 745-52.
Peck, A.J., 1975. Interactions between vegetation and stream water quality in Australia. In: Heady, H.F., Falkenborg, D.H. and Riley, J. Paul (Eds)., Proc. 5th Workshop of the United States/Australia Rangelands Panel, Boise, Idaho, June 15-22. Utah Water Res. Lab., College of Engineering, Utah State Univ., Logan, Utah.
Peck, A.J., 1977. Development and reclamation of secondary salinity. In: Soil Factors in Crop Production in a Semi-Arid Environment, pp. 301-19. (Univ. of Queensland Press, St. Lucia).
Peck, A.J. and Hurle, D.H., 1973. Chloride balance of some farmed and forested catchments in southwestern Australia. Water Resour. Res., 9: 648-57.
Penman, H.L., 1948. Natural evaporation from open water, bare soil and grass. Proc. Roy. Soc., London, A193, 120 pp.
Philip, J.R., 1969. Theory of infiltration. In: Ven Te Chow (Ed.), Advances in Hydroscience, 5: 216-296. Academic Press.
Quinlivan, B.J. and Francis, C.M., 1976. Subterranean clover in W.A. 1. The current situation. J. Agric. Western Australia, 17: 5-9.
Sedgley, R.H., 1979. A simple evapotranspiration model for use in assessing hydrological changes in catchments in response to changing land use. Hydrol. and Water Resour. Symp., Perth, 1979, pp. 123-127. The Inst. of Engrs., (Canberra).
Smith, S.T., 1962. Some aspects of soil salinity in Western Australia. M.Sc. (Agric.) Thesis, Univ. of Western Australia, Perth.
Shachori, A., Rosenzweig, D. and Poljakoff-Mayber, A., 1967. Effects of mediterranean vegetation on the moisture regime. In: Sopper, W.E. and Lull, H.W., (Eds.), Forest Hydrology. Proc. Nat. Sci. Foundation Advanced Sci. Seminar held at the Pennsylvania State Univ., University Park, Pennsylvania., Aug. 29-Sept. 10, 1965. pp. 291-311.
Shuttleworth, W.J., 1979. Evaporation. Rep. No.56, July 1979, Inst. of Hydrology, Wallingford, Oxon.
Specht, R.L., 1957. Dark Island Heath (Ninety-mile Plain, South Australia). IV. Soil moisture patterns produced by rainfall interception and stem-flow. Aust. J. Bot., 5:137-150.

Specht, R.L. and Rayson, P., 1957. Dark Island Heath (Ninety-mile Plain, South Australia) III The Root Systems. Aust. J. Bot., 5: 52-85.
Van Schilfgaarde, J., 1957. Theory of land drainage. In: Luthin, J.N. (Ed.), Drainage of Agricultural Lands. Am. Soc. Agron: Madison, Wisconsin, pp. 79-112.
Walter, C.J. and Barley, K.P., 1974. Depletion of soil water by wheat at low, intermediate and high rate of seeding. Proc. 10th Int. Congr. Soil Sci., (Moscow). 1: 150-158.
Williamson, D.R., 1978. The water balance of deep sands near Meckering, Western Australia. M.Sc. (Agric.) Thesis, Univ. of Western Australia, Perth.
Wit, C.T. de, 1958. Transpiration and crop yields. Versl. Landbouwk. Onderz. 64. 6, Pudoc, Wageningen.
Wolfe, E.C. and Southwood, O.R., 1980. Plant productivity and persistence in mixed pastures containing lucerne at a range of densities with subterranean clover or *Phalaris* spp. Aust. J. Exp. Agric. and Animal Husbandry, 20: 189-196.

SOUTH AUSTRALIA'S APPROACH TO SALINITY MANAGEMENT IN THE RIVER MURRAY.

K.J. SHEPHERD
Engineering and Water Supply Department,
Adelaide, South Australia.

ABSTRACT

Shepherd, K.J., 1981. South Australia's approach to salinity management in the River Murray. Agric. Water Manage., 1981.

According to recent estimates, annual economic losses in South Australia due to River Murray salinity amount to $4 million. At constant prices they would rise to $10 million annually in the absence of remedial action. In recognition of the valley-wide problem, the three River Murray States and the Commonwealth have developed, with the aid of consultants, a co-ordinated programme for controlling drainage and salinity problems. The programme is now being implemented.

The South Australian contribution consists of six measures, varying from the $13.2 million (1980/81 values) Noora Scheme, which will divert saline drainage water to an evaporation basin out of the valley, to agricultural measures such as improving irrigation practices to reduce saline accessions to groundwater and to drainage systems. Development of this set of measures was greatly aided by a public involvement programme.

As a result of these initiatives, South Australia's salinity objectives will be achieved in the immediate future, but unless further control measures are identified and implemented, salinities will continue to increase, and by the year 2000 will again greatly exceed objective levels. An extensive research, monitoring and investigation programme is under way to help identify further opportunities for salinity mitigation and to aid in future decision making. Improvements in on-farm water application efficiency, and management of the whole River Murray, having regard to salinity as well as water quantity objectives, appear to present important opportunities.

1 INTRODUCTION

South Australia vies with Western Australia for the title of the driest State of Australia. Its average rainfall is less than 200 mm(8 inches), with only 12% receiving more than 300 mm(12 inches). The consequence of this low rainfall is that no local river can be relied on for substantial water supplies. The River

Murray is South Australia's most reliable and important water resource, although virtually all of its catchment ($1x10^6km^2$) lies in the eastern States.

In South Australia it is the sole source of irrigation water for about 43 000 ha of a wide variety of crops. It provides water to human settlements ranging in size from isolated homesteads to metropolitan Adelaide (population 920 000), and varying in character and location from the primary production centre of Keith in the south-east to the industrial cities of Whyalla, Port Augusta and Port Pirie on Spencer Gulf. Over 22 000 km of pipelines carry Murray water throughout the State.

In an average season the River Murray supplies 49% of the State's total irrigation, stock, domestic and industrial water consumption. In dry periods, this increases to 62%. The remaining supplies are derived from local streams and groundwater.

In dry years the water resources of the Murray in South Australia are over-committed, as the total irrigation, stock, domestic and industrial demand exceeds the amount of water available, after allowing for losses.

In addition to the importance of the Murray as a source of water, it is a major focus of recreation activities and supports significant and unique flora and fauna.

There are a number of threats to water quality in the river system but in terms of economic and social cost, the major immediate threat to the River Murray is salinity. There has been a steady increase in salinity levels and the situation can become critical in dry years. During periods of high flow, salinity tends to be low and conversely during periods of low flow salinity tends to be high.

The River Murray Commission, a body established under the River Murray Waters Agreement, a compact made in 1915 between Governments of the Commonwealth, New South Wales, Victoria and South Australia, has responsibility for operating and developing the River Murray System. The Agreement specifies rules for the sharing of the resources of the river for diversion by the States, and provides for navigation, but makes few concessions for operating with regard to water quality or other objectives. The parties to the Agreement have, however, recently agreed that the Commission may operate having regard to water quality, pending the acceptance of a new Agreement by the four Governments.

A description of the arrangements that have developed over the years to permit the rational development, use and management of the River Murray has been given in a recent paper by Bromfield (1980), Chairman, State Rivers and Water Supply Commission, Victoria. Further information on resource management and in some of the conflicts that will arise if single purpose operation is changed to multi-objective management, is provided in a paper by Johnson (1980), Executive Engineer,

River Murray Commission.

Fig. 1. shows the section of the river controlled by South Australia for the River Murray Commission.

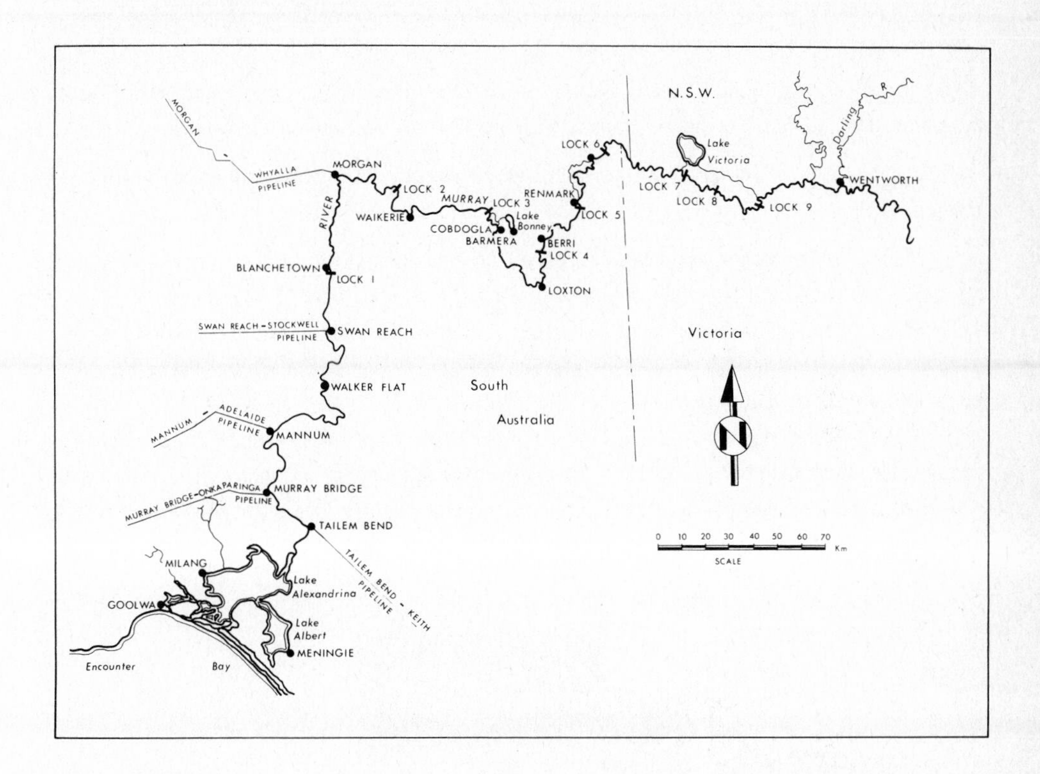

Fig. 1. Section of River Murray controlled by South Australia for River Murray Commission.

2 CAUSES OF SALINITY

Several million years ago a broad gulf of sea extended over much of the area through which the Murray now flows. About one million years ago the sea began to retreat, leaving the surface exposed. The exposed area was highly calcareous and contained vast reserves of residual salt.

Considering this, it is not surprising that groundwaters are generally highly saline in the area through which the River Murray passes in South Australia. The salinity is not very different from that of sea water and is substantially higher in some localities.

Much of this saline groundwater finds its way into the river, the salinity of which increases as it passes downstream.

Salinity enters the River Murray in three ways:

(a) Natural groundwater inflow. In South Australia, groundwater drains towards the river for the greater part of its length. The reserves of saline groundwater are so enormous that natural inflows are expected to continue indefinitely at present rates (unless action is taken to intercept them).

(b) River structures. Locks and weirs on the river raise up-stream water levels by about 3 m, which increases downstream inflows of saline groundwater.

(c) Irrigation. Irrigation increases river salinity in South Australia in three ways:

(i) Groundwater mounds are formed under irrigation areas, increasing hydraulic gradients towards, and hence saline seepage into, the river.

(ii) Some of the evaporation basins, into which drainage is pumped for disposal, induce seepage of saline water into the river as they are held at a higher level than the river.

(iii) The areas of the basins are insufficient to dispose of present and future drainage quantities. It is therefore necessary to release water to the river when basins are full. This is normally done when flows in the river are high, but occasionally releases during relatively low flows are necessary.

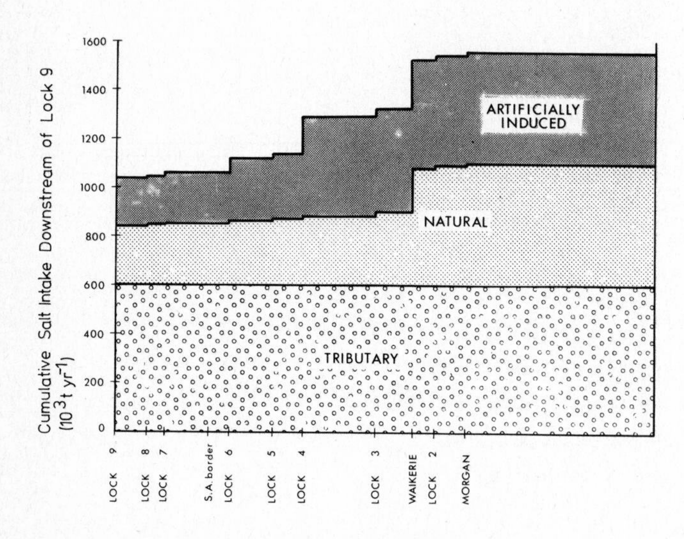

Fig. 2. Cumulative salt intake in S.A.

About 1 100 000 tonnes of salt enters South Australia annually, and about 300 000 tonnes is diverted from the river by water users within the State.

Of the 470 000 tonnes of salt per annum contributed within South Australia,

150 000 tonnes come from natural inflows, 120 000 tonnes from inflow aggravated by river structures and 200 000 tonnes from irrigation drainage. At present approximately half of the salt inflow due to irrigation drainage is due to the evaporation basins. As the quantity of drainage water increases with increased drainage areas, this component is expected to increase by nearly 70% by the year 2010.

Fig. 2. shows the present cumulative salt intake within South Australia.

3 EFFECTS OF SALINITY

High salinities affect virtually all consumers of River Murray water throughout South Australia. The three major affected sectors of the community are:

(a) Agricultural users. Murray irrigators rely totally on river water. High salinities reduce crop yields, limit the choice of crops which can be grown and, at higher concentrations over long periods of time, may kill trees and other crops.

(b) Domestic users. Rural and urban householders are affected by "hardness" which results from salinity. More soap and detergents have to be used or water softeners installed and maintained. Hardness causes scaling in pipes and heaters and increases fabric wear. Domestic gardens are also affected by high salinity, stunting growth and killing some plants.

(c) Industrial users. Water is used by industry for steam generation, processing, for washing, cooling, and for the provision of amenities. Corrosion, scale and poor steam quality are the main problems caused by saline water. Softening and demineralisation are used to deal with these problems but they are costly.

The effects of salinity are distributed through the community in a way which causes considerably greater hardship for some sectors than others.

On an individual basis, irrigators suffer most from the effects of salinity. High salinity can cause substantial loss of yield and may cause tree loss in extreme circumstances. Metropolitan users individually suffer relatively small effects. However, the total aggregated cost is substantially greater for metropolitan water users than for irrigators.

It is only in the last decade that serious attempts have been made to quantify the economic impact of salinity for agricultural, domestic and industrial users. The most significant studies have been carried out in connection with the Colorado River water quality improvement program. The methodology of this programme was used in a study carried out by the Australian Mineral Development Laboratories (AMDEL) for the Engineering and Water Supply Department of South Australia

(Blesing and Tuffley, 1977).

The AMDEL report was restricted to a discussion of economic costs and, for the most part, direct economic costs, whereas the United States study attempted to assess both direct and indirect costs.

Using present and year 2010 predicted average salinities, AMDEL estimated the annual cost at 1978 values to be $4 million at present and $10 million by 2010. Ignoring inflation effects and without discounting future losses, the total cost by 2010 would be $210 million, of which $55 million would be borne by agricultural users, $110 million by domestic users and $45 million by industry.

4 DEVELOPMENT OF THE SOUTH AUSTRALIAN SALINITY CONTROL PROGRAMME

4.1 Investigations

Salinity has been continuously monitored in the South Australian Section of the River Murray since 1932 but investigations directed specifically to the identification of sources of salinity did not commence in earnest until work was initiated in conjunction with the Consultants Gutteridge, Haskins and Davey in 1967. Following the release of the Consultant's report (Gutteridge et al., 1970) a joint programme of investigatory work in the problem areas identified in the report and other areas was initiated by the Engineering and Water Supply Department and the S.A. Department of Mines and Energy.

This involved some 70 detailed salt source surveys, the drilling of 400 observation holes and many thousands of salinity measurements. In some areas the solutions proposed by the Consultants were verified and implemented but in others the improved data revealed that the proposed solutions would not have been effective.

Along with the commencement of field investigations, the Engineering and Water Supply Department developed two computer based simulation programmes aimed at determining the optimal method for the disposal of irrigation drainage effluents from the major areas in the Riverland Region.

The first system simulated the natural flow and salinity conditions in the River Murray, incorporating "SALTRAN" as the central computer programme. The second system simulated the irrigation and drainage effluent system and its interaction with the River Murray. Programme "DISALT" was developed for this purpose. Both systems were established using monthly data, and were used for simulation purposes over the period 1915 to 1972.

Simulation of the River Murray system was carried out in three phases, using "SALTRAN". The first phase involved computation of river discharges from flow, diversion and evaporation loss estimates. Secondly, river water travel times were computed from river volume and flow data. The third phase involved calculation of river salinity from flow data and estimates of salt influx to the river from uncontrolled sources (e.g. naturally occurring groundwater inflows and seepage from groundwater mounds under irrigation areas). The output from "SALTRAN" was then used as input to programme "DISALT".

Computer programme "DISALT" simulated the irrigation drainage system and its interaction with the River Murray. The effect of evaporation basin operation on river salinity was computed by water and salt balance calculations. The effect of salinity control works and drainage disposal schemes was computed in a similar manner.

A further programme "QLOOK" was developed to handle the immense volume of data output from "DISALT", to compute appropriate statistics, and plot results.

In this way the effect on river salinity of alternative disposal schemes was compared, enabling selection of the optimal scheme.

4.2 Public Involvement

Following the release of the Gutteridge, Haskins and Davey Report in 1970, there was considerable public pressure to implement the proposed solutions. Both Government and the Departments concerned came under criticism for their seemingly tardy reaction to the proposals of the Consultants, in spite of the recommendations contained in the report that further detailed field investigations were necessary before implementation. A major public criticism of the Report was the non-consideration of environmental effects of the proposed measures.

Having regard to these criticisms and the firm opinion of some irrigators and horticulture advisors that an engineering organization would give emphasis to engineering solutions to the salinity problems of the river, Government accepted a recommendation from the Engineering and Water Supply Department that a public involvement programme be initiated to ensure that a full community contribution was invited before plans were developed for a salinity control programme in South Australia (Allen and Killick, 1979). In 1976 the Consultants Trojan, Owen and Associates were engaged to assist in conducting this programme.

Its principal objectives were to provide the study team with an understanding of the people who would have to live with the results of the programme and also to disseminate information to prevent misunderstanding and misinterpretation of alternative proposals.

The initial phase was to carry out a survey to determine the community's attitudes to and perceptions of River Murray Salinity control. This involved interviews with individual growers, grower and commercial interests in the Riverland

Region, conservationists, politicians, Local and State Government officers and the media. This indicated that many people were concerned about the salinity problem and that considerable value could be achieved from a programme of community involvement.

A series of six position papers were released in the period April 1977 to January 1978 which summarized current ideas and activities and invited contributions and criticisms. These were released to the media, the general public and about 500 people and organizations who displayed interest in the problems and felt that they could make a contribution to the solutions. Responses were received by mail, personal interviews or by means of a special telephone answering service. All communications were acknowledged, and where appropriate the writers were informed of the Department's reaction to their suggestions.

Additionally, meetings and seminars were arranged with Local Government authorities and broad interest groups to discuss the position papers and test reaction to the aims of the salinity control programme.

Many of the suggestions from the public involvement programme had been taken into account by the Department in developing its own proposals, but two principal points were stressed:

(a) a national approach would produce the best results,

(b) solutions should not only involve drainage disposal, but also drainage reduction.

4.3 Objectives

In determining the objectives of the South Australian programme, account was taken of the overall salinity objective as recommended by the River Murray Working Party (1975), viz, "to maintain the salinity of River Murray water such that established uses are protected".

The Fourteenth Biennial Conference of Engineers representing Authorities controlling Water Supply and Sewerage undertakings serving cities and towns of Australia held in 1969 adopted a salinity criterion of 500 mg L^{-1} (83 m Sm^{-1}) as the highest desirable salinity for potable water, which is in accord with the recommendations of the World Health Organisation. Fig. 5, showing the distribution of salinity at Morgan, indicates that in the absence of remedial measures, the salinity at the offtakes of the major pipelines out of the Murray would exceed this figure for some 25% of the time.

Furthermore, unless appropriate irrigation practices were used, damage to stone fruit plantings could occur when salinities were in excess of 50 m Sm^{-1}, a level which would be exceeded 75% of the time.

Recognising that it would never be possible to completely eliminate high salinities but, on the other hand, it would be unacceptable to operate the River Murray System in a way which keeps salinities at or only just below some fixed maximum level, the following was adopted as a practical overall objective for salinity in the River Murray in South Australia -

"To recommend action which, having regard to the economic, environmental and social consequences, will hold salinity in the River Murray below 80 m Sm^{-1} at Morgan for 95% of the time."

There is an important qualification to this criterion in that it is acceptable only if years of high salinity water are followed by years of low salinity water. Crops will otherwise suffer long-term damage and it is essential, therefore, that future management of the River Murray System should continue to provide adequate periods of low salinity.

4.4 Identification and Assessment of Options

The first stage in identifying the options involved the Government agencies in gaining an understanding of the causes of the salinity problem. This stage began in the late 1960's and consisted of extensive field investigations followed by computer modelling of the river and adjacent hydrogeological systems.

From this basic understanding of the causes, the Government agencies developed a range of possible control options. These were based purely on technical effectiveness. No other criterion, such as environmental impact, was explicitly taken into account. This stage was completed late in 1976.

To ensure that the range of possible options was as extensive as possible, the general public was then involved in the programme, as described in Section 4.2 above.

The public added further salinity control options to those developed by the Government agencies. By the end of 1977, 32 options had been identified.

All options were then assessed in terms of technical, economic, environmental and social feasibility. This assessment was undertaken in two phases. The first was a broad evaluation of all options, which was concerned principally with the technical and economic assessment. Assessing the environmental and social impact of every option would have been extremely time consuming. If an option was not reasonably cost-effective, there was little point in identifying its environmental and social impact. Only if the environmental or social impact of a scheme was obviously unacceptable, was a scheme rejected in the broad evaluation.

The second assessment phase involved a detailed evaluation of the options that "passed" the broad evaluation. Alternatives within these options were also assessed in this second phase. The methods used in assessing each option in each phase are discussed below.

4.5 Assessment Methods

4.5.1 Technical assessment

The principal technical criterion was the salinity reduction achieved by each scheme. This was determined mainly through computer models simulating the river flow and salinity. Other technical aspects which had to be evaluated included the hydraulic, hydrogeological, and engineering feasibility.

4.5.2 Economic assessment

The economic costs of each option were the capital and operating costs expressed as a present value.

The economic benefits for each option were determined from a salinity reduction-economic benefit response function. The benefits were those gained by the irrigators, and domestic and industrial users. These benefits were also expressed as a present value.

4.5.3 Environmental impact assessment

The guidelines for environmental assessment were provided by the South Australian Department for the Environment. The guidelines identified the ecological and social aspects which could be affected by salinity changes or the implementation of any salinity control measures. It also identified the activities which could have adverse environmental effects.

The environmental assessment was undertaken at two levels. Firstly, in the broad evaluation of all options only the obvious deleterious effects were taken into account by the planning engineers. Where necessary, investigations were undertaken to quantify and clarify possible deleterious effects of options that otherwise appeared acceptable. Specialists were brought into these investigations.

In the second, more detailed evaluation phase, the detailed effects of the remaining options were determined by specialists in flora, fauna and social science. This allowed choices to be made between alternative options, as well as alternative actions within the options.

4.5.4 Social impact assessment

This was treated in a similar fashion to the environmental assessment. In addition to specialist advice, comment on the options from the public, especially those affected by the options, was encouraged. This gave the planners a measure of the social acceptability of each option.

The processes of environmental and social assessment resulted in the elimination of an option that in economic terms only would have yielded the highest net benefit. This option was the diversion of the saline drainage waters to a lake used extensively for recreation. It was shown that the lake would have become eutrophic, which was found unacceptable in both the environmental and social assessments.

5 THE SOUTH AUSTRALIAN SALINITY CONTROL PROGRAMME

The results of the South Australian investigations and proposals for salinity control measures were released in January 1979 in a report by the Engineering and Water Supply Department (1978).

This report presented the total salinity problem as the interaction of a number of complex processes - hydrogeology, irrigation and river regulation. The methods of salinity control suggested by the Department and the public during the community involvement programme were evaluated on technical, economic, environmental and social grounds and a combined approach incorporating six salinity control measures was proposed, involving engineering works, improving agricultural practices and optimising river regulation. The measures are:

(i) The Noora Drainage Disposal Scheme - involving the pumping of irrigation drainage water from existing evaporation basins at Berri and Renmark to an out of the valley basin some 20 km east of Loxton (see Fig. 3). The estimated capital cost is $13.2 million.

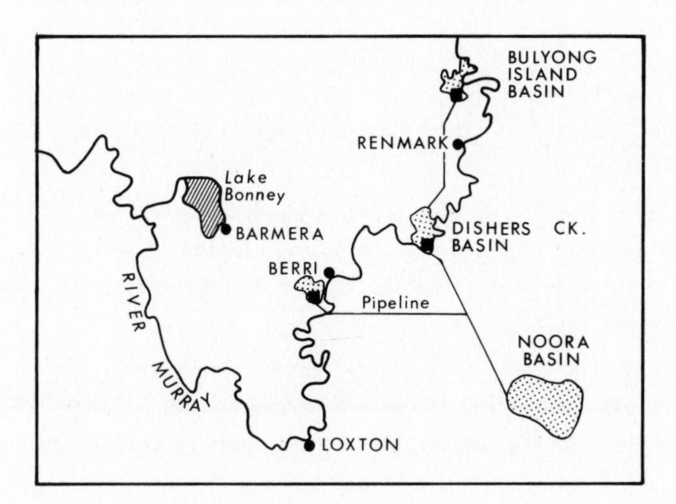

Fig. 3. Noora Drainage Disposal Scheme.

(ii) Upgrading control works on existing evaporation basins - providing for new outlet works in existing basins to allow water levels to be kept lower, reducing the differential head between the basins and the river and also the area of the basins.

(iii) Rufus River Groundwater Interception Scheme - involving the interception of saline groundwater flows which are induced by the differential head between Lake Victoria storage and the river and the pumping of the water to an evaporation basin (see Fig. 4). The estimated capital cost of this scheme is \$2.8 million.

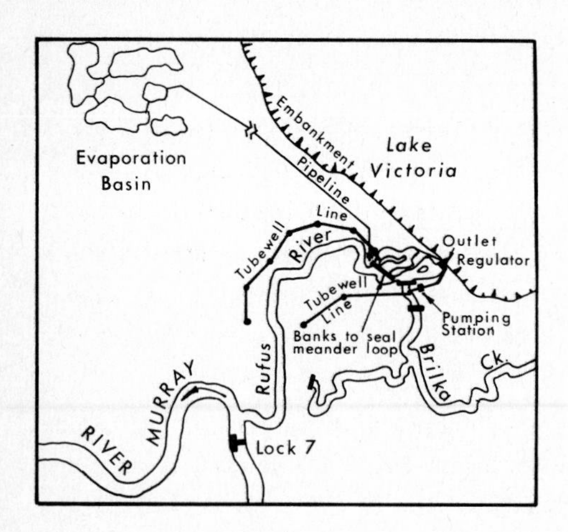

Fig. 4. Rufus River Groundwater Interception Scheme.

(iv) Replacement of furrow irrigation with improved irrigation practices - involving the provision of low interest finance to assist growers in making conversion. This would allow more efficient irrigation and a resulting reduction in drainage and seepage to the river.

(v) Technical support for conversion of overhead to undertree irrigation and for the installation of adequate drainage - reducing the effects of salinity on irrigated crops. Technical rather than financial support is proposed since this conversion would not reduce River Murray salinity.

(vi) Advances on South Australia's entitlement under the River Murray Waters Agreement - continuing present policies which, in certain circumstances, allow requests to the River Murray Commission for advances on monthly entitlements which are used for dilution.

At the time of writing (September, 1980) construction of the Noora Scheme, Measure (i), has commenced. It will be partially operational in 1982, and fully so in 1983. The upgrading of control works on existing evaporation basins,

Measure (ii), is complete. Approval is awaited to enable commencement of construction of the Rufus River Scheme, Measure (iii).

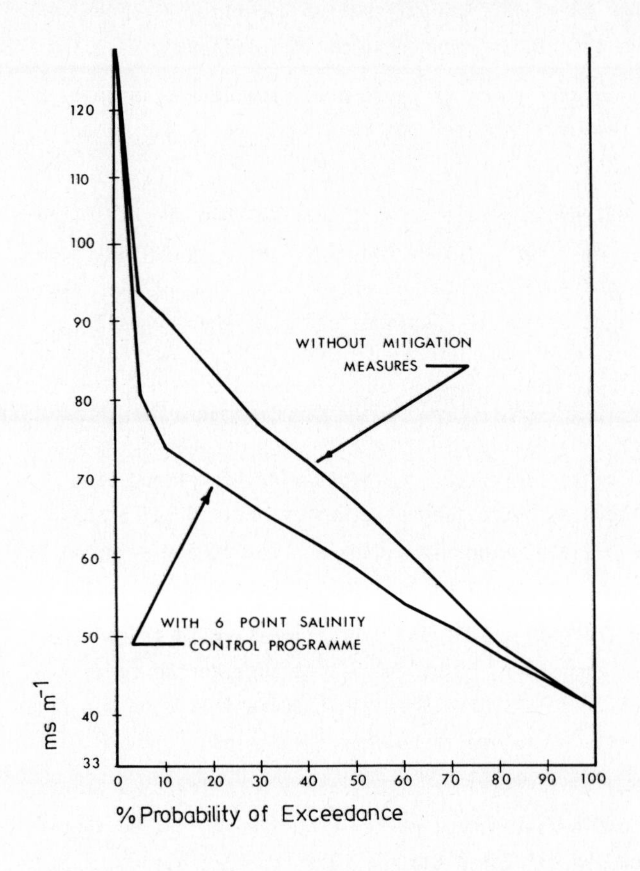

Fig. 5. Predicted salinity at Morgan, 1980 demand on 1915 to 1972 historic flows.

Further progress in the implementation of improved irrigation practices, Measure (iv), has been assisted by higher returns for fruit crops in the last year, and by recent tax concessions. It is planned to commence financial assistance in the year 1981/82.

Technical support for the conversion of overhead sprinklers to undertree irrigation, Measure (v), is a continuing service; and Measure (vi), Advances on entitlement flow, is an option which will continue to be taken up whenever the opportunity arises.

Results of computer modelling show that the effects of the combined six measures would meet the objective. However, salinities would be maintained at objective levels only for a few years. Predictions indicate that by 2010 salinity would be less than 80 m Sm^{-1} at Morgan for only 85% of the time. Although this is a substantial improvement over the "do nothing" situation under which salinity in 2010 would be then less than 80 m Sm^{-1} for only 47% of the time, it highlights the need for continuing efforts to reduce and mitigate salinity.

An important assumption in the derivation of these predicted levels is that there will be no increase in salinity of water entering South Australia after 1980.

The total capitalised cost of the proposals is $23.1 million and the estimated capitalised benefits are $24.3 million. In addition, there are significant non-quantifiable economic, social and environmental benefits, and all benefits taken together are more than sufficient to justify implementation of the proposals.

6 PROGRESS TOWARDS A VALLEY-WIDE APPROACH

Salinity on the River Murray is a valley-wide problem. The irrigated Riverine Plains of New South Wales and Victoria suffer from high water tables and land salinisation. In addition, there is increasing evidence of dryland salinisation on the northern slopes of the Dividing Range in Victoria. Downstream areas - the Sunraysia district and all of South Australia - suffer from high salinities.

Prior to the short, intense drought of 1967/68, virtually all efforts to elicit action by the River Murray Commission, or by the States outside the context of the Commission, to take water quality into account in operations and investigations, foundered due to conflicts of interests between the States.

The publication of the Murray Valley Salinity Investigation (Gutteridge et al., 1970), the first significant basin-wide study, marked a quickening in the intensity of salinity investigations by all three States. A further milestone, the report of the River Murray Working Party of 1975, which resulted in an agreement in principle by the four Governments that the River Murray Commission should take water quality into account in its operations, gave salinity investigations a further impetus.

Nevertheless, by 1977 few proposals to mitigate salinity or drainage problems had reached the implementation stage. In October of that year, a meeting of Ministers was held, which emphasised the need for co-ordinated action and directed that priorities for measures to deal with salinity and drainage problems be developed on a valley-wide basis.

This resulted in a major study report by the Consultants Maunsell and Partners, (1979). This report presented a plan of action for implementation over a 25-year period at a cost of $123 million, including $75 million over the period 1980 to 1985

Works under this programme are now being constructed by all three States and the Commonwealth is contributing 50% of capital expenditure on programmes that were demonstrated by the Consultants to meet criteria of efficiency, effectiveness and priority. From 1977, State proposals were developed in conjunction with this study, including the six-point South Australian Programme.

7 PROBLEMS AND OPPORTUNITIES

These studies, and subsequent South Australian investigations, have drawn attention to a number of difficulties and deficiencies.

(a) Groundwater mounds associated with a number of irrigation areas continue to rise and extend in area. Contributions of saline groundwater and drainage flows to the river will continue to increase.

(b) Further irrigation development on some tributaries, primarily the upper tributaries of the Darling, appears to be in train. This seems likely to diminish flows to storages and to the River Murray, thus increasing salinities in South Australia at certain times.

(c) Intensive investigations have failed to identify economic measures for adequately dealing with saline contributions to the river from the Shepparton and Kerang irrigation areas in Victoria. Proposals for the interception of saline contributions to the river in South Australia proved by geographic accident to be much the cheaper. This gives rise to an equity problem in that, despite a significant Commonwealth contribution, South Australia is required to incur heavy expenditures on works partly required because of salinity contributed upstream of the border.

(d) South Australia's efforts to identify further salinity mitigation measures will be handicapped by the fact that most of the more concentrated discharges that are economic to intercept will already have been dealt with. Further drainage and groundwater interception schemes within South Australia to enable it to meet its salinity objectives are being sought, but it is unlikely that more schemes as economic and effective as the Rufus River Scheme, or with the major beneficial impact of the Noora Scheme, will be found.

(e) For these reasons, measures for combatting salinity problems will be effective in enabling the salinity in the river to meet South Australia's salinity objectives in the short term (5 to 10 years), but not in the longer term.

(f) South Australia's entitlement flow provided under the River Murray Waters Agreement (the flow provided to South Australia during the periods of regulated flow) is insufficient to meet all demands and losses during years of peak demand, despite:

a. a policy in force since 1968 that no further irrigation licences will be issued;

b. an increase in entitlement from the commissioning of Dartmouth Reservoir in 1979;

c. a reduction in total irrigation allocations in 1979;

d. increasing attention to managing the demands of urban and industrial water users, including greater emphasis on pay-for-use pricing;

e. increasing use of more efficient water application methods by irrigators.

(g) While a one-year deficit can be absorbed by drawing down the terminal lakes, as frequently happens, sequences of such years would necessitate restrictions and would lead ultimately to failure of the system in terms of quantity and/or quality. Measures to manage such situations and minimise adverse effects and the probability of failure, are being investigated as a high priority. The salinity problems of the river contribute substantially to this problem.

(h) Progress by farmers in replacing wasteful furrow irrigation with more water-efficient methods which make less contribution to salinity and drainage problems, has been roughly at the rate expected and planned for. The direct Government contribution has been less than planned in 1979, but indirect contributions have been significant due to recently announced tax concessions. It is becoming apparent that opportunities for increased overall community benefit are being missed as a result of the slow rate of conversion. Projections suggest that the present proportion of 70% furrow irrigation could reduce to about 40% by 1990. Reliance is at present wholly placed on farmers volunteering to upgrade their systems, with financial assistance available only to restricted categories of irrigators. Means of providing more widely useful financial assistance are being sought. Attention is also to be given to providing more extension services. Efforts will also be directed at encouraging more attention to watering based on actual plant needs.

(i) There is insufficient information presently available in a number of areas fundamental to future decision making. Accordingly, an $800 000 research programme has been commenced by South Australia, with Commonwealth financial assistance.

It includes research and monitoring on:

a. crop types;

b. the response of crops to changes in salinity;

c. the effects of improved irrigation practices on irrigation requirement, drainage runoff, salt balance, and yield;

d. irrigation techniques and adoption rates of improved irrigation practices;

e. evaluation of irrigation equipment;

f. potential for improved irrigation of former swamp areas;

g. soil degradation;

h. impact of salinity on urban water users;

i. the groundwater regime.

(j) There are opportunities for improved management of the river system on a valley-wide basis, particularly as a result of the failure of intensive investigations to yield economic solutions to the problems of large quantities of saline water entering the river from the Shepparton and Kerang irrigation areas. Investigations are currently being undertaken by Consultants for the River Murray Commission to evaluate the feasibility of water quality modelling of the main stem of the river upstream of South Australia, to link up with the existing South Australian models. If this is proceeded with, it will provide a useful tool to assist in the operation of the river to take more opportunities to meet water quality objectives as well as water quantity objectives. Investigations are also proceeding into the development of possible operating targets for salinity.

(k) There appears to be much scope for salinity management through improved irrigation practices in the eastern States where broad-acre pasture irrigation dominates, as well as in South Australia, where tree crops dominate.

8 CONCLUSIONS

The aggregate salinity reductions resulting from the South Australian Salinity Control Programme, together with other salinity mitigation measures in New South Wales and Victoria, will enable South Australia's salinity objectives to be met in the short term.

The development of the South Australian Salinity Control Programme was greatly assisted by a well-structured public involvement phase and a systematic process of development and evaluation. The public imvolvement programme achieved its objectives of preventing uncontrolled conflict, adequately informing the affected public on a range of technically complex issues and establishing feedback from the public. It contributed materially to the development of a final set of proposals which will meet the needs of the affected community.

In the longer term the salinity objectives for the South Australian section of the river will not be achieved unless further action on a variety of fronts is taken.

While further opportunities are being sought for works to intercept saline drainage and groundwater and dispose of it out of the valley, the greatest challenges and opportunities are most likely to be in the area of management and agricultural solutions rather than engineering solutions.

The measures most likely to yield the greatest benefits to the community will be:

(a) Improved on-farm water application efficiency; and,

(b) Management of the whole of the River Murray having regard to salinity objectives as well as water quantity objectives.

9 ACKNOWLEDGEMENTS

Much material in this paper has been drawn freely from the work of a number of officers of the Engineering and Water Supply Department. Acknowledgement is particularly due to Messrs. R.J. Shannon, J.C. Killick, C.M. Allen and R.M. Ebsary.

This paper is published by permission of Mr. D.J. Alexander, Acting Director-General and Engineer-in-Chief, Engineering and Water Supply Department, Adelaide, South Australia.

10 REFERENCES

Allen, C.M. and Killick, J.C., 1979. Autocracy or Democracy? - Public involvement in water resources planning. The Inst. of Engrs., Australia, Hydrology and Water Resources Symp., Perth, 1979.

Blesing, N.V. and Tuffley, J.R., 1977. Study of potential economic benefits from reduction in salinity of River Murray water. Australian Mineral Development Labs., Adelaide. Rep. No. 1179.

Bromfield, W.E., 1980. Management of the River Murray. Address to Australian Water Res. Found. State Rivers and Water Supply Comm., Victoria.

Engineering and Water Supply Department, 1978. The South Australian River Murray Salinity Control Programme. South Australian Govt., Adelaide. 4 vols.

Gutteridge, Haskins and Davey in association with Hunting Technical Services, 1970. Murray Valley salinity investigation. River Murray Comm., Canberra.

Johnson, K.E., 1980. Management of the water resources of the River Murray - beneficial uses and conflicts. Paper II, The Instn. of Engrs., Australia, Victoria and South Australian Div. Joint Country Convention, Mildura, October, 1980.

Maunsell and Partners, 1979. Murray Valley Salinity and Drainage Report. For Murray Valley Salinity Study Steering Committee, C/- Dept. National Development and Energy, Canberra.

River Murray Working Party, 1975. Report to Steering Committee of Ministers, October, 1975. Australian Govt. Publg. Serv., Canberra.

THE EVOLUTION OF A REGIONAL APPROACH TO SALINITY MANAGEMENT IN WESTERN AUSTRALIA

B.S. SADLER AND P.J. WILLIAMS
Public Works Department, West Perth, Western Australia, 6005.

ABSTRACT

Sadler, B.S. and Williams, P.J., 1981. The evolution of a regional approach to salinity management in Western Australia. Agric. Water Manage., 1981.

A regional approach to managing the effects of salinity on the water resources of south-western Australia has evolved rapidly in recent years. As a rational basis for the evolving policies and strategies, a classification of river basins has been adopted relating land use and water resource potential to zones of increasing salinity hazard. Salinity management measures pertaining to both water quality and agricultural productivity are classified, described and evaluated in strategic terms as perceived by water resources managers. Although a comprehensive regional approach integrating water and agricultural objectives is yet to develop, a few tentative comments are offered in this direction.

1 INTRODUCTION

There has long been concern for management of specific dryland salinity problems in Western Australia. As far back as 1908 engineers concluded that they had induced a salinity rise in the Helena River by clearing native forest to increase streamflow. To rectify this situation the cleared area was replanted to pines and became the first recorded remedial measure responding to stream salinity in south-western Australia.

However, despite this prompt action at the beginning of the century, and although Wood (1924) published a theoretical explanation of the salinity problem, no regional planning for protection of river salinity emerged until half a century later. Throughout that intervening period, the concern for effects of salinity focussed mostly on losses of agricultural productivity through salt land development. Nevertheless there has been limited success in dealing with the agricultural problem.

In view of the extensive effects of salinity caused mostly by agricultural clearing in south-west river basins, growth of concern for salinity as a regional water resource problem is belated. Until relatively recent times the expansion of the regional economy has been very dependent on dryland agriculture.

Furthermore at the end of the second world war, when new land settlement was vigorously encouraged for creating employment and economic development, water utilisation had only grown to a small percentage of regional water resources. Consequently, until the most recent period of the region's history, a pioneering attitude to water and land development prevailed and the community was not ready to accept measures which today it perceives as essential.

Stimulated by the rapid growth of water use in the post-war period public attitudes have changed to a strong concern for the regional management and protection of water resources. With public acceptance that positive action is required a regional approach to management of river salinity has begun to evolve.

Recognising the potential for further serious degradation of water resources and the present imprecisions in predicting the effectiveness of some control measures, a policy of containment and minimisation of risk is being emphasised at this stage.

2 CLASSIFICATION OF RIVER BASINS

For broad definition of the regional variations of salinity hazard, described more fully by Loh and Stokes (1981), the south-west Region has been divided into four zones (Fig. 1). These are the *coastal plain zone* between the coast and the Darling Scarp, the *high rainfall zone* from the scarp eastwards to the 1100 mm rainfall isohyet, the *intermediate zone* with between approximately 1100 and 900 mm of rainfall and the *low rainfall zone* inland from the 900 mm isohyet. The first two regions, for the most part, have little or no salt in storage and do not harbour any significant threat to water quality. In the fourth inland zone, with less than 900 mm of rainfall, salt storage is high and clearing almost invariably causes salt release. The third or intermediate zone is, as the name implies, a zone of transition between the second and fourth zones with highly variable potential for salt release.

The rivers of south-western Australia generally form their most suitable damsites near the Darling Scarp where they begin to descend to the coastal plain from the Western Plateau. The principal land units for strategic planning of salinity control in the region are the river basins commanded by such sites. Downstream of the scarp line the rivers have more limited development potential.

The river basins above the Darling Scarp can be grouped into classes defined according to their salinity hazard and perceived development potential. To a large degree this classification is related to river length because the further inland the drainage extends, the lower is the rainfall, the higher the salt storage, and the more likely that the river drains from cleared agricultural areas situated inland of the State Forest. For more detailed planning these catchments can be further divided into zones of salinity hazard related to the isohyetal zones described above. The river basin groups are described in the following paragraphs and their geographic distribution is illustrated in Fig. 2.

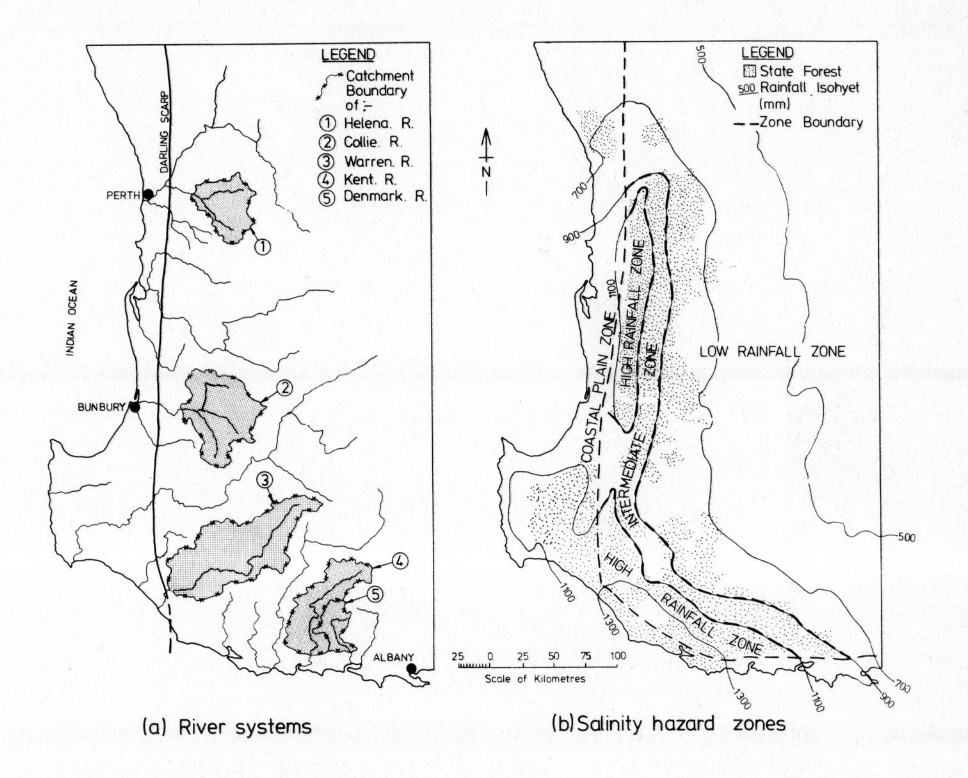

Fig. 1. The south-west Region.

Group 1 - Forested Basins. Forested Basins which are entirely, or almost entirely, in State Forest. Rivers in this group fall into two categories:-

Group 1A - *(Little Salt Hazard)*. These forest basins are wholly in high rainfall areas exceeding 1100 mm annually. The streams are relatively small but fresh and high yielding. These streams drain land having salt storage below the concentrations which are likely to cause significantly increased salinity.

Group 1B - *(Significant Salt Hazard)*. These forest basins extend inland of the 1100 mm isohyet to encompass significant areas where land use changes could cause salinity increase, but where most or all of the sensitive land at present is protected by State Forest. The rivers of Group 1B are fresh and will remain so as long as the State Forests maintain sufficient transpiration in sensitive areas to prevent salts from being mobilised. The group includes most of the important rivers already developed for water supply or irrigation.

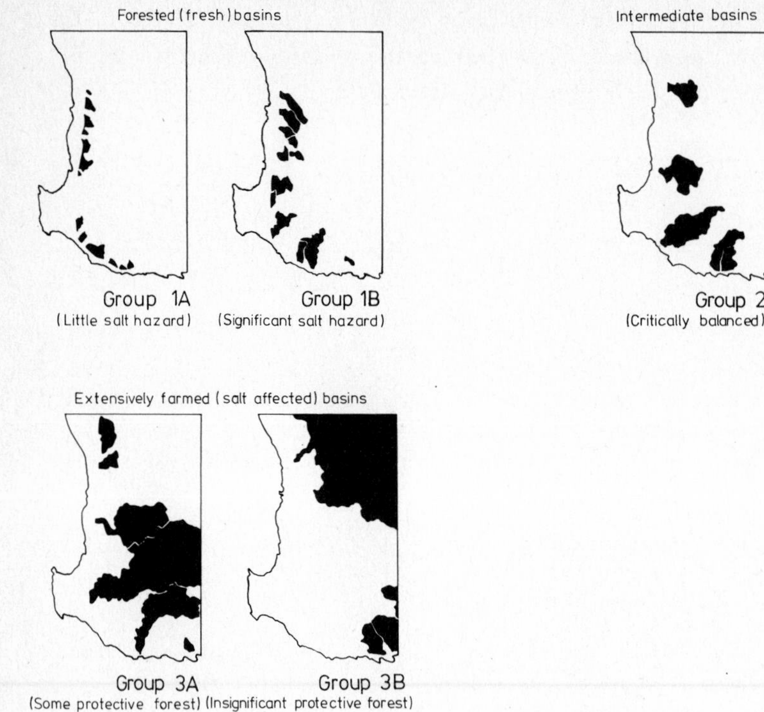

Fig. 2. River basin groups of the south-west Region.

Group 2 - Intermediate Basins. Basins which are predominantly forested but extend inland sufficiently to include moderate areas of land partially cleared for agriculture. The rivers are fresh enough to have important potential for utilisation although they have experienced increases in salinity and are at risk from further clearing. This is the group of basins which has the most immediate risk of further serious salinity increase.

Group 3 - Extensively Farmed Basins. Basins which are predominantly freehold land where agriculture is the most extensive land use and has caused the main river to be brackish or saline. Rivers in this group fall into two categories:-

Group 3A - *(Some Protective Forest)*. These basins, though salt affected, still retain some potential for development of the main stream or at least of forest tributaries.

Group 3B - *(Little or no Protective Forest)*. In these basins agricultural land use is completely dominant with very little forest or other areas yielding fresh water.

Group 0 - Coastal Basins. The catchments of the coastal plain have been designated Group 0. Within this area, clearing has no significant effect on salinity, but the rivers have limited development potential.

Table 1 gives a breakdown of the surface water resources of the south-west Region based on the river basin group classification. It includes the total mean annual flow of rivers in each category and their assessed divertible yield of fresh, marginal, brackish and saline water, together with the yield as developed to date. The proportion (43.8%) of the region's fresh and marginal water resources in the sensitive Groups 1B and 2, and the proportion (33.1%) of the region's potential yield already brackish or saline in Groups 3A and 3B emphasise the importance of salinity control in this region.

TABLE 1

Surface water resources of south-western Australia

River Basin Groups	Total Mean Annual Flow $m^3 x 10^6$	Divertible Yield $m^3 x 10^6 yr^{-1}$					Total Yield Developed To Date $m^3 x 10^6 yr^{-1}$
		Fresh <500 mgL^{-1} TSS†	Marginal 500-1000 mgL^{-1} TSS	Brackish 1000-3000 mgL^{-1} TSS	Saline >3000 mgL^{-1} TSS	Total	
0. Coastal	1510	196	-	-	-	196 (7.5%)	-
1. Forested							
1A. Little Salt Hazard	800	406	-	-	-	406 (15.6%)	81
1B. Significant Hazard	1010	612	3	-	-	615 (23.6%)	185
2. Intermediate	785	50	415	60	-	525 (20.2%)	130
3. Extensive Farms							
3A. Some Forest	1330	-	7	699	-	706 (27.1%)	-
3B. Insignificant Forest.	445	-	-	20	135	155 (6.0%)	-
TOTAL	5880	1264 (48.6%)	425 (16.3%)	779 (29.9%)	135 (5.2%)	2603 (100%)	396*

*Fresh 296 $m^3 x 10^6 yr^{-1}$, Marginal 100 $m^3 x 10^6 yr^{-1}$
†TSS = total soluble salts.

3 SALINITY MANAGEMENT MEASURES

3.1 Classification of Measures

In Table 2 the range of basic measures currently receiving some form of consideration for salinity management is identified and sub-divided into two main groups according to whether the prime objective is water quality management, or the management of salt seeps on developed farmland. The measures are also further classified into groups of management strategies which are either designed to control the release of salts from the sub-soil or to adapt to the salts after they are released.

Table 2 indicates that where management of water supply quality is the prime objective, there is a much wider range of measures available or potentially available than where agricultural productivity is the principal objective. There are two main reasons for this. Firstly, whereas all measures to reduce the area of salt seeps must be based on the land, water managers can also utilise engineering measures designed to adjust to increased salinity in the river systems. Secondly, whereas agricultural objectives can only be satisfied by a limited range of solutions which must allow a large percentage of land to be left under agriculture, catchment management measures which withhold significant areas of land from agriculture can also be considered if water quality is the over-riding objective.

These measures are broadly evaluated in turn in Sections 4 to 9 where their strategic relevance in regional salinity management is related to the basin groups defined in Section 2.

TABLE 2

Salinity management measures in south-western Australia

Management Strategies		Management Measures	
		Primary Objective-Water Quality	Primary Objective-Salt Free Farmland
Controlling the Release of Salts	Prevention or Limitation	Halting alienation of uncleared Crown Land. Protective management of forest Control of clearing. Partial clearing.	Partial clearing.
	Restoration or Amelioration	Reforestation complete or partial Modification of crops and cultivation. Drainage of saline groundwater.	Partial reforestation. Modification of crops & cultivation. Drainage of saline groundwater. Shallow or surface drainage. Comprehensive saltland treatment.
Adapting to the Release of Salts	Salt Disposal	Reservoir management. Seasonal withdrawal. Saline diversion. Desalination.	Not applicable.
	Dilution	Blending fresh & brackish waters. Increasing runoff in salt free areas.	Not applicable
	Adaptation of Land and Water Use	Adapting water use to higher salinity. Separate brackish supplies.	Adapting land use by establishing salt tolerant plants.
	Substitution	Abandonment of undeveloped sources. Tributary development. Abandonment of developed sources for alternative.	Abandonment of seeps.

3.2 Basis for Evaluating Management Measures

When concerned with management on a river basin scale, it is normally a combination of measures which is appropriate. None of the possible measures listed in Table 2 have any universality in application and most of those directed at controlling the release of salt are not yet developed to their potential level of refinement. Evaluation of the relative merits of alternative salinity management measures depends very much on the circumstances and seriousness of the problem for which they are being considered. In evaluating alternative salinity management measures it is important to clearly distinguish between practices which are sufficiently developed for use on contemporary problems and those which only offer prospects for use in the future.

To be usable in dealing with present day problems it is clearly necessary that the measures are practically or theoretically demonstrated to be effective and that workable techniques have been developed for field implementation. However it is also essential that adequate administrative and legal machinery exists to encourage, regulate or authorize the prescribed actions and this must be supported by social acceptance. Other very important practical requirements are that the consequences of particular measures must be reasonably predictable to allow some rational process of planning and efficient implementation. Furthermore, the time required for the salinity benefits to take effect must be acceptable in the particular problem situation.

One criterion which has not yet been met by any developed measure listed in Table 2 and which represents a potentially important objective for research is that the measure substantially satisfies the joint objectives of water quality and maximizing productive farmland.

Finally it is essential that the benefits expected from a measure justify its cost and, although this may seem a self-obvious statement, it does in some circumstances imply complex issues of resource economics.

4 CONTROLLING THE RELEASE OF SALTS - PREVENTION

4.1 General

The most basic strategy available for managing river salinity is that of preventing or limiting the release of salt from the landscape by retaining all or part of the natural vegetation. By maintaining high levels of evapotranspiration in areas of high salt storage discharge of saline groundwater will be minimized.

The policy underlying recent preventive actions in key river basins has been to minimize risk of further deterioration where doubt exists, and in most cases this has involved maintenance of full vegetative cover at least until some other measure can be proven as safe. Measures of this nature are much less relevant where agricultural production is the prime objective of salt land management.

4.2 Halting Alienation

More than 36% of the region's divertible surface water resources fresher than 1000 mg L^{-1} TSS are in Group 1B river basins which are wholly forested, but which have areas of high salt storage. In these basins protection of forested Crown Land is the sole action necessary to prevent the release of salts into the stream.

In Group 2 basins, the success of other measures to manage water quality would be nullified if forested Crown Land was not retained in potentially saline areas. These Group 2 basins include another 27.5% of the region's divertible resource fresher than 1000 mg L^{-1} TSS.

In Group 3 basins, there are potable tributary streams draining potentially saline areas where salinity is controlled by forested Crown Land. Throughout all of the foregoing areas, alienation of forested Crown Land is now severely restricted either by: dedication as State Forest; executive policy of State Government; or by administrative processes for legally proclaimed water catchments. This policy, designed to withhold from clearing the maximum possible area of forested Crown Land in areas of high salt storage, is the single most significant water quality control measure in the region. Although the policy represents a virtual exclusion of conventional agriculture, less than 1% of the total land area of Western Australia is reserved as State Forest whereas agricultural and pastoral activity are now very extensive. The retention of forest land is therefore a policy having its own intrinsic merits.

4.3 Protective Management of Forest

In a large area of the forest of this region the dominant tree species is jarrah *(Eucalyptus marginata)*. The jarrah forest has been seriously affected by dieback disease. Caused by a microscopic fungus *(phytophthora cinnamomi)* which destroys the fine root structure, the disease is fatal to jarrah and to many other forests species. Dieback, has been spread by vehicles, logging equipment and by downhill movement along drainage paths. This disease severely reduces the forest canopy and consequently is expected to cause a reduction in evapotranspiration.

The dieback disease is more widespread in the western high rainfall area where a significant increase of salinity is not expected. In the lower rainfall areas of the forest, where substantial canopy reduction might cause a more serious salinity increase, the disease has not been as widely distributed and a large portion of the forest has been quarantined to allow time for recent infections to develop and accurate mapping to be undertaken. When such mapping is completed it will be possible to devise more effective forest hygiene which may allow some renewed access. Measures for controlling and treating dieback are being actively studied and recent research has uncovered promising possibilities.

The effects on stream salinity which would result from widespread incidence of dieback in a catchment with low rainfall cannot yet be predicted with any confidence. Adhering to a policy of minimising serious risks in situations of uncertainty, the forest quarantine and dieback research programmes have a strong policy support from water authorities.

Forest operation in this region is based on multiple land use policies in which priority uses are assigned to land units and then secondary and tertiary uses are admitted according to their compatibility with the higher priority uses (Forests Department W.A., 1980). Throughout the areas of forest with high salt storage and draining into potentially usable streams, catchment protection has been given a priority status and other uses are only admitted if compatible with catchment protection.

The most radical land use activity which has been introduced into the forests is bauxite mining, a form of surface mining in which the laterite capping covering upland areas is removed. After removal of the laterite ore, the land is revegetated by rehabilitation measures which are still in early years of development. Mineral leases have been granted covering the whole of the northern forest. However mining is presently confined to the high rainfall zones of the forest where salt storage is low and there is no risk of significant salinity increase. On present rates of mining activity, the industry is assured of 25 years or more of operation in this low salinity zone.

The mining company must prove there is no risk of significant increases in stream salinity consequent on its operation before being permitted to move into areas of high salt storage. This will require proof that rehabilitation is effective and that salinity problems will not be a secondary consequence through the spread of dieback. Given the time requirement for maturation of trees and the inherently long time scale of hydrologic processes, there is not a long lead time in which to conclude effective research.

The routing of service corridors through forests has been a matter of some concern, not so much because of the trees removed, but because of possible distribution of dieback along transects of the forest. Establishment of new corridors is now the subject of more vigorous constraints but cannot be wholly restricted.

In southern forests, the recent development of wood-chipping operations involving clearfelling has been the subject of public controversy and potential salinity problems have been suggested as arguments for restricting these operations. In response to these concerns, as a temporary precaution until research conclusively determines whether salinity is a matter of concern, the woodchip operation has been excluded from areas of high salt storage. However because this forest operation establishes vigorous regeneration, the effects on salinity of a river basin are expected to be temporary and insignificant.

4.4 Clearing Control on Crown and Freehold Land

In the 1970's water resources planning focussed attention on the continuing salinity deterioration of Group 2 rivers, and identified a need for urgent action in these basins (Public Works Department W.A., 1979a). Also during this period as a result of education and debate, public attitude to the water quality problem developed strongly in favour of firm control measures.

By the latter half of the 1970's, quantitative understanding had developed sufficiently at catchment scale, to allow very approximate predictions of the salinity increases which would result from clearing. Predictions were made that the salinity of Group 2 rivers would rise significantly (Table 3) if all uncleared private land was cleared for agriculture. In these circumstances, with 20% of the divertible resources at risk of serious degradation, legislation was passed in 1976 by State Parliament, imposing clearing controls on the catchment of the Collie River. Subsequently in 1978 controls were extended to the Denmark, Helena, Kent and Warren Rivers (Fig. 1). The legislation provided for compensation payments to land owners refused clearing licences.

TABLE 3

Salinity predictions for Group 2 rivers.

River	Catchment Area km^2	Land Uses in Catchment			Mean Annual Salinity Level mg L^{-1} TSS		
		State Forest & Reserves %	Private Land		At Present	As a Result of Clearing To Date*	If All Private Land Were Cleared*
			Uncleared %	Cleared %			
Collie	2830	65	12	23	750	1100	1700
Helena	1470	95	2.5	2.5	360	360	700
Denmark	650	79	5	16	570	640	850
Warren	3890	54	14	32	725	1000	1400
Kent	1650	45	15	40	1100	1500	2500

* Ultimate figures allowing for natural time lags.

It is an objective in the administration of this legislation to seek the development of partial clearing and any other management techniques which will help minimise the impact of these controls on land owners. However although such developments are being actively pursued, it is an over-riding policy that the introduction of such refinements must be based on sound evidence and not involve risk taking in relation to stream salinity.

As a basis for administering clearing controls, the Group 2 river basins have been divided into four zones which are a simple refinement of those illustrated in Fig. 1. These zones range from the inland zone of highest salinity potential where controls are most stringent, through two zones of reducing salinity potential where some relaxations of the controls are progressively admitted, to a zone of very low salinity potential where up to 90% of a land holding may be licensed for clearing.

At this stage, the principal relaxations allow clearing of small areas in the intermediate zones where the number of land holders who benefit is relatively large and the potential compensation costs and regulatory difficulties so avoided are substantial. In the longer term, prescriptions for partial clearing, and other such refinements will probably also be defined zone by zone.

In the Group 2 basins of this region, clearing control is the least costly proven measure available for controlling water quality. Although in some basins the introduction of clearing control may have occurred too late to be adequate without supplementary action, it will nevertheless maintain quality within close range of the acceptable limits. Clearing control is an essential prerequisite to the partial reforestation policy adopted in two Group 2 basins and holds open the option for partial reforestation in other basins.

In monetary terms the cost of clearing control is small in relation to the cost of developing other effective alternatives. However, the costs may be incurred decades before the protected water resource is utilised. In terms of loss of agricultural potential the clearing controls on the five Group 2 basins have affected approximately 1200 km^2, which is only a very small portion (0.7%) of the 180 000km^2 of farmland within private holdings in the agricultural areas of south-western Australia.

4.5 Partial Clearing

Water balance studies and observations of changes in groundwater following clearing have indicated that in zones of highest salinity hazard the difference in annual use between the original forest cover and agricultural crops or pastures is only about 10% or 15%. This relatively small difference between agricultural and forest consumption of water suggests that the original water and salt balance might be maintained by leaving comparatively small areas of freely transpiring native vegetation. Measures for partial clearing, if effective would at least permit some modified form of agricultural practice and achieve a reduction of constraints on agriculture in Group 2 basins.

Whereas partial clearing has relevance in areas where water quality control is an objective, it is much less likely to be of economic interest to farmers seeking to maximize areas of productive land. Salt seeps, on average, amount to 2% or less of developed farmland (Malcolm and Stoneman, 1976). However, it is expected that the amount of uncleared land which would need to be withheld from development to prevent the occurrence of salt seeps would at least be an order of magnitude greater.

In the deeply weathered profiles of the salt prone areas of this region, the pallid clays holding the accumulated salt, overlay the weathered bedrock which acts as a confined aquifer. Recharge of the aquifer in the weathered bedrock is believed to occur near the ridges through the lateritic soils, and on the slopes particularly by preferred pathways such as old root channels in the pallid clay. Salt discharge commonly occurs as a result of this groundwater emerging under pressure through the salt laden clays near the valley bottom. Shallow underground water also commonly moves downslope above the pallid clay in a perched water table which discharges fresh water near the valley bottom.

There are two broad aspects of water movement to which clearing control measures are directed and on which emphasis is varied in alternative patterns of partial clearing. These water control objectives are:

(i) providing an umbrella effect to control the descent of water through the root zone and thereby limit the recharge to saline groundwater;

(ii) accepting that recharge occurs, and using deep-rooted trees to control saline discharge from groundwater in the vicinity of seep areas.

Two patterns of partial clearing, on which development effort is being focussed, are parkland clearing and strip clearing.

The concept of parkland clearing is to thin out the natural forest, planting pasture between the remaining trees which are distributed uniformly. Parkland clearing will limit the recharge passing through the root zone. However at the present time no minimum effective tree density prescriptions have been developed and proven. Consequently in the river basins subject to clearing control regulations, licences for parkland clearing are at present limited to special cases and to field trials. Apart from the need for further technical development it will also be essential, before parkland clearing is accepted, to resolve practical and administrative problems associated with formulating prescriptions, legally defining a basis for long-term maintenance and regeneration, and monitoring compliance with licence prescriptions.

Proposals for strip clearing envisage that strips of natural vegetation will be retained:

(i) near the valley bottom within, or immediately upslope of, potential seep areas; or

(ii) along the ridges and other possible upslope areas of recharge.

The effectiveness of strips of vegetation low in the valley will be dependent on the ability of trees to control discharge from the deeper saline groundwater. Apart from the problem of lack of salt tolerance which may particularly affect trees in the poorly drained valley bottoms, there is a possibility that trees will preferentially draw fresher water draining through the perched water table. These detailed water and salt balances have not yet been sufficiently defined and quantified for a sound design basis to be established. For strips of vegetation retained upslope, there has not yet been sufficient description of recharge processes to determine the location and extent of vegetation required to control recharge.

Trials and research for evaluation, and ultimately for design of strip clearing are in progress. Strip clearing, in contrast to parkland clearing, retains the possibility of cropping. Licensing of strip clearing would also be administratively simpler than licensing of parkland.

For partial clearing to be acceptable under the clearing controls imposed in Group 2 basins, research must successfully establish a basis of design which will ensure that no significant salt release occurs.

5 CONTROLLING THE RELEASE OF SALTS - RESTORATION OR AMELIORATION

5.1 General

Several broad groups of restorative measures are listed in Table 2. The first of these, reforestation, is the one measure which at present is considered of major consequence.

Reforestation is currently being applied in two Group 2 basins and may be relevant in other basins of this Group. Reforestation measures are to be researched for restoring the natural water and salt balance after bauxite mining enters the drier areas of several Group 1B basins. Also, as a long term objective, it is hoped that agroforestry techniques will be developed for the joint benefit of water quality and agriculture in some Group 3A basins as well as in basins of Group 2.

The search for a restorative measure which will effectively and economically control salt seeps on farmlands in this region has proved most difficult. None of the measures listed for restoration of agricultural land has yet proved of more than limited benefit in the West Australian situation.

5.2 Reforestation

Although there are indications of reforestation having affected underlying groundwater, there is as yet no direct experimental confirmation that reforestation will control salinity. However if the discharge of deeper groundwater can be controlled there is reason to expect that reforestation should restore, or partly restore, the original salt balance. Also, for the same reasons that partial clearing may prevent salinity release, it is considered likely that close to full reversal might be achieved with partial reforestation of treated areas.

At the catchment scale in this region, the rise of stream salinity as a consequence of clearing typically occurs after a time lag of ten years or more. It is expected that similar or longer time lags will occur between reforestation and any resultant salinity improvement. These time lags present major practical and economic difficulties in use of reforestation as a restorative measure.

In considering the efficiency of reforestation for water quality restoration it is convenient to define effectiveness as follows:-

$$\text{Effectiveness of Reforestation} = \frac{\text{Ultimate drop in annual salt release after reforestation}}{\text{Ultimate rise in annual salt release after Clearing.}}$$

On any land unit, the effectiveness of reforestation will be determined by site conditions and by the design of reforestation in relation to such matters as: planting pattern; percentage of the land unit covered by replanting; and selection of tree species. In translating these effects from small land units to catchment scale, the main determinants of the amount by which river salinity is reduced will be the fraction of cleared land which can be included in the scheme, and the relative amounts of land rehabilitated in various salinity zones.

Some useful observations relating to design and implementation of reforestation can be derived by using the Collie River as an illustration. The Collie River catchment covers 2830 km^2 and has been developed for irrigation and water supply purposes. Clearing was halted in 1976 by which time the total cleared area was 645 km^2. Because of natural time lags the full effect of recent clearing has yet to be felt, but it is predicted that salinity will rise to 1100 mg L^{-1} TSS from an original figure of near to 200 mg L^{-1} TSS. A programme of reforestation commenced in 1979 with the objective of restoring the resource to a more acceptable salinity level.

A much higher quantity of salt is released per unit area of cleared land in the inland portion of the Collie catchment between the 900 and 650 mm rainfall isohyets than is released in the western high rainfall zone (Loh and Stokes, 1981). Furthermore, 85% of the cleared land lies between the 900 mm and 650 mm rainfall isohyets. This lower rainfall zone has contributed an estimated 98% of the increase of salinity in the Collie River, and it is in this zone that reforestation is being implemented.

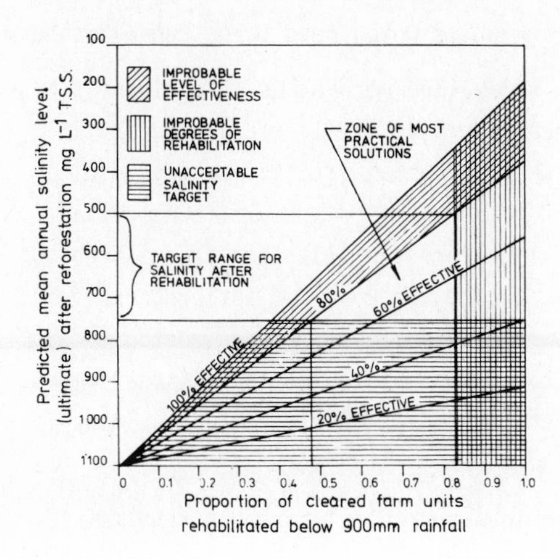

Fig. 3. Collie River reforestation - design diagram for catchment wide planning.

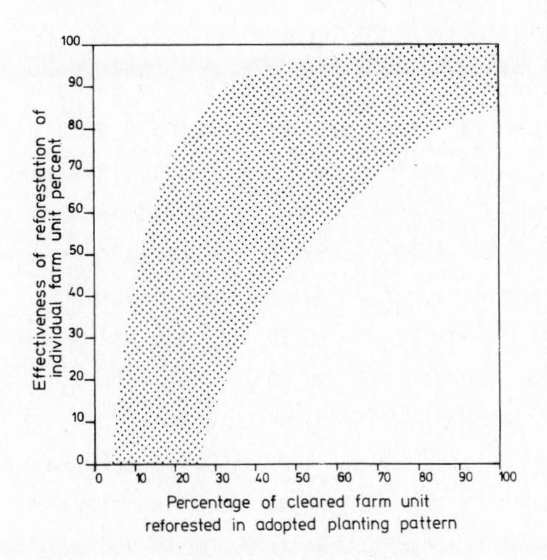

Fig. 4. Collie River reforestation - assumed working envelope - effectiveness of reforestation on individual farm units.

A major design problem in reforestation is in estimating the effectiveness with which salt discharge from individual farm units will be controlled after they have been rehabilitated by some pattern of partial reforestation. As a starting point therefore, Fig. 3 has been constructed to enable the design process to consider the implications of achieving different levels of effectiveness on treated farms and also to determine a general target range which must be set for this parameter and for the proportion of farm units which need to be rehabilitated.

Fig. 3 relates alternative design levels of average percentage effectiveness in treatment of individual farm units, and the proportion of cleared farm units in the 650 mm to 900 mm rainfall zone which are to be rehabilitated, to the resultant equilibrium salinity after reforestation. Because it is difficult to assert a selective priority over the acquisition of land for rehabilitation it has been assumed that rehabilitation proceeds randomly through the zone and not progressively from the area of highest to the area of lowest salt release. Superimposed on Fig. 3 are zones of unacceptable salinity reduction, improbable levels of effectiveness and nominal constraints on the proportion of cleared farm units which could be included within the partial reforestation scheme.

The constraints superimposed on Fig. 3 enclose a group of possible results of reforestation which would be acceptable for their salinity reduction. The diagram shows that, to be moderately effective in rehabilitation of this catchment, reforestation must achieve an average effectiveness on treated farm units of better than 50%. Also the proportion of all cleared farm units inland of the 900 mm isohyet which must be rehabilitated is relatively high, between 50% and 80%.

The costs of reforestation are directly related to the number of farm units included in the scheme, the planting pattern, and the percentage of farm unit which is reforested. To link Fig. 3 with reforestation costs and to achieve an efficient design for the reforestation scheme, a relationship is needed between the percentage of any cleared farm unit which is reforested and percentage effectiveness. The confident definition of such relationships for various planting patterns is beyond the presently developed "state of the art", but the water manager must make a working judgement in order to plan and implement the scheme. Fig. 4 is a subjective judgement of the envelope of likely relationships between effectiveness and percentage cover for present plantings in the Collie reforestation project. The figure reflects the present uncertainty confronting planners.

For the initial plantings of the Collie River reforestation scheme, strips of trees situated low in the valley, and on the ridges, are being designed for 50% cover of cleared farm units. The intention with this initial plan of partial reforestation has been to design conservatively and give a high expectation of achieving the desired levels of effectiveness indicated as necessary in Fig. 3.

Such designs might involve unnecessarily high levels of capital cost and greater than optimum contraction of farmland. Consequently, if the technical basis can be established, an important aim in future development of this programme will be to design for a lower percentage cover.

Research to assess the most effective planting patterns and quantify the likely effectiveness for different percentages of cover is crucial to further practical development of reforestation and partial clearing. Research in this area deserves the highest level of priority.

5.3 Modification of Crops and Cultivation

Conceptually a most attractive measure for controlling the release of salts would be one which achieved a satisfactory water and salt balance by modifying crops and crop management. However, although such measures as continuous cropping and use of deep-rooted perennial crops have been used with some success in North America, they are not immediately adaptable, and no crops have been found which will continue growth and re-establish the transpiration of the original bushland under the dry summer conditions of south-western Australia.

If in the long-term, improved cropping measures could be developed which achieved some control of salinity, albeit with a low percentage effectiveness, the measures would be of considerable strategic significance in several Group 2 basins and might even lead to re-evaluation of the water resource potential of some Group 3A basins. If agriculturally attractive measures could be developed, which farmers would adopt voluntarily or with only small incentives, there could be merit in introducing them speculatively as an addition to the range of management strategies presently being followed in Group 2 basins. From the viewpoint of a water resource planner, the possible benefits of improved cropping systems justify expanded research effort.

5.4 Drainage of Saline Groundwater

A potential method for managing and re-directing the release of salts, which has found little application in the region because of its expense, is the use of sub-surface drainage specifically planned to lower a saline water table or reduce the head in a saline leaky-confined aquifer.

As a possible measure for reclaiming salt seeps on agricultural land there have recently been some limited development trials of saline groundwater drainage in south-western Australia. However these drainage measures are normally regarded as uneconomic for farm management in this region. Furthermore the drainage would normally be disposed of by discharge into stream channels and this is not likely to be acceptable in catchments where water management is a primary objective.

In special cases where the costs are justified, drainage of saline groundwater may have some potential for controlling salt seeps on agricultural land. In Group 3 basins such measures could not be restricted by water authorities, although extensive use of such measures might cause concern and require review in some Group 3A basins.

In Group 2 basins, it is likely that drainage of saline groundwater into water courses could not be permitted. For such measures to be acceptable, suitable disposal must be established which will not find its way into the river. One possible means of disposing of the saline drainage would be discharge into evaporating basins and, on the river Murray in eastern Australia, salt lakes have been used for this purpose. In Western Australia salt lakes forming potential evaporating basins are not available in most Group 2 basins but there may be parts of the Kent River basin in which such measures are physically possible. The general concept may warrant further evaluation in specific cases.

5.5 Shallow or Surface Drainage

A theory that one of the main causes of salt affected land is movement of water through the shallow surface soils has gained some popular support in the south-west region in recent years. Measures have been advocated by a group of farmers for the interception of perched water moving laterally in shallow soil by cutting a drain or ditch down to the clayey sub-soil. A number of farmers believe these *interceptor drains or ditches* have achieved some success in control of land salinity.

In a report on an experimental trial of interceptor drains for water quality management, (Public Works Department, W.A., 1979b) it was observed that, on the site selected, the shallow sub-surface water carried only small quantities of salt which could not account for the observed salinity. It was concluded that to manage the discharge of salts it would be more appropriate to concentrate on controlling the more saline confined groundwater. A review by Holmes (1979) has supported this conclusion and further concluded that shallow interceptor drains have little prospect, at any acceptable spacing, of significantly lessening recharge to the confined groundwater.

Locally it must be anticipated that the debate on interceptor drains will continue. However from this debate there has emerged a more detailed discussion and recognition of the variations from site to site in the predominant paths of salt and water movement.

5.6 Comprehensive Saltland Treatment

The Department of Agriculture (1978) encourages the use of a combination of farm management practices aimed at using surplus water and controlling capillary rise. These include practices such as contour cultivation, control of stocking on seep areas, maintenance of pastures in spring, mulching seep areas, and drainage of waterlogged areas. These combined measures have met with good success (Nulsen, personal communication, 1980) in assisting the establishment of salt tolerant vegetation and may exercise some limited control over salt seepage. However only in mild salt land situations are they claimed to arrest salt discharge. The continuing managerial commitment of these measures has limited their acceptance by farmers.

6 ADJUSTING TO THE RELEASE OF SALTS - SALT DISPOSAL

6.1 General

There are several engineering measures which can be utilized to control salt after it has entered the water course by separating and disposing of a brackish component of streamflow. Apart from desalination, which is possibly the most obvious salt disposal measure, these measures all take advantage of the areal and temporal variations of stream salinity within a river basin to separate out a saline component. All of these disposal measures involve loss of water resources in the process of rejection or blowdown although, in the most efficient cases, this loss may be relatively small. Depending upon how the rejected water is disposed of, these measures also commonly involve some problem of degradation of other water courses.

At present the main interest in application of these measures is in Group 2 basins to supplement a range of preventive measures. Ultimately they may be adopted for gaining some utilisation of Group 3A basins. Current policies for Group 3A basins are therefore based on the premise that a salt disposal measure may eventually prove effective if presently forested zones of the catchment are protected.

6.2 Reservoir Management

A low cost method of salt disposal which has been utilized in the region, makes use of the temporal variation in stream salinity and of stratification phenomena in reservoirs (Patterson et al., 1978).

In this region, on salt affected catchments, the early winter flows wash down the surface encrustations of salt which formed during the previous dry summer and therefore tend to be considerably more brackish or saline than flows of late winter. These early winter flows also tend to be colder and denser than water stored in the reservoir which would have been well mixed by an autumn turnover. This colder brackish water underflows the stored water forming a brackish wedge at the base of the reservoir. By opening the bottom scours of the dam before mixing occurs, the brackish wedge can be released to waste, thereby improving the average salinity of water within the reservoir at the cost of reducing the quantity available.

This technique has been researched, developed and become standard procedure for Wellington Reservoir on the Collie River, a Group 2 basin with a critical salinity problem which has been aggravated by an extended run of dry winters. The resulting benefits to the average salinity level of water stored in Wellington Reservoir have been small, being generally in the range of a 5% to 10% improvement for releases in the order of 10% of the normal annual draw from the reservoir. The effects have been small because the contrasts between salinities of fresh and brackish inflows to the reservoir are not great, having been reduced by instream mixing of saline headwaters with fresher waters further downstream towards the reservoir. However significant operational benefits have also been obtained by holding the wedge below the level at which it affects the water supply offtake.

One of the serious limitations of this measure which has become apparent in Wellington Reservoir of recent years is that the greatest need for scouring arises in the driest years. At such times there is a shortage of water and scouring can only be carried out at the cost of a further diminution of available supplies. Controlled releases from storage reservoirs can only be considered as a supplementary salinity management measure in this region.

6.3 Seasonal Withdrawals

Another measure which has been considered for utilisation of temporal variations of salinity is seasonal withdrawal by pumping, at times when water quality is acceptable, into a water supply system or offstream storage. However in brackish rivers the period when water quality is acceptable may be limited to the low flow period in autumn and a period of two months or less in late winter. A high rate of pumping would be necessary to gain a supply of any significance and in dry years the supply of potable water may be very small or non-existent.

Seasonal withdrawal has limited potential in the South-West Division. As is the case with reservoir management, this measure is limited by the extent to which instream mixing of saline and fresh water takes place.

6.4 Saline Diversions

Use of the areal variations of salinity offers more positive means of separation of saline and fresh water components of river flow than do measures based purely on temporal variation. Typically in south-western Australia it is the headwater tributaries which drain land of high salt storage and which have saline discharges as a result of clearing. Schemes taking advantage of spatial variation of salinity therefore would involve diversion of saline headwaters enabling conventional development of the fresher water resources of the lower catchment.

The feasibility of this measure in any basin is largely determined by topographic features determining the ease with which saline headwaters can be collected and discharged to waste. This disposal may be by diversion past the freshwater damsite, by diversion to another river which acts as a drain, or by diversion to a salt lake.

Feasibility studies have recently been completed for a scheme on the Warren River in topographically favourable circumstances (Public Works Department, W.A., 1980). It was demonstrated in this study that a scheme diverting only 10% of mean annual flow could reduce an average annual salinity of 1200 mg L^{-1} TSS to approximately 500 mg L^{-1} TSS. This degree of effectiveness could be achieved because the average salinity of the headwater diversion would be as high as 7000 mg L^{-1} TSS. By comparison, experience indicates that because of instream mixing, the mean salinity of water which could be scoured from a stratified reservoir, if the Warren River was dammed at its main development site, would only be in the order of 1500 to 2500 mg L^{-1} TSS.

The high salinity of diverted water which causes such an effective salinity reduction also creates a major disposal problem. In the case investigated, the saline diversion was to be discharged into an adjacent Group 3A river system. The scheme was being investigated as a possible means for relaxing clearing controls but, because of the effects on the quality of the receiving river and the relatively high engineering costs the proposal was not adopted.

As a measure for reclaiming a water resource affected by headwater clearing, saline diversion has the advantage of being more immediate and more accurately predictable in its effects than is reforestation. In the long-term saline diversion is a measure likely to gain acceptance in specific circumstances for basins of Group 3A and possibly Group 2 categories. However for the present it has been accepted that there is a need for caution in making irreversible decisions committing a Group 3A river to act as a drain for saline diversion or allowing clearing to turn a Group 2 river brackish on the assumption that it can be reclaimed by subsequent headwater diversion.

6.5 Desalination

Desalination remains a very costly and energy intensive process to which must be added the cost of pre-treatment, conventional source development, delivery, and disposal of blowdown. Nevertheless, techniques such as reverse osmosis have improved the economics of brackish water desalination. Although, in the foreseeable future, desalination is not likely to be utilised as a salinity management strategy in south-western Australia, it does provide an *off the shelf* solution to extreme water quality management problems and is an upper bound figure for reclamation costs against which alternative measures can be evaluated.

7 ADJUSTING TO THE RELEASE OF SALTS - DILUTION

7.1 Blending Fresh and Brackish Water

Blending of fresh water and marginal or brackish water within the supply system is a common water supply practice in this region, particularly in borefield development. For this dilution process to be effective in a particular situation the acceptable salinity of the blended water must not be greatly below the mid range of the salinities of the fresh and brackish sources or a large proportion of the water must be drawn from the fresh source leaving only a small proportion of the brackish water to be utilised. In consequence, blending is most successful as a measure supplementing other controls and dealing with waters whose quality is only marginally impaired or as a measure where a moderately high salinity can be accepted for the blended water.

7.2 Increasing Runoff in Salt Free Areas

A measure which has been proposed at various times for improving the salinity of salt affected streams is to thin the forest in low salinity hazard areas of catchments and increase the fresh component of the total stream flow. This suggested measure encounters the basic difficulty of all dilution measures namely, that unless there is a large injection of good quality water the consequent lowering of salinity would be small. Furthermore in a catchment having serious salinity problems only a portion of the annual runoff would be from zones of low salinity hazard which are suitable for thinning. Consequently to produce an overall increase of 10% to 15% in catchment runoff may require increases in the order of 50% in the runoff from the low salinity zone. If a catchment like that of the Collie River, whose average annual salinity of discharge is 1100 mg L^{-1} TSS, was managed by thinning in the high rainfall zone so that the mean annual flow of the total catchment was increased by 15% through an inflow of water at 200 mg L^{-1} TSS, the resulting average annual salinity would only be improved to 980 mg L^{-1} TSS.

Catchment management for increased runoff in salt free areas is of long-term interest in water resources management. However, in terms of salinity improvement the benefits are only likely to be marginal.

8 ADJUSTING TO THE RELEASE OF SALTS - ADAPTATION OF LAND & WATER USE

8.1 Adapting Water Use to Higher Salinity

As water development proceeds in this region one of the necessary strategic considerations is some measure of adaptation to a higher average salinity of water supply. Because most water consumption is from reticulated public water supplies and public irrigation systems this possibility resolves largely into a basic question of what salinity is acceptable for such systems.

In this region, water systems for both public water supplies and for irrigation are based on maintaining salinity below a preferred upper limit of 500 mg L^{-1} TSS. However, salinities up to 1000 mg L^{-1} TSS are accepted in difficult circumstances. Where necessary to get the best use of a resource, supplies will be blended up to one or other of these accepted limits and, in the event of a drought or other short-term difficulty, supplies will be accepted for a short period at salinities in excess of the limit. The predominant ions in surface waters of the south-west are sodium and chloride, and at the accepted limits of total soluble salts concentration, no other constituents are present at concentrations approaching levels considered toxic to humans.

One possible example of adaptation of use is to consider providing separate brackish supplies to consumers able to utilise this water and thus help conserve the available fresh resources. However, in practical economic terms the development of separate brackish supplies is dependent on finding large industrial consumers who can utilise brackish water in place of fresh water and who can dispose o the waste water where it will not affect fresh resources. Situations of this kind present themselves occasionally and are actively followed as part of the regional water strategy. However, as with many other management measures this strategy achieves only small benefits in relation to the total problem.

The question of supply authorities accepting or designing for increases in salinity is a relatively complex matter. Value judgements with respect to quality involve socio-political issues which, in the short-term, may be more important in decision making than the technical and economic considerations. For longer-term planning and future cost-benefit decisions in salinity management, involving either raising or lowering of salinity levels, there is a case for a systematic and comprehensive assessment of the regional costs associated with changes in salinity.

8.2 Adapting Land Use with Salt Tolerant Vegetation

The farmer's means of adaptation to patches of incurable salt land is the planting of salt tolerant vegetation with the aim of achieving some form of agricultural production. A range of suitable shrubs, forbs and grasses have been identified which offer good prospects for grazing, or alternatively trees may be planted to improve appearance (Department of Agriculture, W.A., 1978). M. Smith (personal communication, 1980) is working on a research programme investigating the use of tissue culture to develop salt tolerant lines of pasture legumes.

9 ADAPTING TO THE RELEASE OF SALTS - SUBSTITUTION

As an alternative to the measures outlined in Sections 4 to 8 for control of salinity or for adaptation to its presence, there remains the option of abandonment of a resource and substitution of an alternative. In some cases abandonment may be the only measure which is physically, economically and socially feasible. However, in south-western Australia the development history has effectively seen some undeveloped water resources within Group 3 basins abandoned by default rather than by considered decision.

Recent decisions for Group 3A basins have established de facto a policy of avoiding for the present any new decisions which further reduce the possibility of these resources later being utilised. Thus, land alienation has been halted in forested portions of Group 3A basins and diversion of saline flows from Group 2 rivers into Group 3A rivers has been rejected for the present. At some future time there may be a necessity to degrade the role of certain Group 3A rivers to act as salt drains for adjacent salt affected rivers. However in the foreseeable future such a decision is not required.

10 SUMMARY OF SALINITY MANAGEMENT MEASURES

10.1 General

A moderately wide range of measures are listed in Table 2 for their potential in water quality management and a large proportion of these are being, or could be, utilised. However most of these could only be classed as supplementary or limited measures in the sense that either; their effects are minor; their costs or tradeoffs are high; or there are only limited sets of circumstances in which they are applicable. Nevertheless, there is a small number of measures which have the potential to achieve a major benefit in river basin management at an acceptable cost and with a reasonably wide field of application.

By contrast the managers of agricultural land have a more restricted range of measures at their disposal which are either more limited in their effectiveness, unfavourable in their benefit/cost relationships, or extremely limited in their applicability. Consequently there is an essential difference in the research and development needs of these two groups.

The most immediate need of the water manager in research and development is for refinement of preventive and restorative techniques of catchment management including a quantitative appreciation of their effectiveness. Understanding and management of the effects of jarrah dieback in the intermediate and lower rainfall zones, and development of effective rehabilitation methods for the future movement of bauxite mining into those zones, are also perceived as important areas of research and development. On the other hand, for advances in the management of agricultural salt land which seek to control salt seeps rather than adapt to their presence with salt tolerant pasture, there would appear to be a need, in this region, for some new breakthrough in research and development.

10.2 Water Quality Control Measures

In Table 2 the broad range of salinity control measures under consideration has been separated into two basic groups namely;

(i) those directed to controlling the release of salts from the landscape; and

(ii) those adapting to the release of salts.

The first group, particularly the basic preventive measures now implemented, form the essential foundation of the regional water management strategy. Of the restorative measures, partial reforestation is being pursued optimistically as an important measure for use with clearing control in some Group 2 and possibly Group 3A basins. Partial reforestation has been implemented operationally in a programme aimed at rehabilitating a 2830 km^2 Group 2 river basin. This programme has been initiated because of the critical water problems of the basin, but is also being developed as a full scale pilot study aimed at progressively addressing and resolving the practical problems of reforestation in a situation of freehold agricultural land use. However, the potential of reforestation for wide regional use will be dependent on refinements that enable retention of the maximum possible areas of land under agriculture and develop the maximum economic return from tree plantings in zones which are not commercially favourable for conventional forestry.

The second group of measures which in popular terms might be classed as the *technical fixes* are mostly engineering measures involving developed technology. With the exception of desalination, these measures are virtually all techniques of separation or blending which can be effective only where substantial fresh resources have been protected. The second group of measures are therefore seen as measures which can only supplement and not replace a regional strategy of catchment protection. Nevertheless, in partly cleared river basins having salinity problems, these measures are important and may be applied in various combinations for the tactical management of water supply quality. In the future, as regional water development proceeds, the adoption of saline diversion as a means of developing the forested catchment areas of Group 3A River basins may be justified.

10.3 Agricultural Land Management Measures

The measures most commonly applied for agricultural salt land management in this region are:

(i) management of plants, stock and cultivation to increase water use where it falls;

(ii) management of grazing and soil condition to minimize capillary rise;

(iii) surface drainage to control water logging; and

(iv) use of salt tolerant pastures in seep areas.

The benefits achieved by these measures have been perceived as limited by a large proportion of farmers. Although the W.A. Department of Agriculture's comprehensive salt land treatment techniques have met with success in establishment of salt tolerant pastures, they have not been popular because of the high managerial input required. As a consequence, there has been some speculative experimentation by farmers, particularly in the use of several variations of large contour banks and ditches.

Partial reforestation, deep-rooted vegetation and deep drainage have all been proposed as possible control measures but none have yet been developed to provide economic control of salt seeps.

11 FURTHER DEVELOPMENT OF A REGIONAL STRATEGY

A regional strategy adopting a range of measures for river basin management has evolved over the past decade through a progression of decisions and actions both small and major. Of necessity this strategy has been built only on those salinity management measures which, on present knowledge, are able to be applied with reasonable confidence.

In the development of a regional strategy for managing stream salinity, the management of salt land for improved agricultural production has been a separate objective. These two objectives have only been loosely linked in planning, partly because of the lack of measures which are effective in satisfying both objectives, but also because the greater agricultural problem is in Group 3A and 3B basins whereas the main thrust of water management has been in Group 1B and Group 2 basins. As techniques present themselves, and particularly if more effective agricultural measures are developed, it will be desirable to seek a more comprehensive strategy in Group 3A and some Group 2 basins.

In terms of land management the evolving salinity strategy has, of necessity, recognised multiple land use as an established policy in this region. Land uses, in addition to agriculture and water catchment include timber production, bauxite mining, recreation and conservation of flora and fauna. Within State Forests multiple land use objectives are incorporated in statutory forest planning by assigning primary, secondary and tertiary priorities to designated management areas. For the areas of alienated land no such formal land use priority system exists, although agriculture is established as the dominant use. However, recent clearing control legislation has overlaid water catchment as the management priority in a small proportion of this alienated land.

The first priority of regional water management strategy has been the establishment of a broad programme of preventive measures designed to halt any further avoidable salinity deterioration in potable water resources. Progressively these preventive actions have been extended to establish a regional coverage. They have been applied within the region according to the status of individual river basins and to the areal variations of salinity hazard and land uses within basin boundaries. In situations where any technical doubts have existed in prescribing these measures it has been a consistent policy to err on the side of least risk to water quality leaving room for some relaxation at a later time.

By sub-dividing the region into river basin groups which reflect resource potential and the potential for salinity degradation, and then overlaying these basins with zones of progressively increasing groundwater salinity a regional basis has been established for mapping salinity control strategies and relating these to other objectives of multiple land use planning in the region.

Following from the definition of broad regional strategies the more detailed planning for salinity control in rivers is developing progressively within individual river basins. According to circumstances, this more detailed planning encompasses adaptive measures as well as measures to control the release of salts. At this more detailed and basin orientated level of planning, regional or inter basin considerations remain important in determining such matters as the value of resources, the economic and social benefits and costs of action, and the acceptability or otherwise of salt drainage into neighbouring brackish rivers.

At this time detailed basin planning is concentrated on the Group 2 basins where the need for action is greatest and where the economic and social benefits of refinements will be most apparent in the short-term. Progressively however, over the next two decades, it will be necessary to evolve more specific basin plans to accommodate land uses such as bauxite mining in the lower rainfall zones of Group 1B basins. Also there may be growing pressure from rising water demand and from other land use interests to be more definitive in the control strategies presently applied in Group 3A basins. A major research project (Bennett and Thomas, 1980) investigating the applications of systems analysis to planning the use of land in a Group 3A basin has provided useful insights into the complexities of decision making for salinity management within these larger basins.

Although this paper has concentrated on the physical aspects of salinity control, enough has been said for it to be apparent that socio-political, economic, environmental and practical administrative considerations will be major factors in future decision making. Particularly significant among these considerations are:-

(i) the social and economic impact of salinity control on other land uses, particularly on agriculture;

(ii) the perceived costs to the community of increased salinity of urban and irrigation water supplies;

(iii) the value the community attaches to protection of undeveloped water resources for benefit of future generations;

(iv) the perceived environmental costs of increased stream salinity;

(v) the social and administrative feasibility of implementation.

In the next few decades as water use continues to increase it must be anticipated that the community will further increase the value it places on protection of water resources and also increase its concern for loss of agricultural production through dryland salting. Nevertheless the benefits and tradeoffs of future decisions in water quality management appear likely to be far more complex to evaluate than has been the case for determining strategies of the last decade. Consequently the needs for research and development should not be perceived solely in terms of the scientific aspects of salinity control but also in decision analysis and in practical machinery for implementation.

12 REFERENCES

Bennett, D. and Thomas, J.F., 1980 (Eds.) On rational grounds - planning the use of land and water resources in a salt affected catchment in Western Australia. CSIRO Div. of Land Resour. Manage., Perth.

Department of Agriculture of Western Australia, 1978. Saltland and what to do about it. Bull. 4048.

Forests Department of Western Australia, 1980. Land use management plan - northern jarrah forest management priority areas.

Holmes, J.W., 1979. The Whittington Interceptor Drain Trial. Report to the Public Works Department of Western Australia.

Loh, I.C. and Stokes, R.A., 1981. Predicting stream salinity changes in South-western Australia. Agric. Water Manage., 4: 227-254

Malcolm, C.V. and Stoneman, T.C., 1976. Salt encroachment - the 1974 saltland survey. J. Agric. West Aust., 17: 42-9.

Patterson, J., Loh, I.C., Imberger, J. and Hebbert, R., 1978. Management of a salinity affected reservoir. Proc. Inst. of Engrs. Aust., Hydrology Symp., Canberra.

Public Works Department of Western Australia, 1979a. Clearing and stream salinity in the South West of Western Australia. Doc. No. MDS 1/79.
Public Works Department of Western Australia, 1979b. Investigations of the Whittington Interceptor System of salinity control. Prog. Rept.
Public Works Department of Western Australia, 1980. Tone River and Upper Kent River diversion study. Prog. Rept. Doc. No. MDS 1/80.
Wood, W.E., 1924. Increase of salt in soil and streams following the destruction of the native vegetation. J. Roy. Soc. West Aust., 10: 35-47.

DRYLAND MANAGEMENT FOR SALINITY CONTROL

J. VAN SCHILFGAARDE

United States Salinity Laboratory, U.S.D.A., California, U.S.A.

ABSTRACT

Van Schilfgaarde, J., 1981. Dryland management for salinity control. Agric. Water Manage., 1981.

The objective of managing the lands in a watershed to maintain or enhance a dependable water yield of low salinity differs fundamentally from that of enhancing agricultural production in situ. The challenge is to devise strategies compatible with both.

Vegetative management to increase evapotranspiration reduces salt emissions; it also reduces water yield and, if achieved by forestation, agricultural production. However, U.S. experience indicates that crop selection to increase water use in recharge areas is an effective practice to ameliorate downslope saline seeps.

It appears the physico-chemical principles that control salt and water flow through geologic systems, and the effects of vegetation thereon, are well established. This is true, at least, for systems where the predominant salt is NaCl derived from deposition in rainfall. The mathematical tools to make use of these principles are also adequate. The data base, however, frequently is not sufficient to describe the system, nor is our ability to make the necessary field measurements at a reasonable cost.

Aside from economic considerations, potential solutions for dryland salinity problems must be related to the specific site conditions. They may include interception drainage, drainage of water from perched water tables, reduction of hydrostatic pressure in artesian systems, as well as soil and crop management systems. The viability of these (or other) solutions can only be assessed after adequate delineation of the site conditions, including identification of the recharge area, description of the subsurface conditions with evaluation of the hydraulic properties of the aquifer materials traversed by the flux, and sufficient information to derive the flow paths. In addition, the time dependence of the flow system must be considered. Whereas flow problems have most often been solved in terms of potential distributions, it will be helpful to pay more explicit attention to velocity fields and transit times. Examples of specific situations, real or imagined, will be used to illustrate the points made above. A parallel will be drawn with similar problems under irrigated agriculture.

1 INTRODUCTION

In assessing management strategies for salinity control, the objective must be clearly perceived. One strategy may be best suited for maintaining or enhancing in situ agricultural production, while another leads to increased yield of good quality water. These two objectives may well be in conflict with each other. One must also clearly distinguish among various problem situations, as solutions feasible in one circumstance may not apply in another.

I shall briefly outline several typical situations encountered in the United States and describe the implications of various management strategies. Although these circumstances may differ drastically from those found in Western Australia, principles should apply. Finally, I wish to stress some specific points of concern.

2 SALINITY IN NORTH AMERICA

In the terminology preferred by Peck (1978), salt seeps are areas where saline seepage is of "recent" origin, maintained by local recharge and resulting from lateral flow of the infiltrated water through more or less horizontal layers or lenses of permeable material that may truncate near the surface, or flow over less pervious layers that outcrop on hillsides. Such relatively shallow flow systems seem typical of the salt seeps in North Dakota (Doering and Sandoval, 1976), Montana (Halvorson and Black, 1974) and similar areas of the Northern Great Plains (Van der Pluym, 1978). The excess of infiltrated precipitation over that evapotranspired at the recharge site provides the seepage water. The salts, derived from the mantle and of geologic origin, tend to be calcium, magnesium and sodium sulfates; they become concentrated at the seep by evaporation. Changes in vegetation from the original grassland, especially the introduction of grain-fallow rotations, together with somewhat wetter than average weather cycles, have accelerated the development of salt seeps. The primary concern in the Northern Great Plains is the loss of agricultural production and solutions seek prevention of new and reclamation of existing salt seeps.

Two strategies can be pursued. One can manage the recharge site to change the water balance, or deal with the problem downstream. In principle, prevention at the source is preferred to correction after the fact. The simplistic answer to the former is to plant a deep-rooted perennial crop in place of small grain and fallow. As detailed by Black at this conference, far more sophisticated crop management schemes have been developed recently that make use of as much of the available water as possible by a flexible cropping sequence. Solving the problem after recharge has taken place required interception of the flow. Doering and Sandoval (1976) demonstrated that subsurface interceptor drains, in the right circumstances, can be effective in this regard. They are expensive, however, especially if an outlet can only be found at some distance; disposing of the drain flow can also encounter regulatory problems. A possible alternative, not pursued to my knowledge in practice, might be an evaporation pond.

A different situation prevails in the Red River Valley of North Dakota. There, some 160 000 ha of agricultural lands have saline water tables near the surface. A sandstone aquifer about 20 m thick is overlain with 35 m of glacial till and/or alluvium. The aquifer, fed hundreds of km to the west, is under pressure and saline. Numerous experiments have been conducted dealing with tile drainage systems, changes in soil surface topography, tillage, mulching and cropping sequences to find ways to "live with" the salinity problem, none very successful. Pump tests demonstrated that wells penetrating the aquifer could be pumped to reduce the pressure and thus reverse the direction of the artesian flow. A total of 11 wells pumping at a combined rate of 0.35 m^3 s^{-1} (about 7 mm yr^{-1}) would alleviate the problem on the entire 160 000 ha. This flow, at a salinity of 4400 mg L^{-1}, would have to be discharged in the Red River; it amounts to 0.5% of the long-term average river flow. This discharge would, of course, increase the river salinity; this increase would be substantial at some times of the year unless temporary storage were provided. (Benz et al., 1976). A larger number of shallower wells terminating in the glacial till should also provide a solution. A 3-year test of such a shallow well gave disappointing results. The deep-well solution, originally proposed and demonstrated in 1968, has not been implemented and it probably won't be. About 100 km north of the problem area, the Red River flows into Canada, and the objectives of the farmers in North Dakota's Red River Valley differ from those of the water users downstream.

In the Lower Rio Grande Valley of Texas, some 40 000 ha of cultivated, non-irrigated soils have salinity problems. Salty patches are often interspersed with non-saline areas in the same field. The salty areas tend to be slightly higher in elevation and their soils have a higher clay content. The area is underlain by a shallow sand aquifer whose top ranges in depth from 1.5 to 9 m below the surface. The water table generally is well above the aquifer, ranging from 0.5 to 2 m deep. Rektorik et al. (1976) found that differential leveling agriculture, one has control over water inputs; in dryland agriculture, the same objective must be met by choosing the appropriate cropping sequence.

A final example, taken from California, concerns use of brackish water in contrast to disposal (Rhoades et al., 1980). In some parts of the Central Valley, drainage water contains as much as 6000 mg L^{-1} of salt; disposal by gravity would require construction of a drain to the San Francisco Bay, technically and politically expensive. Use of such water to irrigate cotton produces a full crop and reduces the volume of drainage water to one third or less. Thus, the disposal problem is lessened but not eliminated; to maintain a viable system of production, there must be export of salt. In this area plagued by water shortages, we advocate appropriate use of low quality water, if not for salt tolerant crops, then to produce halophytes for other end uses. Such use offers a responsible means for conserving water while reducing a waste disposal problem. One must recognize, however, that plants are not an effective sink for salts; they effectively extract water from the solution, but remove only negligible amounts of salt.

3 SALINITY IN WESTERN AUSTRALIA

The nature of the salt seep problem in Western Australia apparently is similar to that of the Northern Great Plains. Excess water percolation due to changes in vegetation results in increased lateral movement through semi-confined aquifers. There are also important differences: The native vegetation in Australia was forest rather than grassland. The salts are dominantly chlorides deposited in rainfall and, a key point, more emphasis is placed on downstream water quality than on agricultural production. As in the Great Plains, a change in vegetation to increase evapotranspiration in the recharge area provides a viable option to alleviate the problem. Reforestation, however, will adversely affect farm income and, as I understand it, introduces some knotty institutional problems as well. This is in contrast to the situation in Montana, where farm income can be maintained or even increased. Neither does it address the problem of *phytophthora*, an area concerning which I plead complete ignorance. Finally, reforestation reduces water yield.

For correcting saline seeps "after the fact", one again is reduced to interception of the seepage water and its disposal. It is my understanding, undocumented, that a number of farmers in Australia have successfully reduced their problem areas by cutting shallow surface drains into less pervious layers, thus intercepting lateral flows. Although the on-farm problem is resolved, returning such water to a receiving stream goes counter to the objective of better downstream water quality. Combining such drains with evaporation ponds may be a feasible compromise. An alternative approach, sometimes advocated, is the planting of of the fields, with the salt affected areas about 0.1 m lower than the nonsaline areas, was effective in enhancing leaching by rainwater and thus in increasing total grain yield. Alternatively, under similar conditions, Rektorik (1976) was successful in providing effective drainage by installing a series of shallow well points that were connected with PVC pipes to one self-priming centrifugal pump. On a smaller scale, this system is similar to that proposed by Benz et al. (1976) for the Red River Valley, but it doesn't suffer the institutional problems associated with conflicting interest groups.

An extreme situation dealing with salinity control is found in the Paradox Valley of Colorado. Brine inflow into the Dolores River from a 4.5 km deep salt dome that is crossed by the river contributes about 200 000 tonnes of salt per year. Tentative plans are to intercept the upward flow of water and salt with a number of wells before it enters the river, and to dispose of it in an evaporation pond. This pond will be at an elevation of more than 600 m above the river, and the pipeline (with a number of pumping plants) will be 33 km long. The pumping rate is still uncertain. The expected reduction in river salt load is 180 000 tonnes, which in turn should reduce the salinity of the lower Colorado River by 15 mg L^{-1}. By one estimate, 7 x 10^6 m^3 yr^{-1} of water will be evaporated (U.S. Bureau of Reclamation, 1974). In this instance, (dis)benefits from the salinity at the site are negligible and the project is based strictly on downstream water quality improvement, primarily of benefit to the metropolitan Los Angeles area.

The Paradox Valley project is one of a number of endeavors being planned or implemented in a program to reduce the salinization of the Colorado River. Since 37% of the salt in the lower river is ascribed to return flow from irrigated areas, considerable emphasis has been placed on irrigation management and irrigation water distribution systems in the development of this program (USBR, 1974). Although salt management of irrigated lands is outside the scope of this conference, a brief reference may be useful in illustrating some parallels.

In all instances, the emphasis is on increasing irrigation efficiency; expressed differently, the amount of water applied is matched as closely as possible to crop need, thus reducing the volume of drainage and/or seepage water to a minimum compatible with crop salt tolerance. Reducing drainage reduces salt displacement from aquifers, salt dissolution from soils, and volume of drainage water needing disposal or treatment. As stressed in work at this Laboratory (van Schilfgaarde et al.,1974), reducing the deep percolation may also induce salt loss from the irrigation water by mineral precipitation and thus reduce the total mass of dissolved salts in the system. Which of these processes is operative depends on local circumstances. In all cases, however, the key objective is to reduce deep percolation by matching water applied to water used by the crop. In irrigated salt-tolerant plants in or just above a seep area. Although such plantings no doubt should increase evapotranspiration temporarily, they do not seem to offer a viable, long-term solution. As noted above with reference to drain water disposal in California, plants do not remove salt from the system; the accumulation of salts would soon make it impractical to grow any plants in the interceptor area.

An important consideration is that seeps result from changes in vegetation, or from the disturbance of an equilibrium. Comparing the steady state situation on a forested watershed with a cleared watershed would favor the cleared watershed, with a lower salt concentration and higher water yield in the latter. Unfortunately, Peck and Hurle (1973) have calculated that, for typical watersheds, the time required to reach the new equilibrium is likely to vary from 20 to 400 years. Clearly, we have but limited interest in solutions that require centuries to come about. Yet it is also suggested by Peck and Hurle's findings that it may be reasonable to identify some specific watersheds where improvements in salinity can be expected in a few decades (10 to 30 years) and on these even to accelerate the water throughput by cropping adjustments. As long as the percentage of streamflow derived from these areas is small enough, such a strategy could have interesting advantages.

4 SALINITY AND SOIL SCIENCE

The foregoing discussion has restated the obvious, without any profound new ideas. Its purpose was to put into perspective the nature of the problems under discussion and potential solutions. It also served to introduce some observations of a somewhat different nature. Collis-George (1979) recently opined that most major conceptual advances have been made in the field of soil science and that future advances should be expected in learning to apply existing concepts to field problems and in making appropriate measurements. He may be proven a false prophet. Yet there is little question that the greatest immediate challenge we face is to learn to better apply our general understanding to specific, real world situations. With reference to the subject at hand, the fundamentals of water and solute transport are understood well enough (Peck, 1978), but it is less clear how to use these principles to delineate the problems in the field, or to make the appropriate measurements needed for quantitative interpretations.

Rhoades and various co-workers adapted the principles of electrical conductivity measurements to develop rapid techniques for the delineation of areas affected by saline seepage or sites of incipient seeps (e.g. Rhoades and Halvorson, 1977). A resistivity probe was also developed, now available commercially, to more readily measure salinity as a function of depth (Rhoades and van Schilfgaarde, 1976). More recently, Rhoades has had tentative success in adapting electromagnetic measurements to the evaluation of near-surface salinity (unpublished). This recent activity - and substantial progress - in developing instrumentation for salinity investigation is evidence of a long existing need. Laborious methods of sample collection and laboratory analysis were often too costly to permit adequate field investigations.

Among the soil properties most important for the evaluation of water and salt fluxes is the hydraulic conductivity (K). Numerous methods are available to measure this parameter; hundreds of papers have been written on the subject. There still are many difficulties with such measurements, as well as with their interpretation. It is the latter that is of primary concern. For example, Peck et al. (1980), using a slug test technique, verified observations by others that K was log-normally distributed; they found that K varied, within a given area, over several orders of magnitude. An indication of the difficulties associated with a correct and accurate quantitative evaluation of K is found in their closing statement that "... then it must be concluded that the bulk conductivity is low in each area of this study ". Such wide variability is not restricted to K and, in recent years, substantial attention has been paid to the spatial and temporal variability of other parameters as well. For example, Dagan and Bresler(1979) developed a theoretical framework for describing solute dispersion in heterogeneous, unsaturated soil and used it to compute the distribution of concentration under some simple flow conditions. Warrick and Amoozegar-Fard (1980) recently compiled a very useful interpretation of methods to describe spatially variable flow conditions and consequences of these methods with reference to water and salt transport in the field. In my humble opinion, tremendous progress has indeed been made in understanding the consequences of variability, but we have not yet reached the point where we can use this knowledge for predictive purposes.

There are, of course, numerous situations where the conceptual understanding suffices to draw important conclusions without the detailed knowledge of hydraulic properties. Peck and Hurle's (1973) work on salt and water balances falls in that category. In other instances, more detailed knowledge is indeed needed. When changes in water quality are the primary item of interest, the historically well-developed application of potential theory is more useful when interpreted in terms of the stream function rather than the potential function. Raats (1978) made a significant contribution along this line, emphasizing flow lines and travel times in contrast to equipotential lines. Unfortunately here, also, detailed knowledge of the properties of the porous media is frequently needed.

5 CONCLUSION

In summary, I would paraphrase Collis-George that the greatest challenge in the field of soil water flow is the development of more useful field measurement techniques and methods for interpreting these measurements. Dryland salinity problems are as diverse in character as they are widespread. Their solutions are equally varied. They depend greatly on the objectives as well as on the local physical conditions. These conditions may not be too difficult to describe qualitatively but frequently the detailed, quantitative description poses a great challenge.

6 REFERENCES

Benz, L.C., Sandoval, F.M., Doering, E.J. and Willis, W.O., 1976. Managing saline soils in the Red River of the North. U.S. Dept. Agric. ARS-NC-42, 54 pp.

Collis-George, N., 1979. The contribution of soil science to the agricultural industry in Australia in the 21st. century. J. Aust. Inst. Agric. Sci. : 106-108.

Dagan, G. and Bresler, E., 1979. Solute dispersion in unsaturated heterogeneous soil at field scale. Soil Sci. Soc. Am. J. 43: 461-467.

Doering, E.J. and Sandoval, F.M., 1976. Hydrology of saline seeps in the Northern Great Plains. Trans. ASAE 19: 856-861, 865.

Halvorson, A.D. and Black, A.L., 1974. Saline-seep development in dryland soils of northeastern Montana. J. Soil and Water Conserv. 29: 77-81.

Peck, A.J., 1978. Salinization of non-irrigated soils and associated streams: A review. Aust. J. Soil Res. 16: 157-168.

Peck, A.J. and Hurle, D.H., 1973. Chloride balance of some farmed and forested catchments in southwestern Australia, Water Resources Research 9: 648-657.

Peck, A.J., Yendle, P.A. and Batini, F.E., 1980. Hydraulic conductivity of deeply weathered materials in the Darling Range, Western Australia. Aust. J. Soil Res. 18: 129-138.

Raats, P.A.C., 1978. Convective transport of solutes by steady flows I. General theory. Agric. Water Manage. 1: 201-218.

Rektorik, R.J., 1976. Field drainage with manifold well points. Trans. ASAE 19: 81-84.

Rektorik, R.J., Allen, R.R. and Lyles, L., 1976. Surface modifications for water management and salinity control in a non-irrigated area. Trans. ASAE 19: 699-703.

Rhoades, J.D. and Halvorson, A.D., 1977. Electrical conductivity methods for detecting and delineating saline seeps and measuring salinity in Northern Great Plains soils. U.S. Dept. Agric. ARS W-42, 45 p.

Rhoades, J.D. and van Schilfgaarde, J., 1976. An electrical conductivity probe for determining soil salinity. Soil Sci. Soc. Am. J. 40: 647-651.

Rhoades, J.D., Rawlins, S.L. and Phene, C.J., 1980. Irrigation of cotton with saline drainage water. ASCE Convention and Exposition, Portland, Oregon, April 1980: Preprint 80-119.

U.S. Bureau of Reclamation, 1974. Colorado River Quality Improvement Program Status Report. 125 p.

Van der Pluym, H.S.A., 1978. Extent, causes and control of dryland saline seepage in the Northern Great Plains region of North America. Dryland-Saline-Seep Control Symposium, ISSS, Edmonton, Canada: 1.48-1.58.

van Schilfgaarde, J., Bernstein, L., Rhoades, J.D. and Rawlins, S.L., 1974. Irrigation management for salt control. J. Irrig. Drain. Div. ASCE 100 (IR3): 321-338.

Warrick, A.W. and Amoozegar-Fard, A., 1980. Areal prediction of water and solute flux in the unsaturated zone. Report to EPA, Rob S. Kerr Env. Res. Lab., Ada, Okla. In press.

AUTHOR INDEX

Bergatino. R.N. 115
Black. A.L. 295
Bresler. E. 35
Brown. P.L. 295
Donovan. J.J. 115
Ghorbanzadeh. A. 275
Henschke. C.J. 173
Hillman. R.M. 11
Holmes. J.W. 19
Imberger. J. 225
Jenkin. J.J. 143
Johnston. C.D.......... 83
Konikow. L.F. 187
Loh. I.C. 227
Miller. M.R............ 115
Nulsen. R.A. 173
Orlob. G.T. 275
Peck. A.J. 83
Raats. P.A.C. 63
Sadler. B.S. 353
Sanderegger. J.L. 115
Schmidt. F.A. 115
Sedgley. R.H. 313
Shepherd. K.J. 335
Smith. R.E. 313
Stokes. R.A. 227
Talsma. T. 103
Tanji. K.K. 207
Tennant. D. 313
Van Schilfgaarde. J. ... 383
Williams. P.J. 353
Williamson. D.R. 83
Wronski. E.B. 19